AF361854

Multispectral Remote Sensing Satellite Data for Mineral and Hydrocarbon Exploration: Big Data Processing and Deep Fusion Learning Techniques

Multispectral Remote Sensing Satellite Data for Mineral and Hydrocarbon Exploration: Big Data Processing and Deep Fusion Learning Techniques

Editors

Amin Beiranvand Pour
Omeid Rahmani
Mohammad Parsa

MDPI • Basel • Beijing • Wuhan • Barcelona • Belgrade • Manchester • Tokyo • Cluj • Tianjin

Editors

Amin Beiranvand Pour
Institute of Oceanography &
Environment (INOS)
University Malaysia
Terengganu (UMT)
Kuala Nerus
Malaysia

Omeid Rahmani
Department of Natural
Resources Engineering and
Management, School of
Science and Engineering
University of Kurdistan
Hewlêr (UKH),
Erbil
Iraq

Mohammad Parsa
Department of Earth Sciences,
University of New Brunswick
Fredericton, NB,
Canada

Editorial Office
MDPI
St. Alban-Anlage 66
4052 Basel, Switzerland

This is a reprint of articles from the Special Issue published online in the open access journal *Minerals* (ISSN 2075-163X) (available at: www.mdpi.com/journal/minerals/special_issues/MRSSDMHE).

For citation purposes, cite each article independently as indicated on the article page online and as indicated below:

LastName, A.A.; LastName, B.B.; LastName, C.C. Article Title. *Journal Name* **Year**, *Volume Number*, Page Range.

ISBN 978-3-0365-6795-2 (Hbk)
ISBN 978-3-0365-6794-5 (PDF)

Contents

About the Editors . . . . **vii**

Amin Beiranvand Pour, Omeid Rahmani and Mohammad Parsa
Editorial for the Special Issue: "Multispectral Remote Sensing Satellite Data for Mineral and
Hydrocarbon Exploration: Big Data Processing and Deep Fusion Learning Techniques"
Reprinted from: *Minerals* **2023**, *13*, 193, doi:10.3390/min13020193 . . . . **1**

**Abdallah M. Mohamed Taha, Yantao Xi, Qingping He, Anqi Hu, Shuangqiao Wang and
Xianbin Liu**
Investigating the Capabilities of Various Multispectral Remote Sensors Data to Map Mineral
Prospectivity Based on Random Forest Predictive Model: A Case Study for Gold Deposits in
Hamissana Area, NE Sudan
Reprinted from: *Minerals* **2023**, *13*, 49, doi:10.3390/min13020193 . . . . **7**

**Saad S. Alarifi, Mohamed Abdelkareem, Fathy Abdalla, Ismail S. Abdelsadek, Hisham
Gahlan, Ahmad. M. Al-Saleh and Mislat Alotaibi**
Fusion of Multispectral Remote-Sensing Data through GIS-Based Overlay Method for
Revealing Potential Areas of Hydrothermal Mineral Resources
Reprinted from: *Minerals* **2022**, *12*, 1577, doi:10.3390/min12121577 . . . . **39**

**Monera Adam Shoieb, Haylay Tsegab Gebretsadik, Syed Muhammad Ibad and Omeid
Rahmani**
Depositional Environment and Hydrocarbon Distribution in the Silurian–Devonian Black
Shales of Western Peninsular Malaysia Using Spectroscopic Characterization
Reprinted from: *Minerals* **2022**, *12*, 1501, doi:10.3390/min12121501 . . . . **59**

Li He, Pengyi Lyu, Zhengwei He, Jiayun Zhou, Bo Hui, Yakang Ye, et al.
Identification of Radioactive Mineralized Lithology and Mineral Prospectivity Mapping Based
on Remote Sensing in High-Latitude Regions: A Case Study on the Narsaq Region of Greenland
Reprinted from: *Minerals* **2022**, *12*, 692, doi:10.3390/min12060692 . . . . **75**

Aref Shirazi, Ardeshir Hezarkhani and Amin Beiranvand Pour
Fusion of Lineament Factor (LF) Map Analysis and Multifractal Technique for Massive Sulfide
Copper Exploration: The Sahlabad Area, East Iran
Reprinted from: *Minerals* **2022**, *12*, 549, doi:10.3390/min12050549 . . . . **97**

**Stephen E. Ekwok, Ogiji-Idaga M. Achadu, Anthony E. Akpan, Ahmed M. Eldosouky, Chika
Henrietta Ufuafuonye, Kamal Abdelrahman and David Gómez-Ortiz**
Depth Estimation of Sedimentary Sections and Basement Rocks in the Bornu Basin, Northeast
Nigeria Using High-Resolution Airborne Magnetic Data
Reprinted from: *Minerals* **2022**, *12*, 285, doi:10.3390/min12030285 . . . . **117**

**Stephen E. Ekwok, Anthony E. Akpan, Ogiji-Idaga M. Achadu, Cherish E. Thompson,
Ahmed M. Eldosouky, Kamal Abdelrahman and Peter András**
Towards Understanding the Source of Brine Mineralization in Southeast Nigeria: Evidence from
High-Resolution Airborne Magnetic and Gravity Data
Reprinted from: *Minerals* **2022**, *12*, 146, doi:10.3390/min12020146 . . . . **133**

Timofey Timkin, Mahnaz Abedini, Mansour Ziaii and Mohammad Reza Ghasemi
Geochemical and Hydrothermal Alteration Patterns of the Abrisham-Rud Porphyry Copper
District, Semnan Province, Iran
Reprinted from: *Minerals* **2022**, *12*, 103, doi:10.3390/min12010103 . . . . **151**

Mastoureh Yousefi, Seyed Hassan Tabatabaei, Reyhaneh Rikhtehgaran, Amin Beiranvand Pour and Biswajeet Pradhan
Application of Dirichlet Process and Support Vector Machine Techniques for Mapping Alteration Zones Associated with Porphyry Copper Deposit Using ASTER Remote Sensing Imagery
Reprinted from: *Minerals* **2021**, *11*, 1235, doi:10.3390/min11111235 . . . . **175**

Behnam Mehdikhani and Ali Imamalipour
ASTER-Based Remote Sensing Image Analysis for Prospection Criteria of Podiform Chromite at the Khoy Ophiolite (NW Iran)
Reprinted from: *Minerals* **2021**, *11*, 960, doi:10.3390/min11090960 . . . . **201**

Roberto Bruno, Sara Kasmaeeyazdi, Francesco Tinti, Emanuele Mandanici and Efthymios Balomenos
Spatial Component Analysis to Improve Mineral Estimation Using Sentinel-2 Band Ratio: Application to a Greek Bauxite Residue
Reprinted from: *Minerals* **2021**, *11*, 549, doi:10.3390/min11060549 . . . . **219**

Mohamed Abd El-Wahed, Basem Zoheir, Amin Beiranvand Pour and Samir Kamh
Shear-Related Gold Ores in the Wadi Hodein Shear Belt, South Eastern Desert of Egypt: Analysis of Remote Sensing, Field and Structural Data
Reprinted from: *Minerals* **2021**, *11*, 474, doi:10.3390/min11050474 . . . . **237**

About the Editors

Amin Beiranvand Pour

Amin Beiranvand Pour is a well-experienced and internationally established researcher in the field of geological remote sensing and mineral exploration. He was listed among the World's Top 2% Scientists by Stanford University for the years 2020 and 2021. He has a full academic background in applied geology, economic geology, remote sensing and mineral exploration. He was the project leader of numerous mineral exploration research projects using satellite remote sensing technology in arid and semi-arid terrains, Antarctic, Arctic and tropical environments. His experience ranges widely, from mineral exploration to environmental issues such as geo-hazards, structural mapping, geothermal and geomorphic and coastal geology investigations. He has published more than 170 research papers in the field of geological remote sensing, mineral exploration and geohazard modeling.

Omeid Rahmani

Omeid Rahmani has a Ph.D. in Petroleum Engineering from Universiti Teknologi Malaysia (UTM). He has been working as a Senior Researcher Fellow at the University of Kurdistan Hewlêr (UKH), Hewlêr, the Kurdistan Region. Omeid is currently serving as the Academic Editor of PLOS ONE (IF=3.752), Topical Advisory Panel of Minerals (IF=2.818), and Editor of Carbonate & Evaporates (IF=1.492), and as an active reviewer on the board of the Journal of Petroleum Science and Engineering (IF=5.168) and other prestigious journals in Elsevier, ACS, RSC, Springer, MDPI, Emerald, and Wiley. Omeid has also accomplished an exceptional academic record with solid R&D activities by publishing high-impact papers in prestigious journals, such as *Energy, J CO2 Utilization, Construction & Building Materials, Ultrasonics, Remote Sensing, Colloids & Surfaces A, J Petroleum Science & Engineering, Marine & Petroleum Geo., Molecules, Energy & Fuels, J Geo Chem. Exploration, RSC Advances, Int. J Earth Sciences, Processes, Energies, Minerals, Plos ONE, ACS Omega,* and *Sustainability.* He has delivered a diverse range of projects in: CO2 sequestration using industrial wastes and natural minerals; enhanced oil recovery (EOR) by adsorption of surfactant and nanoparticles from aqueous solutions in porous media and ultrasonic simulated water-flooding conducted on a long unconsolidated sand pack; waxy crude oil, drilling fluids, wellbore instability; source rocks evaluation, organic geochemistry; reservoir characterizations, rock typing, and energy conservation.

Mohammad Parsa

Mohammad Parsa is an exploration geologist with expertise in exploration geochemistry and GIS-based mineral prospectivity mapping. He received his Ph.D. in mineral exploration engineering from the Amirkabir University of Technology in 2018. Currently, he is working as a postdoc fellow for the University of New Brunswick, Canada. He has published about 20 peer-reviewed papers in the fields of geochemistry and mineral exploration. He has also reviewed many peer-reviewed publications for prestigious journals. Currently, he is serving as the guest editor for some Special Issues, including Special Issues in the *Journal of Geochemical Exploration, Minerals, and Mining.*

 *minerals*

Editorial

Editorial for the Special Issue: "Multispectral Remote Sensing Satellite Data for Mineral and Hydrocarbon Exploration: Big Data Processing and Deep Fusion Learning Techniques"

Amin Beiranvand Pour [1,*], **Omeid Rahmani** [2] **and Mohammad Parsa** [3]

[1] Institute of Oceanography and Environment (INOS), Universiti Malaysia Terengganu (UMT), Kuala Nerus 21030, Terengganu, Malaysia

[2] Department of Natural Resources Engineering and Management, School of Science and Engineering, University of Kurdistan Hewlêr (UKH), Erbil 44001, Kurdistan Region, Iraq

[3] Mineral Exploration Research Centre, Harquail School of Earth Sciences, Laurentian University, Sudbury, ON P3E 2C6, Canada

* Correspondence: beiranvand.pour@umt.edu.my; Tel.: +60-9-6683824; Fax: +60-9-6692166

Citation: Pour, A.B.; Rahmani, O.; Parsa, M. Editorial for the Special Issue: "Multispectral Remote Sensing Satellite Data for Mineral and Hydrocarbon Exploration: Big Data Processing and Deep Fusion Learning Techniques". *Minerals* **2023**, *13*, 193. https://doi.org/10.3390/min13020193

Received: 24 January 2023
Accepted: 27 January 2023
Published: 28 January 2023

Using multispectral remote sensing data in cooperation with big data processing and deep fusion learning techniques provides a new approach for mineral and hydrocarbon exploration. Regional-scale mineral exploration in metallogenic provinces and hydrocarbon exploration in inaccessible and harsh areas are challenging due to the difficulty of processing remote sensing big data and the variety of remote sensing datasets needed for different applications [1–5]. Nowadays, spaceborne remote sensing big data sources are available, appropriate and range from free to low-cost for mineral and hydrocarbon exploration projects. Landsat data series, Satellite Pour l'Observation de la Terre (SPOT) data series, Worldview-3 data series, Advanced Land Imager (ALI) data, Phased Array type L-band Synthetic Aperture Radar (PALSAR) data and Advanced Spaceborne Thermal Emission and Reflection Radiometer (ASTER) data have been successfully and continuously used for regional-scale mineralogical-lithological-structural mapping in metallogenic provinces and hydrocarbon exploration [6–8].

Numerous image processing algorithms and Geographic Information System (GIS) modeling can be used for extracting spectral information related to alteration minerals, ore-related lithological units and microseepage-related geochemical alterations. The fusion of extracted information using deep fusion learning techniques has been developing progressively and is crucial for unraveling several image processing challenges [9–11]. Although the techniques are specific to scientific interest in remote sensing in the mineral and hydrocarbon exploration community, a generic implementation is in the initial stages. This Special Issue focused on the recent developments in the applications of multispectral remote sensing satellite data for mineral exploration in metallogenic provinces, onshore oil slick detection, offshore oil spill monitoring and hydrocarbon exploration. We were interested in innovative solutions for deep fusion learning techniques for remote sensing data processing and difficulties. The submission of manuscripts was encouraged for a broad range of related mineral and hydrocarbon exploration themes. Researchers were encouraged to submit novel research or case studies, including: (i) innovative methods for fusing multispectral remote sensing satellite data for prospectivity mapping in vast metallogenic provinces; (ii) multispectral remote sensing satellite data for both macroseepage direct detection and microseepage indirect detection; (iii) mapping ophiolite complexes to understand mineral carbonation and CO_2 sequestration using novel remote sensing approaches and modeling; and (iv) Geographic Information System (GIS) modeling for integrating different remote sensing datasets and geophysical and geochemical techniques for mineral exploration.

A total of 19 manuscripts were submitted to this Special Issue, which were evaluated by professional Guest Editors and reviewers. Subsequently, 12 papers attained the level of quality and novelty expected by *Minerals*, and consequently, were revised, accepted and published. The accomplishments of the articles in this Special Issue are summarized in the following paragraphs.

Abd El-Wahed et al. [12] used Landsat-8, ASTER and PALSAR satellite remote sensing datasets to map geological contacts, lithologies and structural elements controlling gold-bearing quartz veins in the Wadi Hodein Shear Belt, South Eastern Desert of Egypt. Several image processing techniques such as band combinations, band math, Principal Component Analysis (PCA), decorrelation stretch and mineralogical indices were executed and integrated with fieldwork data. The data layers were exported to the GIS environment and, subsequently, fused to produce a potentiality map for shear-related gold mineralization in the study area. The results demonstrated that the gold mineralization was typically restrained to steeply dipping strike-slip shear zones in the marginal parts of the shear belt. Gold-mineralized zones cut heterogeneously deformed ophiolites and metavolcaniclastic rocks and attenuate inside and around granodioritic intrusions. The gold mineralization event was evidently epigenetic in the metamorphic rocks and was likely attributed to rejuvenated tectonism and the circulation of hot fluids during transpressional deformation. Accordingly, it was recommended that the concurrence of shear zones, hydrothermal alteration and crosscutting dikes creates high potential zones for new gold targets in the study area.

Bruno et al. [13] utilized a spatial component analysis to improve mineral estimation using the ferrous iron oxide (4/11) band ratio of Sentinel-2 in a Greek bauxite residue. This study proposed a model to map the iron concentration as the strategic metal within a bauxite residue in Greece. Due to the probability of substituting in a co-kriging system, the whole band ratio information with only the correlated components was analyzed. The approach represented three estimation alternatives: ordinary kriging, co-kriging and component co-kriging. Firstly, only direct samples from the site were used (ordinary kriging—OK estimation). Then, the band ratio (4/11) known for iron detection was used as additional information to map the iron variability within the bauxite residues (co-kriging—CK estimation). For map accuracies, a new method (component co-kriging—CCK estimation) was described to reconstruct the coregionalization model between the sample data and the band ratio information by exploiting the possibility of extracting a specific component from the Sentinel-2 data and using it in the coregionalization models. Lastly, all three models and their products were compared to check the enhancement given by the proposed model for iron estimation maps. Results indicated how utilizing the most correlated component reduces the estimation variance and improves the estimation results. Generally, when a good correlation with ground samples exists, co-kriging of the Sentinel-2 (4/11) band ratio component improves the reconstruction of the mineral-grade distribution, and consequently, affects the selectivity.

Mehdikhani and Imamalipour [14] evaluated mapping chromite-bearing mineralized zones within the Khoy ophiolite complex in NW Iran by analyzing spectral bands (VNIR+SWIR) of the ASTER satellite sensor. The optimum index factor (OIF), band ratio (BR), spectral angle mapper (SAM) and PCA analysis were used for lithological mapping. A specialized OIF, the RGB (8, 6, 3) color composite, was generated for discriminating lithological units in the study area. The RGB color composition of $(4 + 2)/3$, $(7 + 5)/6$ and $(7 + 9)/8$ band ratios showed good performance for identifying ophiolite complex lithology units. The SAM and PCA analysis were able to map harzburgite and dunite as the host units of the chromite lens. The results showed that the integration of information extracted from ASTER data is very useful for chromite prospecting and lithological mapping in ophiolitic zones located in mountainous and remote regions around the world. Yousefi et al. [15] applied Dirichlet Process (DP) and Support Vector Machine (SVM) techniques for mapping alteration zones associated with the Zefreh porphyry copper deposit in the Urumieh-Dokhtar Magmatic Arc (UDMA) of central Iran using ASTER remote data. The DP process was utilized to specify the training data, where alteration zones were detected by using spectral mapping methods such as Relative Band Depth (RBD), Linear Spectral Unmixing (LSU), Spectral Feature Fitting (SFF) and Orthogonal Subspace Projection (OSP). Executing the SVM and SAM methods on the ASTER data helped with identifying phyllic, argillic, propylitic and iron oxide alterations at a regional scale. This study demonstrated the use of the SVM algorithm for mapping hydrothermal alteration zones associated with porphyry copper deposits in metallogenic provinces worldwide.

Timkin et al. [16] used geochemical and remote sensing techniques to identify blind mineralization (BM) and zone-dispersed mineralization (ZDM) in the Abrisham-Rud porphyry copper deposit, Semnan province, Iran. Sentinel-2 and ASTER data were utilized for mapping lineaments and alteration zones in the study area using Automatic Line Extraction, Logical Operator and PCA algorithms. The zonality method was applied to separate geochemical anomalies and to calculate erosion levels. Using the zonality method, the geochemical maps of multiplicative haloes were produced. The K-nearest neighbor (KNN) algorithm was implemented to fuse rock units, faults and alterations as a geological layer. The results of both methods correspond to each other in the southern part of the study area, indicating a high potential zone. Ekwok et al. [17] used high-resolution airborne magnetic (HRAM) and gravity data to understand the genesis of brines in southeast Nigeria. The result of the analytic signal exposed the locations and spatial distribution of short- and long-wavelength geologic structures associated with igneous intrusions. Low-pass filtering, upward continuation and 2D modeling procedures indicated critical synclinal structures, which corresponded with the location of brine fields. The study discloses that igneous intrusions and associated hydrothermal fluids are responsible for brine generation. Ekwok et al. [18] used HRAM data to evaluate the thicknesses of sedimentary series in the Bornu Basin, northeast Nigeria, utilizing three depth approximation techniques including source parameter imaging, standard Euler deconvolution and 2D GM-SYS forward modeling methods. The maximum sediment thickness values from the various depth estimation methods used in this study correlate relatively well. Additionally, the anomalous depth zone exposed using the 2D forward model overlaps with the locality of the thick sedimentation revealed via the source parameter imaging and standard Euler deconvolution (St-ED) methods.

Shirazi et al. [19] fused a lineament factor (LF) map analysis and multifractal technique for massive sulfide copper exploration in the Sahlabad Area, Eastern Iran. The rose diagram analysis, Fry analysis, LF map analysis and multifractal technique were implemented using geological and geophysical data. Aeromagnetometric data were analyzed to determine the presence of intrusive and extrusive masses associated with structural systems. The results showed that the NW–SE fault systems control the host rock's lithology for copper mineralization. Therefore, the NW–SE fault systems are consistent with the main trend of lithological units related to massive sulfide copper mineralization in the area. He et al. [20] identified radioactive mineralized lithology and mineral prospectivity mapping techniques using Worldview-2 and Landsat-8 TIRS thermal infrared data in the Narsaq Region of Greenland. Employing a weight-of-evidence analysis technique that combines machine-learned lithological classification information with information on surface temperature thermal anomalies, the prediction of radioactive element-bearing deposits in the study area was performed. Shoieb et al. [21] evaluated the hydrocarbon functional groups, aromaticity degree, and depositional environment in the Silurian–Devonian Kroh black shales of western peninsular Malaysia using Fourier transform infrared spectroscopy (FTIR). The existence of humic acid and the enrichment of aromatic hydrocarbons in the Kroh shales confirmed that the organic matter in these shales contains plant-derived hydrophilic minerals with terrestrial origin. These discoveries may offer evidence on the depositional and thermal maturation of organic matter for the exploration efforts into the pre-Tertiary sedimentary successions of peninsular Malaysia. Alarifi et al. [22] used multiple criteria inferred from Landsat-8, Sentinel-2 and ASTER data using a GIS-based weighted overlay multicriteria decision analysis approach to construct a model for the delimiting of hydrothermal mineral deposits in the Khnaiguiyah district, Saudi Arabia. Mohamed Taha et al. [23] investigated the proficiencies of several multispectral remote sensing data to map mineral prospectivity using the random forest predictive model for gold deposits in the Hamissana area, NE Sudan.

The comments provided by the reviewers helped improve each of the papers published in this Special Issue, which was possible only because they were willing to volunteer their time and attention. We hope that the investigations published in this Special Issue will assist the mineral exploration communities and mining and hydrocarbon exploration companies to apply and integrate multispectral remote sensing satellite data and big data processing and deep fusion learning techniques for mineral and hydrocarbon exploration.

Author Contributions: A.B.P. writing; O.R. and M.P. editing. All authors have read and agreed to the published version of the manuscript.

Funding: This research received no external funding.

Acknowledgments: The Guest Editors would like to thank the authors who contributed to this Special Issue and the reviewers who helped to improve the quality of the Special Issue by providing constructive recommendations to the authors. We would like to express our appreciation to all authors and reviewers who contributed their time, research and special knowledge to this Special Issue. We extend our sincere gratitude to the MDPI Editorial Team for supporting the Guest Editors in efficiently processing each manuscript.

Conflicts of Interest: The authors declare no conflict of interest.

References

1. Pour, A.B.; Park, Y.; Park, T.S.; Hong, J.K.; Hashim, M.; Woo, J.; Ayoobi, I. Regional geology mapping using satellite-based remote sensing approach in Northern Victoria Land. *Antarctica. Polar Sci.* **2018**, *16*, 23–46. [CrossRef]
2. Pour, A.B.; Hashim, M.; Park, Y.; Hong, J.K. Mapping alteration mineral zones and lithological units in Antarctic regions using spectral bands of ASTER remote sensing data. *Geocarto Int.* **2018**, *33*, 1281–1306. [CrossRef]
3. Pour, A.B.; Park, T.S.; Park, Y.; Hong, J.K.; Zoheir, B.; Pradhan, B.; Ayoobi, I.; Hashim, M. Application of multi-sensor satellite data for exploration of Zn-Pb sulfide mineralization in the Franklinian Basin, North Greenland. *Remote Sens.* **2018**, *10*, 1186. [CrossRef]
4. Pour, A.B.; Hashim, M.; Hong, J.K.; Park, Y. Lithological and alteration mineral mapping in poorly exposed lithologies using Landsat-8 and ASTER satellite data: North-eastern Graham Land, Antarctic Peninsula. *Ore Geol. Rev.* **2019**, *108*, 112–133. [CrossRef]
5. Pour, A.B.; Park, Y.; Park, T.S.; Hong, J.K.; Hashim, M.; Woo, J.; Ayoobi, I. Evaluation of ICA and CEM algorithms with Landsat-8/ASTER data for geological mapping in inaccessible regions. *Geocarto Int.* **2019**, *34*, 785–816. [CrossRef]
6. Sekandari, M.; Masoumi, I.; Pour, A.B.; Muslim, M.A.; Hossain, M.S.; Misra, A. ASTER and WorldView-3 satellite data for mapping lithology and alteration minerals associated with Pb-Zn mineralization. *Geocarto Int.* **2022**, *37*, 1782–1812. [CrossRef]
7. Fossi, D.H.; Dadjo Djomo, H.; Takodjou Wambo, J.D.; Pour, A.B.; Kankeu, B.; Nzenti, J.P. Structural lineament mapping in a sub-tropical region using Landsat-8/SRTM data: A case study of Deng-Deng area in Eastern Cameroon. *Arab. J. Geosci.* **2021**, *14*, 2651. [CrossRef]
8. Ishagh, M.M.; Pour, A.B.; Benali, H.; Idriss, A.M.; Reyoug, S.S.; Muslim, A.M.; Hossain, M.S. Lithological and alteration mapping using Landsat 8 and ASTER satellite data in the Reguibat Shield (West African Craton), North of Mauritania: Implications for uranium exploration. *Arab. J. Geosci.* **2021**, *14*, 2576. [CrossRef]
9. Aali, A.A.; Shirazy, A.; Shirazi, A.; Pour, A.B.; Hezarkhani, A.; Maghsoudi, A.; Hashim, M.; Khakmardan, S. Fusion of Remote Sensing, Magnetometric, and Geological Data to Identify Polymetallic Mineral Potential Zones in Chakchak Region, Yazd, Iran. *Remote Sens.* **2022**, *14*, 6018. [CrossRef]
10. Shirazi, A.; Hezarkhani, A.; Pour, A.B.; Shirazy, A.; Hashim, M. Neuro-Fuzzy-AHP (NFAHP) technique for copper exploration using Advanced Spaceborne Thermal Emission and Reflection Radiometer (ASTER) and geological datasets in the Sahlabad mining area, east Iran. *Remote Sens.* **2022**, *14*, 5562. [CrossRef]
11. Maleki, M.; Niroomand, S.; Rajabpour, S.; Pour, A.B.; Ebrahimpour, S. Targeting local orogenic gold mineralization zones using data-driven evidential belief functions: The Godarsorkh area, Central Iran. *All Earth* **2022**, *34*, 259–278. [CrossRef]
12. Abd El-Wahed, M.; Zoheir, B.; Pour, A.B.; Kamh, S. Shear-Related Gold Ores in the Wadi Hodein Shear Belt, South Eastern Desert of Egypt: Analysis of Remote Sensing, Field and Structural Data. *Minerals* **2021**, *11*, 474. [CrossRef]
13. Bruno, R.; Kasmaeeyazdi, S.; Tinti, F.; Mandanici, E.; Balomenos, E. Spatial Component Analysis to Improve Mineral Estimation Using Sentinel-2 Band Ratio: Application to a Greek Bauxite Residue. *Minerals* **2021**, *11*, 549. [CrossRef]
14. Mehdikhani, B.; Imamalipour, A. ASTER-Based Remote Sensing Image Analysis for Prospection Criteria of Podiform Chromite at the Khoy Ophiolite (NW Iran). *Minerals* **2021**, *11*, 960. [CrossRef]
15. Yousefi, M.; Tabatabaei, S.H.; Rikhtehgaran, R.; Pour, A.B.; Pradhan, B. Application of Dirichlet Process and Support Vector Machine Techniques for Mapping Alteration Zones Associated with Porphyry Copper Deposit Using ASTER Remote Sensing Imagery. *Minerals* **2021**, *11*, 1235. [CrossRef]

16. Timkin, T.; Abedini, M.; Ziaii, M.; Ghasemi, M.R. Geochemical and Hydrothermal Alteration Patterns of the Abrisham-Rud Porphyry Copper District, Semnan Province, Iran. *Minerals* **2022**, *12*, 103. [CrossRef]
17. Ekwok, S.E.; Akpan, A.E.; Achadu, O.-I.M.; Thompson, C.E.; Eldosouky, A.M.; Abdelrahman, K.; András, P. Towards Understanding the Source of Brine Mineralization in Southeast Nigeria: Evidence from High-Resolution Airborne Magnetic and Gravity Data. *Minerals* **2022**, *12*, 146. [CrossRef]
18. Ekwok, S.E.; Achadu, O.-I.M.; Akpan, A.E.; Eldosouky, A.M.; Ufuafuonye, C.H.; Abdelrahman, K.; Gómez-Ortiz, D. Depth Estimation of Sedimentary Sections and Basement Rocks in the Bornu Basin, Northeast Nigeria Using High-Resolution Airborne Magnetic Data. *Minerals* **2022**, *12*, 285. [CrossRef]
19. Shirazi, A.; Hezarkhani, A.; Pour, A.B. Fusion of Lineament Factor (LF) Map Analysis and Multifractal Technique for Massive Sulfide Copper Exploration: The Sahlabad Area, East Iran. *Minerals* **2022**, *12*, 549. [CrossRef]
20. He, L.; Lyu, P.; He, Z.; Zhou, J.; Hui, B.; Ye, Y.; Hu, H.; Zeng, Y.; Xu, L. Identification of Radioactive Mineralized Lithology and Mineral Prospectivity Mapping Based on Remote Sensing in High-Latitude Regions: A Case Study on the Narsaq Region of Greenland. *Minerals* **2022**, *12*, 692. [CrossRef]
21. Shoieb, M.A.; Gebretsadik, H.T.; Ibad, S.M.; Rahmani, O. Depositional Environment and Hydrocarbon Distribution in the Silurian–Devonian Black Shales of Western Peninsular Malaysia Using Spectroscopic Characterization. *Minerals* **2022**, *12*, 1501. [CrossRef]
22. Alarifi, S.S.; Abdelkareem, M.; Abdalla, F.; Abdelsadek, I.S.; Gahlan, H.; Al-Saleh, A.M.; Alotaibi, M. Fusion of Multispectral Remote-Sensing Data through GIS-Based Overlay Method for Revealing Potential Areas of Hydrothermal Mineral Resources. *Minerals* **2022**, *12*, 1577. [CrossRef]
23. Mohamed Taha, A.M.; Xi, Y.; He, Q.; Hu, A.; Wang, S.; Liu, X. Investigating the Capabilities of Various Multispectral Remote Sensors Data to Map Mineral Prospectivity Based on Random Forest Predictive Model: A Case Study for Gold Deposits in Hamissana Area, NE Sudan. *Minerals* **2023**, *13*, 49. [CrossRef]

Article

Investigating the Capabilities of Various Multispectral Remote Sensors Data to Map Mineral Prospectivity Based on Random Forest Predictive Model: A Case Study for Gold Deposits in Hamissana Area, NE Sudan

Abdallah M. Mohamed Taha, Yantao Xi *, Qingping He, Anqi Hu, Shuangqiao Wang and Xianbin Liu

School of Resources and Geosciences, China University of Mining and Technology, Xuzhou 221116, China
* Correspondence: xyt556@cumt.edu.cn; Tel.: +86-516-8359-0106

Abstract: Remote sensing data provide significant information about surface geological features, but they have not been fully investigated as a tool for delineating mineral prospective targets using the latest advancements in machine learning predictive modeling. In this study, besides available geological data (lithology, structure, lineaments), Landsat-8, Sentinel-2, and ASTER multispectral remote sensing data were processed to produce various predictor maps, which then formed four distinct datasets (namely Landsat-8, Sentinel-2, ASTER, and Data-integration). Remote sensing enhancement techniques, including band ratio (BR), principal component analysis (PCA), and minimum noise fraction (MNF), were applied to produce predictor maps related to hydrothermal alteration zones in Hamissana area, while geological-based predictor maps were derived from applying spatial analysis methods. These four datasets were used independently to train a random forest algorithm (RF), which was then employed to conduct data-driven gold mineral prospectivity modeling (MPM) of the study area and compare the capability of different datasets. The modeling results revealed that ASTER and Sentinel-2 datasets achieved very similar accuracy and outperformed Landsat-8 dataset. Based on the area under the ROC curve (AUC), both datasets had the same prediction accuracy of 0.875. However, ASTER dataset yielded the highest overall classification accuracy of 73%, which is 6% higher than Sentinel-2 and 13% higher than Landsat-8. By using the data-integration concept, the prediction accuracy increased by about 6% (AUC: 0.938) compared with the ASTER dataset. Hence, these results suggest that the framework of exploiting remote sensing data is promising and should be used as an alternative technique for MPM in case of data availability issues.

Keywords: remote sensing; mineral prospectivity mapping; machine learning; random forest; gold mineralization; Sudan

Citation: Mohamed Taha, A.M.; Xi, Y.; He, Q.; Hu, A.; Wang, S.; Liu, X. Investigating the Capabilities of Various Multispectral Remote Sensors Data to Map Mineral Prospectivity Based on Random Forest Predictive Model: A Case Study for Gold Deposits in Hamissana Area, NE Sudan. *Minerals* **2023**, *13*, 49. https://doi.org/10.3390/min13010049

Academic Editors: Amin Beiranvand Pour, Omeid Rahmani and Mohammad Parsa

Received: 27 November 2022
Revised: 24 December 2022
Accepted: 24 December 2022
Published: 28 December 2022

1. Introduction

The prediction of mineral prospectivity is one of the substantial practices in mineral exploration, which is used to fulfill the growing demand for mineral resources in industrial development countries [1,2]. Mineral prospectivity mapping (MPM), also known as mineral prospectivity modeling, is a multivariable decision-making tool that aims to delimit and prioritize high-potential zones for exploring a particular type of mineral in unexplored regions [2–4]. Model-based MPM is a vital but challenging process that essentially attempts to establish a function for integrating a collection of geological features (input variables) with the presence of the targeted mineral (output variables) [5]. Establishing this integration function is carried out by analyzing the spatial relationships between input variable features and known mineral occurrences through different numerical algorithms [6]. Hence, it is essential to select a convenient algorithm that is capable to learn the complex relationships between variables (input/output) to obtain an accurate prediction [7]. In practice, the most critical procedure in prospectivity modeling is the selection of evidential features that represent the spatial representatives of ore-controlling factors, which can be extended to combine available multi-source exploration datasets such as geological, geophysical, geochemical, and remote sensing data [8,9]. Based on the ease of implementation and the availability of data and software tools, prospectivity modeling can be categorized into two types: (i) knowledge-driven models that depend on expert knowledge to heuristically estimate the parameters of the models using the given information of mineral deposits in the given geological setting [10,11]; (ii) data-driven models that depend on the quantitative measures of the spatial associations between evidential features and targeted deposit locations to empirically estimate the parameters [6,12–14].

Remote sensing data have been successfully and extensively employed in mineral exploration since they can detect and delineate geological and structural features that aid in identifying new areas of mineralization [15–18]. Remotely-sensed images with the proper spatial and spectral resolution, including multispectral and hyperspectral satellite imagery allow to identify rocks and minerals based on their spectral signatures in the visible-near-infrared (VNIR) and the shortwave infrared (SWIR) regions [19–22]. In specific, multispectral satellite imagery with a high spatial resolution (10–30 m) and coarse spectral resolution such as Landsat-8, Sentinel-2, and ASTER, have been widely utilized to map and remotely sense fault/fracture zones and/or hydrothermal alteration zones associated with ore mineralization [23–28]. Nevertheless, the remote-sensing approach in mineral exploration applications has often been exclusive to classification models and knowledge-driven regression models. Whereas classification models aim to classify different hydrothermal alteration zones (argillic, phyllic, and propylitic) or minerals associated with alteration (e.g., iron-bearing and hydroxyl-bearing minerals) [29–31]. On the other hand, in the GIS-based knowledge-driven method, each remote sensing predictor layer is assigned a weight reflecting its importance in the modeling process. Subsequently, producing a map with a continuous prospectivity score indicates the likelihood of the targeted mineral [15,32].

In recent years, the development of machine learning (ML) and deep learning (DL) methods have boosted the regression models of mineral prospectivivty, which achieve better predictive performance than traditional statistical techniques and empirical explorative models [2,3,33]. Some of the most commonly used supervised learning models include Random Forest (RF) [4,34], Support Vector Machine (SVM) [35], and Artificial Neural Network (ANN) [36], which have been efficiently applied for MPM. RF is well known to be the first choice for data-driven predictive modeling of MPM, considering its accuracy in the delineation of prospective areas and sensitivity of parameter configuration [7]. Furthermore, RF performance is very stable in the case of: (i) the sufficiency of the number of known locations of mineral occurrences [34]; (ii) and the sensitivity of using different training sets of non-occurrence locations [4]. Another advantage that makes RF a great objective tool for data assessment is its capability to measure and rank the importance of evidential features to the training process [37].

Although remote sensing data were utilized in several studies for mapping mineral potential using supervised learning models, they have not been used as the main core for the derivation of the evidential features to train ML data-driven predictive models (e.g., RF, SVM, and ANN). For instance, Mansouri et al. [38] processed ASTER data with a multivariate regression analysis method to map iron mineral resources in the Sarvian area, Iran. Moreover, three multispectral data, namely Landsat-7 ETM+, Landsat-8, and ASTER were utilized by Bolouki et al., [39], to produce several predictor maps (evidential features), then were fused together to train Naïve Bayes (NB) classifier for producing a map showing the probability of gold occurrence in Ahar-Arasbaran area, NW Iran. On the other hand, remote sensing imagery was integrated with other sources of data to train various ML predictive models. As an example of that, two band ratios (BR) images of Landsat TM (5/7 and 3/1) were integrated with other predictor variables such as three geochemical survey maps and a couple of geophysical maps. Rodriguez-Galiano et al., used these two BRs as an indication for ore-related hydroxyl and iron oxide alteration to train RF model for gold MPM in Rodalquilar area, Southern Spain [40].

Considering the global development in GIS and ML fields, data availability is still an issue for conducting a comprehensive MPM study in third-world countries. Carranza [33] reported that from 2006 to 2016 about 116 MPM studies were exclusive only to 25 countries such as Iran, Australia, China, Canada, etc. Whereas countries such as Sudan have almost no research about ML applications in the mineral exploration field, even though Sudan is the third largest gold-producing country in Africa and among the 20 countries in the world gold mine production in 2019 [41]. Therefore, it is worth noting that since the data of several multispectral satellite sensors are free, a comprehensive study of the capability of various remote sensing data for training multiple predictive models is needed.

In this study, mineral prospectivity modeling was performed for delineating gold prospective regions in west Hamissana, northeast Sudan. The present research aims at investigating the potential of Landsat-8, Sentinel-2, and ASTER for mapping mineral potential. Specifically, (i) remote sensing datasets were utilized to identify geological features and hydrothermal alteration zones associated with gold mineralization in the study area; (ii) spatial analysis methods and remote sensing enhancement techniques were applied to produce different thematic layers, which were afterward assessed based on their contribution to the prediction process; (iii) all datasets were also combined into another dataset to investigate the synergy of various data for developing a comprehensive scheme of MPM in the study area; and (iv) random forest algorithm was used as a tool of comprehensive comparison to obtaining the optimal dataset for accurate prediction.

2. Study Area

The study area comprises approximately 1379 km^2, which is situated between latitudes (20°22′ N and 20°50′ N) and longitudes (34°00′ E and 34°45′ E). It is located in Wadi Edom to the west of Hamissana, Red Sea State, Sudan (Figure 1). Topographically, the area studied is in the northern part of the Red Sea hills, which rises almost 2000 m (≈6561 ft) above sea level. The area is characterized by a dry climate, with very poor vegetation cover. The highest temperature reaches 46 °C in the summer (October–march). The Winter season is relatively short, from November to February, with an average temperature of around 25 °C in the daytime [42].

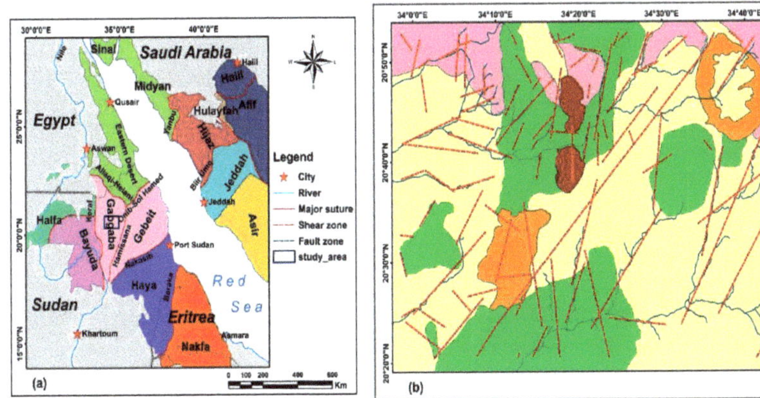

Figure 1. Location of the study area. (**a**) Sketch map of the main terranes, major suture, and shear zones of the Arabian-Nubian shield in the Red Sea Hills Region of Sudan (modified after Bierlein et al. [43]); (**b**) geological map of the study area (modified after Mohamed et al. [42]).

The geological setting of the Hamissana area forms a part of the Arabian Nubian Shield (ANS) (Figure 1a). ANS covers the eastern part of Sudan and broad areas of other countries such as Saudi Arabia, Ethiopia, Eritrea, Yemen, and Egypt [15]. During the Pan-African tectonic event, the collision and the accretion of Neoproterozoic island arcs to the Nile craton formed ANS [44]. The evolution of the shield including the complete orogenic cycle between 900 and 550 Ma is documented, where the island arcs characterized by basic to acid metavolcanics, and metasediments of the Proterozoic age are exposed [44,45]. Different arc assemblages are separated by ophiolitic-decorated suture zone forming five terranes in Sudan, while subduction-related calc-alkaline I-type granodiorites (older granites) intruded these assemblages [45,46]. The entire sequence is intruded by post-orogenic alkali A-type granitoid (younger granites) [44,47]. A NE-SW suture zone named after the study area forms a transition terrane between Bayuda craton terrane and Gebiet island arc terrane, called Gabgaba terrane. The main exposed lithological units in the study area are predominant with metavolcanic, syn-orogenic (older intrusion), and post-orogenic (younger intrusion) rocks [42]. The metasediments represent the oldest rock unit in the study area, which has E-W linear trending and is composed of quartzite and marble. Metavolcanics are generally composed of gray meta-acid volcanic and dark metatrachyte. Granite and coarse to medium-grained granodiorite, form the older intrusions. Younger intrusions are non-foliated and consist of porphyritic microgranite, highly sheared and dark granodiorite, and quartz feldspar porphyry. Several low outcrops of sediments and superficial deposits are scattered in the region. Structural trends of faults and dykes are NW-SE, NE-SW, and E-W, while most faults are in the form of strike-slip faults [42].

3. Materials and Methods

3.1. Data and Data-Preprocessing

Mohamed et al. [42] integrated important geological information about the study area. They established a comprehensive geodatabase containing the updated geological map, primary faults/fractures map, and locations of gold occurrences. All these geological datasets were digitized from paper maps of the published study [42]. Intrusion rock units were extracted separately as shapefile of polygons, while faults with different azimuth directions were saved as line shapefiles. 25 locations of gold occurrence were carefully selected from the overall 34 locations that constituted the database, where the minimum distance between each location corresponds to the grid size of 30m. The preparation of these geological datasets was carried out using ArcGIS 10.6.1 software.

The satellite remote sensing data employed in this study are Landsat-8, Sentinel-2A, and ASTER. All three types of multispectral imagery were freely downloaded from the U.S. Geological Survey's Earth Resources Observation and Science (EROS) Centre, using the USGS earth explorer website (https://earthexploere.usgs.gov). In addition, the user must also register on the National Aeronautics and Space Administration website (NASA) to obtain ASTER data (https://earthdata.nasa.gov). In this study, one scene of Landsat-8, two scenes of Sentinel-2, and four scenes of ASTER were obtained on different dates to cover the study area. All scenes have (0%–2%) cloud coverage and (>0.05) maximum Normalized Different Vegetation Index (NDVI), which suit the basic requirements for geological investigation. Table 1 shows the technical properties of different sensors and the characteristics of various scenes used in this investigation.

Table 1. Technical characteristics and dataset attributes of different remote sensing data.

Satellite	Bands	Wavelength (μm)	Spatial Resolution (m)	Scene ID	Date and Time of Acquisition	Other Info
	Band 1-(coastal/aerosol)	0.435–0.451	30			
	Band 2-Blue	0.452–0.512	30			
	Band 3-Green	0.533–0.590	30			
	Band 4-Red	0.636–0.673	30			
	Band 5-(NIR)	0.851–0.879	30			
Landsat-8	Band 6-(SWIR) 1	1.566–1.651	30	LC817304620	26 December 2021	Path = 173
	Band 7-(SWIR) 2	2.107–2.294	30	21360LGN00	08:08:23	Row = 46
	Band 8-Panchromatic	0.503–0.676	15			
	Band 9-Cirrus	1.363–1.384	30			
	Band 10-(TIRS) 1	10.60–11.19	100 * (30)			
	Band 11-(TIRS) 2	11.50–12.51	100 * (30)			

Table 1. *Cont.*

Satellite	Bands	Wavelength (µm)	Spatial Resolution (m)	Scene ID	Date and Time of Acquisition	Other Info
Sentinel-2	Band 1-(coastal/aerosol)	0.421–0.457	60	S2A_MSIL1C _20211203T08 1321_N0301	3 December 2021 09:28:49	Orbit No.: 78 Tile No.: T36QXJ
	Band 2-Blue	0.439–0.535	10			
	Band 3-Green	0.537–0.582	10			
	Band 4-Red	0.646–0.685	10			
	Band 5-Red edge	0.694–0.714	20			
	Band 6-Red edge	0.731–0.749	20			
	Band 7-Red edge	0.768–0.796	20			
	Band 8-NIR	0.767–0.908	10			
	Band 8A-Narrow NIR	0.848–0.881	20	S2A_MSIL1C _20211203T08 1321_N0301	3 December 2021 09:28:49	Orbit No.:78 Tile No.: T36QXH
	Band 9-Water vapour	0.931–0.958	60			
	Band 10-Cirrus	1.338–1.414	60			
	Band 11-SWIR	1.539–1.681	20			
	Band 12-SWIR	2.072–2.312	20			
ASTER	Band 1-VNIR (green/yellow)	0.520–0.60	15	ASTL1A 070331082541 0010269001	31 March 2007 08:22:08	ASTER Scene ID: (173, 129, 4)
	Band 2-VNIR (red)	0.630–0.690	15			
	Band 3N-VNIR	0.760–0.860	15			
	Band 3B-VNIR	0.760–0.860	15			
	Band 4-SWIR1	1.600–1.700	30	ASTL1A 070331082550 0010269001	31 March 2007 08:22:07	ASTER Scene ID: (173, 130, 4)
	Band 5-SWIR2	2.145–2.185	30			
	Band 6-SWIR3	2.185–2.225	30			
	Band 7-SWIR4	2.235–2.285	30			
	Band 8-SWIR5	2.295–2.365	30	ASTL1A 06122508250 0010269001	25 December 2006 08:24:03	ASTER Scene ID: (173, 129, 5)
	Band 9-SWIR6	2.360–2.430	30			
	Band 10-TIR1	8.125–8.475	90			
	Band 11-TIR2	8.475–8.825	90			
	Band 12-TIR3	8.925–9.275	90	ASTL1A 061225082515 0010269001	25 December 2006 08:24:03	ASTER Scene ID: (173, 130, 5)
	Band 13-TIR4	10.250–10.950	90			
	Band 14-TIR5	10.950–11.650	90			

In this study, the spatial resolution of various multispectral data was resampled to 30m using nearest neighbor technique. Since ASTER scenes were obtained on different dates, the Thermal Infrared (TIR) bands of ASTER and Landsat-8 were excluded to avoid unfavorable changes in surface thermal emission. Moreover, the coastal and cirrus bands of Landsat-8 and Sentinel-2 were designed for atmospheric correction. Therefore, they were not used in the analysis, as well as the panchromatic band (band 8) of Landsat-8 and water-vapor band (band 9) of Sentinel-2. Landsat-8 level 1 terrain corrected (L1T) data and ASTER level 1 Precision Terrain Corrected Registered At-Sensor Radiance (AST_L1A) data are radiometrically calibrated and geometrically corrected [27]. Both datasets were atmospherically corrected using the FLASH (Fast Line of Sight Atmospherics Analysis of Hypercubes) algorithm provided by ENVI 5.2 software. The FLASH algorithm was applied to ASTER data after implementing a cross-track illumination correction to the short waves infrared (SWIR) bands. Dark Object Subtraction (DOS) method in the semi-automatic classification plugin provided by QGIS 3.16.7 software, was employed to automatically atmospherically correct Sentinel-2 data. All atmospherically corrected datasets were georeferenced to the Universal Transverse Mercator (UTM) coordinate system in zone 36 N.

3.2. Random Forest (RF)

RF is an ensemble learning algorithm that is developed based on the concept of Decision Trees (DTs) [48]. The accumulation of multiple classification or regression DTs is employed to obtain repeated predictions of the target phenomenon represented by the training dataset [40]. These trees are grown based on random selection from the original training datasets using a procedure known as "bootstrap bagging" [49]. This sampling method increases the diversity of the trees by generating training subsets (bag samples) using about two-thirds of the training features for prediction, whereby the left out of the training samples (out-of-bag (OOB) samples) are used to validate the prediction accuracy [34].

To overcome the overfitting issue of the DT, RF attempts to grow trees in a way that maximizes the reduction in purity by searching through the optimal feature/split node, which varies from pruning trees according to discriminative conditions in the standard DT [50]. In other words, RF generates a tree using the best variable within bag samples, which reduces the correlation between the trees and minimizes the generalization error [48]. RF uses the Gini index to ensure the best split selection based on the comparison of the information purity of the leaf nodes with that of their root nodes. The Gini index used in this study is shown in the following equation [50]:

$$I_G(f) = \sum_{i=1}^{n} f_i(1 - f_i) \tag{1}$$

where f_i is the probability of class n, which can be calculated by dividing (m_j) the number of samples belonging to class j, by (m) the total number of samples in a specific node. The ultimate decision of RF is made by combining the votes of every DT, then averaging the results as shown in Equation (2) [7]:

$$f_{rf}^K(x) = \frac{1}{K} \sum_{K=1}^{K} T(x) \tag{2}$$

where $T(x)$ represents the result of DTs using x input vector, while K denotes the number of DTs that are grown to obtain RF results (f_{rf}^K) [2].

It is important to mention that RF has another essential advantage besides the unbiased estimation of the generalization error, which is the ability to measure and sort the importance of different predictor variables [51,52]. This is achieved internally by using the OOB samples, which originally are used to calculate the number of classified trees. Variables' importance is measured by randomly permutating each variable including OOB samples and then sending down these permuted OOB cases to the trees again. Calculating the correctly classified cases and subtracting them from the original correctly classified cases derived from non-permuted data, allows measuring the importance of that variable [53]. In other words, RF measures the marginal effect of a specific variable by holding all other predictor variables constant [4]. This asset is vital for multi-source data that are characterized by high dimensionality, where it is significant to grasp the influence of each predictor on the prediction performance [7,37,54].

3.3. Induction of RF Predictive Model

The process of inducting data-driven predictive machine learning modeling consists of three main steps, which directly affect the model's outcome. These three steps are: (i) the preparation of the input training dataset, which is considered the most important and critical step in the MPM field; (ii) specifying the suitable configuration of parameters in each model, also known as "hyperparameters tuning"; (iii) assessing predictive model performance [7,37]. Figure 2 shows the technical flowchart of this study's overall methodology and different stages to completely train RF predictive model. As shown in the figure, the preparation of input data includes generating predictor variables (also called feature predictors) and target variables. Predictor variables are thematic maps derived from integrating muti-source data and guided by a deep understanding of the gold mineral system. These predictor maps represent the critical stipulations for generating a desirable prediction of mineral potential [3]. On the other hand, target variables are the ground truth data of the studied phenomena. Unlike classification tasks where the target is defined by categorical data that are presented as labeled classes, predictive models (regression tasks) use continuous data as target variables to predict a continuous quantity of specific phenomena. In the case of MPM, mineral occurrence and non-occurrence are given as binary values (1 and 0, respectively) to predict continuous output representing the likelihood of gold value. In this study, the generation of different input datasets was accomplished by using ENVI and ArcGIS software. Meanwhile, Python 3 was implemented to train different RF by using "Scikit-Learn" library.

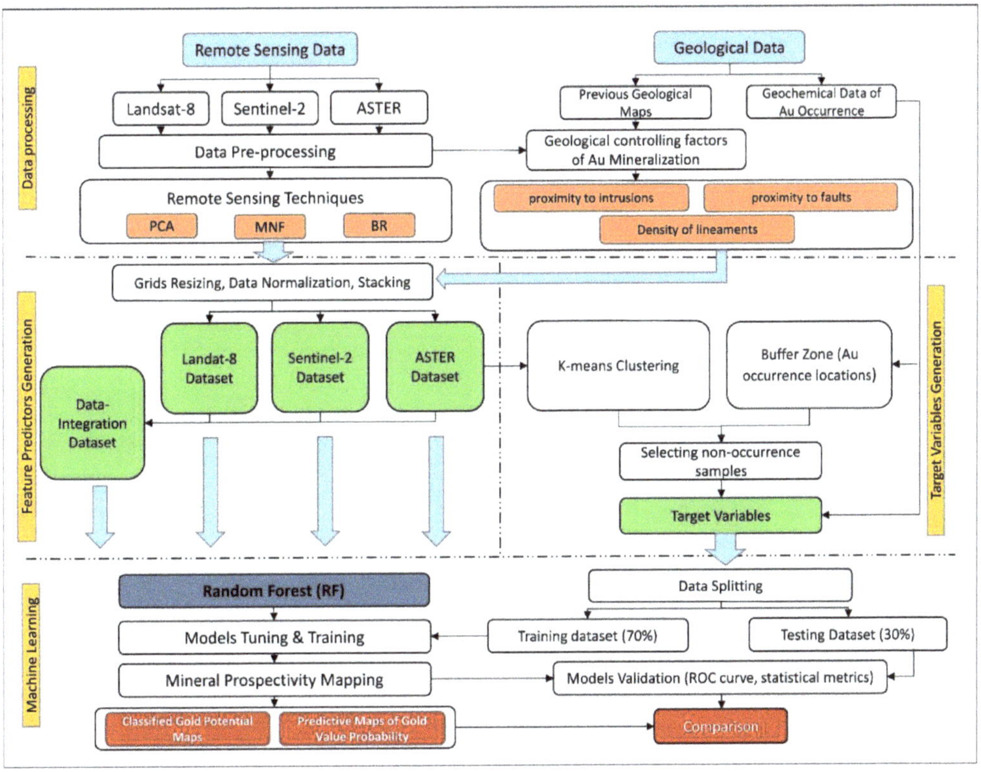

Figure 2. A technical flowchart shows the study's overall methodology to delineate MPM.

In common practice, leave-out and cross-validation methods are utilized to assess model performance [55]. The leave-out approach is achieved by randomly splitting the target variables into training and testing subsets. Data split usually takes different portions according to the user's definition when it typically is carried out at 75:25 or 80:20. On the other hand, the cross-validation approach, namely K-Fold cross-validation method, employs all the target data in the training and testing process simultaneously. This is achieved by splitting the data into k subsets, where each subset serves once as a testing set while the remaining sets are used to train the model. This process is repeated k times until all of the target data appear in the training and testing set. This method, thereafter, averages the scores of the prespecified accuracy metric from each k fold performance. Since the study focuses on regression task, the mean square error (MSE) was utilized to measure the average squared difference between the trained model predicted result ($\hat{y}_i$) and the true value of each sample (y_i). This can be formalized as follow:

$$MSE = \frac{1}{N} \sum_{i=1}^{N} (\hat{y}_i - y_i)^2 \tag{3}$$

where N is the number of samples in the test dataset.

This study uses both approaches for assessing performance and reducing overfitting. The train-test split method was utilized to introduce possible bias since there is limited target data. Moreover, this method aids in comparing the performance of various outputs of RF by measuring accuracy metrics from the testing dataset. On the other hand, the purpose of employing five K-Fold cross-validations is to reduce overfitting and obtain optimal parameters for training each dataset. The possibility to find an optimal combination of parameters varies with different input datasets. Therefore, an objective grid search method known as "GridSearchCV" was used for hyperparameter tuning. This method is provided by the Scikit-Learn library (https://scikit-learn.org). As shown in Figure 3, this method searches through all possible combinations of parameters using k iteration for each combination. The user defines a dictionary of the possible set of values for each parameter whether they are categorical or numerical (e.g., number of trees). Although this process has a high computational cost, it is vital to measure the influence of model configuration on prediction performance. In the present study the range of the number of trees was set between 50 and 500 at intervals of 50, and the number of features between 2 and 12 at 2 intervals [3,7,8,40].

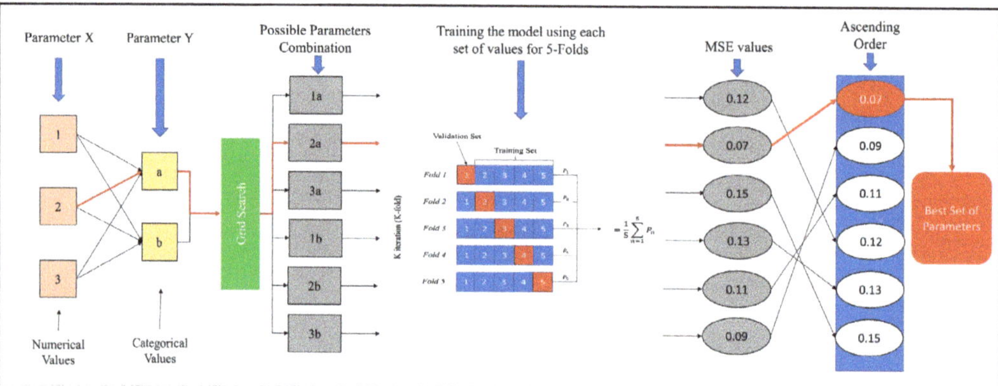

Figure 3. A plot demonstrates the GridSearchCV method: in this case using two parameters (numerical/categorical) and 5-folds cross-validation.

3.4. Predictor Variables

As mentioned before, the input datasets (input-feature vectors) of MLA are the set of information derived from combining different thematic layers at each grid location. In this regard, different layers combination represents a unique input dataset. Four different datasets are employed in this study from integrating geological data with various multi-spectral remote sensing data. In addition to the geological predictor maps in each dataset, predictor maps processed from data of a specific sensor are appended. Therefore, dataset-1, dataset-2, and dataset-3 are formed by Landsat-8, Sentinel-2, and ASTER data, respectively, while the fourth dataset is composed of synergy from the three datasets.

3.4.1. Geological-Based Predictor Maps

According to the primer understanding of gold mineralization controlling factors, and geological data availability as well, we produced four geological-based predictor maps by using GIS spatial analysis methods. Identifying permissive lithologies, structures, and hydrothermal alteration zones is the main criterion of exploration. From prior literature about the Red Sea Hills, it is well known that mineralization zones have the same linear structures and exist in the acid meta-volcanic rocks [42,44,56]. Faults/fractures are favored channels for fluid migration, which represent the main ore-controlling factor in shear zone-related gold deposits. Therefore, two maps of distance to NE- and NW- faults were generated by using the Euclidean distance method (Figure 4a,b). The contact zone of the intrusive rocks (older and younger) lies in meta-sediments and metavolcanics, which may indicate the spatial agreement with gold mineralization in Hamissana area. Moreover, younger intrusions in the study area are highly sheared and contain several dykes. Hence, the proximity to outcropped intrusions was employed as a predictor map (Figure 4c).

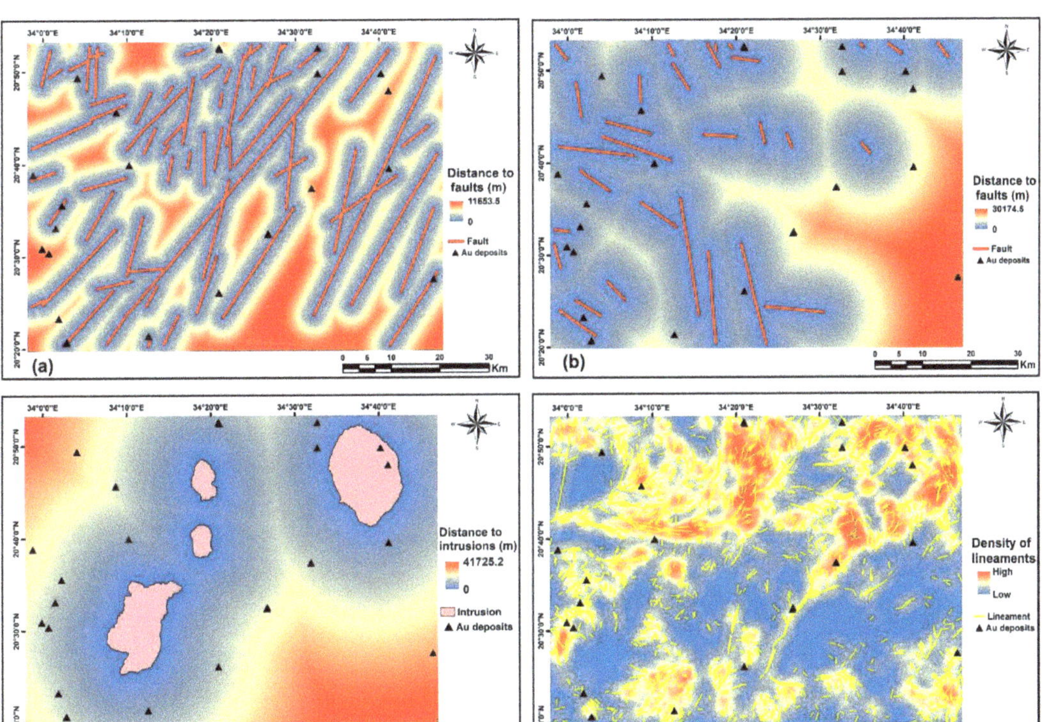

Figure 4. Geological evidential features of the study area used as predictor variables: (**a**) Proximity to NE-SW faults; (**b**) proximity to NW-SE faults; (**c**) proximity to intrusions; (**d**) density of lineaments.

Since the valleys and drainage in the study area are structurally controlled by the shear zone, we automatically extracted lineaments as an indication of structural weakness, faults, fractures, or lines that separate different formations [57,58]. In mineral exploration, excessive lineaments are often localized close to mineralogical deposits, which may correspond to the main conduits for carrying hydrothermal solutions [15,25,58,59]. Therefore, these lineaments are adequate to be an indirect indicator of mining potential. Sentinel-2 has a higher spatial resolution than Landsat-8 and ASTER, which makes it more suitable for lineament extraction. Prior literature reported that Principal Component Analysis (PCA) has a better capability for automated lineament extraction compared with the original remote sensing data and other enhancement techniques [58,60]. Using PCI Geomatica software, lineaments were automatically extracted from PC5 of Sentinel-2. (Figure 4d) shows the concentration of lineaments distribution as a density map, which was employed as the fourth geological-based predictor map.

3.4.2. Remote Sensing-Based Predictor Maps

Remote sensing data provides significant information about different geological objects, such as mineral assemblages, lithological units, and hydrothermal alteration zones. Studying the existence of different alteration zones was another exploration key criterion since economic mineralization is often associated with these alteration zones. Multispectral data such as Landsat-8, Sentinel-2, and ASTER can be utilized to detect surface alteration zones using various remote sensing enhancement techniques. The main objective of these processing techniques is to interpret the remote sensing spectral signature of different alteration zones (Argillic, Phyllic, Propylitic) or minerals that are associated with hydrothermal alteration (iron oxides, clay, and hydroxyl bearing minerals). To generate different thematic layers of different alteration zones, this study employs different enhancement technique methods, such as Band Ratio (BR), PCA, and Minimum Noise Fraction (MNF).

BR is one of the most applicable techniques which aims to reduce the shadow effects of topography [15,61,62]. This method improves the spectral characteristic of specific alteration minerals (e.g., iron oxide, alunite, kaolinite, or chlorite) or alteration zones by dividing the digital number (DN) value of one band by the DN of another band [27,39,63]. On the other hand, Relative Absorption Band Depth (RBD) is another method that attempts to detect the typical absorption of targeted minerals, but it uses three bands to formalize the ratio (the sum of two bands is divided by the absorption band) instead of two bands [39]. Since ASTER sensor was developed particularly for geological investigations, several mineralogical indices were developed using bands in SWIR and TIR regions [64–66]. Table 2 lists all selected BR, RBD, and mineralogical indices, which were suggested by previous studies [15,39,61,66–70] to map targeted alteration minerals and zones. It is important to point out that the BR image for mapping ferric oxide was excluded in the case of Landsat-8 and ASTER datasets because the range of the data values (histogram width) of the generated imagery is very low, which may affect the output of the MLA models.

Table 2. Selected BR, RBD, and mineralogical indices of each sensor to map targeted minerals.

Method	Target	Landsat-8	Sentinel-2	ASTER
BR	Hydroxyl-bearing	6/7	11/12	4/6
	Ferric iron	4/2	4/3	2/1
	Ferrous iron	(7/5) + (3/4)	(12/8a) + (3/4)	(5/3) + (1/2)
	Ferric oxide	6/5 (Excluded)	11/8a	4/3 (Excluded)
	Alunite	-	-	4/5
	Calcite	-	-	4/7
RBD	Argillic (RBD1)	-	-	(4 + 6)/5
	Phyllic (RBD2)	-	-	(5 + 7)/6
	Propylitic (RBD3)	-	-	(6 + 9)/(7 + 8)
Mineralogical Indices	Hydroxyl-bearing (OHI)	-	-	(7/6) * (4/6)
	Kaolinite (KLI)	-	-	(4/5) * (8/6)
	Alunite (ALI)	-	-	(7/5) * (7/8)
	Calcite (CLI)	-	-	(6/8) * (9/8)

"-" represents that there is no mathematical formula for the specific satellite data to map targeted mineral.

PCA and MNF are transformation methods, which have been successfully utilized to enhance remote sensing imagery. Both statistical methods are employed for spectral data reduction by transforming the information in the original remote sensing data into a new set of data. In the PCA procedure, the new dataset (PC components) has less variance, since each component is extracted based on uncorrelated linear combinations of values (also called eigenvector loadings). These eigenvectors are calculated in a matrix called covariance matrix (Eigen matrix), which comes across the statistical relation between all the PCs. On the other hand, MNF method also uses the covariance matrix to rescale and segregate noise in the data. In the new dataset, the noise is reduced and whitened in a descending way based on the eigenvalue of each MNF component.

Since the eigenvector loadings (sign and magnitude) are linked to the spectral feature (absorption and reflectance) of objects, they can be utilized to detect the existence of a specific alteration mineral. For this purpose, the selective PCA technique (also known as

Crosta technique) was developed to extract features of the specific object as bright or dark pixels in the PCs. This method is applied to VNIR+SWIR bands, where bands are selected (mostly 3 or 4 bands) based on the prior knowledge of the spectral behavior of an alteration mineral. One of the PCs will have two strong loadings with opposite signs that indicate the reflection and absorption bands of that alteration mineral. If the loading has a positive sign in the reflection band, the PC enhances the targeted mineral in bright pixels. In the meantime, this PC could also enhance that mineral in dark pixels, if the sign is negative in the reflectance band [29,39,46,71]. In this study, all selected bands from different sensor data to map different hydrothermal alteration zones and minerals, are illustrated in Table 3.

Table 3. Selected bands of each sensor's data to perform PCA transformation for mapping defined targets.

Dataset	Target	Selected Bands
Landsat-8	Hydroxyl-bearing	2, 5, 6 and 7
	Iron oxides	2, 4, 5 and 6
Sentinel2	Hydroxyl-bearing	2, 8a, 11, and 12
	Iron oxides	2, 4, 8a, and 11
ASTER	Hydroxyl-bearing	1, 3, 4 and 6
	Iron oxides	1, 2, 3 and 4
	Argillic	1, 4, 6 and 7
	Phyllic	1, 3, 5 and 6
	Propylitic	1, 3, 5 and 8

Unlike PCA method, MNF technique is less interpretable and very subjective. MNF results are only statistics and do not indicate specific mineral occurrences. However, separating and rescaling the noise process helps MNF to identify differences inside the image in the first few bands, while the latest few bands subsequently convey more noise [24,72]. Therefore, we visually assessed all MNF bands in each dataset, then for each dataset (Landsat-8, Sentinel-2, ASTER), we carefully selected three MNF that have a spatial agreement with different hydrothermal alteration minerals.

3.4.3. Data Preparation

At this stage, different predictor maps are generated from multisource data, so the numeric range of each input data is different. This variance in the range gives a chance for more domination to those inputs with a greater range than those with a smaller one. This issue directly affects the outputs of RF and brings numerical obstacles during the models' execution [73]. In this regard, each input was normalized in the range of [0, 1] using the following equation:

$$x_{norm} = \frac{x - x_{min}}{x_{max} - x_{min}} \tag{4}$$

where x is the input data, x_{max} and x_{min} donate to the maximum and minimum values of the original data respectively. After normalizing each predictor map, they were stacked to form four distinct datasets as shown in Table 4.

Table 4. Input layers of each dataset to conduct data-driven MPM.

Dataset	Remote Sensing-Based			Geological-Based	No. of All Input Layers
	BR	PCA	MNF		
Landsat-8 (Dataset-1)	3	2	3	4	12
Sentinel-2 (Dataset-2)	4	2	3	4	13
ASTER (Dataset-3)	12	5	3	4	24
Data integration (Dataset-4)	19	9	9	4	41

3.5. Target Variable

The target binary variables, corresponding to the gold occurrence and non-occurrence location, are used to train and validate the performance of supervised predictive models. A set of 25 occurrence locations are given a score of 1. In the meantime, the non-occurrence locations corresponding to the score of 0, were selected based on prespecified criteria. The selection of non-occurrence samples was achieved according to (i) a clustering procedure similar to the one proposed by Torppa [74]; (ii) several other criteria that were defined in previous literature [2,3,6,35]. Unsupervised classification (clustering) is utilized to describe the spatial distribution of gold occurrence using several clusters. By classifying these clusters using known occurrences, we can delineate geologically similar areas of occurrence and non-occurrence. In this study, we employed k-means as a clustering method to generate some clusters that do not exceed the number of known occurrences. Hence, 20 clusters were produced by applying this method to ASTER dataset, since ASTER dataset has more input layers than those in Landsat-8 and Sentinel-2. Thereafter, we divided those clusters into six prsopectivity classes (very high, high, moderate, low, very low, non-occurrence), by visually counting the frequency of known occurrence in each cluster. Non-occurrence samples were then selected from low, very low, or non-occurrence classes according to the following criteria:

1. The number of non-occurrence samples must be equal to the number of mineral occurrences.
2. Non-occurrence samples should be spatially distributed randomly.
3. The selection of non-occurrence locations should be distal from any known gold occurrences. Here, we applied a 10 km buffer zone around known occurrences.

By following these requirements, we generated a full set of target variables, which contains 50 points of samples. Furthermore, we randomly split these variables into training and testing datasets. 70% of target variables were assigned to the training dataset (35 points), and the remaining 30% were employed as the testing dataset.

3.6. Model Assessment

The performance of the trained RF predictive model was comprehensively assessed by various statistical measurements, including the prediction and classification performance. Classification, here, means labeling the floating value (0, 1) at each cell as prospective or non-prospective (barren region) by using a 0.5 threshold value. A confusion matrix can be successfully utilized to evaluate and explain the classification performance of predictive models using the following categories: (i) true-positive "TP", when there is an agreement between the model and the reality about mineral occurrence; (ii) true-negative "TN", when there is an agreement between the model and reality about mineral non-occurrence; (iii) false-positive "FP" when the model incorrectly classified a non-occurrence sample into mineral occurrence; and (iv) false-negative "FN", when the model mistakenly classified a mineral occurrence as non-occurrence [2,37,75]. These four situations are used to calculate six statistical metrics, namely overall accuracy (OA), sensitivity, specificity, positive predictive value (*PPV*), negative predictive value (*NPV*), and Kappa [76,77]. These statistical matrics can be formalized as follow [3,78]:

$$\text{Sensitivity} = \frac{TP}{TP + FN} \tag{5}$$

$$\text{Specificity} = \frac{TN}{TN + FP} \tag{6}$$

$$PPV = \frac{TP}{TP + FP} \tag{7}$$

$$NPV = \frac{TN}{TN + FN} \tag{8}$$

$$OA = \frac{TP + TN}{TP + TN + FP + FN} \tag{9}$$

$$K = \frac{TP + TN - [(TP + FN)(TP + FP) + (FP + TN)(FN + TN)]/(TP + FP + TN + FN)}{(TP + FP + TN + FN) - [(TP + FN)(TP + FP) + (FP + TN)(FN + TN)]/(TP + FP + TN + FN)} \quad (10)$$

Furthermore, the overall predictive performances of different datasets were compared using the success-rate curve and receiver operating characteristic (ROC) curve [4,7]. The success-rate curve was created by plotting the percentage of correctly classified gold (true positive rate "TPR") against the area percentage of prospective regions that are generated by reclassifying MPM using moving threshold values [79]. Subsequently, the optimal goal of the model is to capture as many mineral occurrences as possible in the smallest possible prospective area. This method is very useful in delineating or classifying different prospective regions (high, moderate, low), by identifying the change in curve slope. Since the success-rate curve only depends on the TPR, the ROC curve was created to also consider the false positive rate (FPR). In the ROC curve, TPR and FPR are plotted against each other on the y-axis and x-axis, respectively. In addition, the predictive performance can be measured by calculating the area under ROC curve (AUC), where the better model performance is indicated by how closer the curve can be to the upper left corner [7,80,81].

4. Results

4.1. Generating Remote Sensing-Based Predictor Maps

Figure 5a–c illustrate hydroxyl-bearing minerals derived from BR 6/7 of Landsat-8, BR 11/12 of Sentinel-2, and BR 4/6 of ASTER, respectively. The distribution of these minerals (Al-OH and Fe, Mg-OH) is shown as cyan pixels. As seen in the figure, the spatial distribution of hydroxyl-bearing minerals is similar in all three images. However, Landsat-8 and Sentinel-2 ratio images show these minerals in association with drainage channels. On the other hand, ASTER BR image extensively mapped these minerals in the southeastern part. Another method employed to map hydroxyl-bearing minerals is OH bearing altered minerals index (OHI = 6/7 * 4/6). As displayed in Figure 6a, the spatial distribution of these minerals is relatively similar to ASTER BR image. OH-bearing altered minerals are concentrated in the metavolcanic rocks and have a spatial agreement with known gold occurrence.

Iron minerals, including ferrous iron Fe+2 and ferric iron Fe+3, are shown in Figure 5d-i. The light orange color in Figure 5d–f depicts ferrous iron minerals, which were produced by using BRs (7/5) + (3/4) of Landsat-8, BRs (12/8a) + (3/4) of Sentinel-2, and BRs (5/3) + (1/2) of ASTER. The distribution of ferrous iron minerals in the three images is concentrated in the northeastern part. However, it can be seen in other parts of ASTER image, while it almost disappeared in the western part of Landsat-8 image. On the other hand, Figure 5g–i show ferric iron minerals as dark orange pixels, derived from BR 4/2 of Landsat-8, BR 4/3 of Sentinel-2, and BR 2/1 of ASTER, respectively. Unlike the spatial distribution of ferrous minerals, ferric iron minerals are significantly detected in the drainage areas around younger intrusion in the northeast and the lower middle parts of the three BR images. This distribution of these minerals has less association with documented gold occurrences compared with ferrous iron minerals. Moreover, another BR using Sentinel-2 data was used to map ferric oxide minerals, which is 11/8a. In this imagery, iron oxide minerals are illustrated by red pixels in Figure 5j. By using this ratio image, ferric oxide minerals were mapped in a very extensive way that cover most of the outcrops, including younger and older intrusions, metasediments, and metavolcanics rock units. This distribution relatively matches the density of lineaments features.

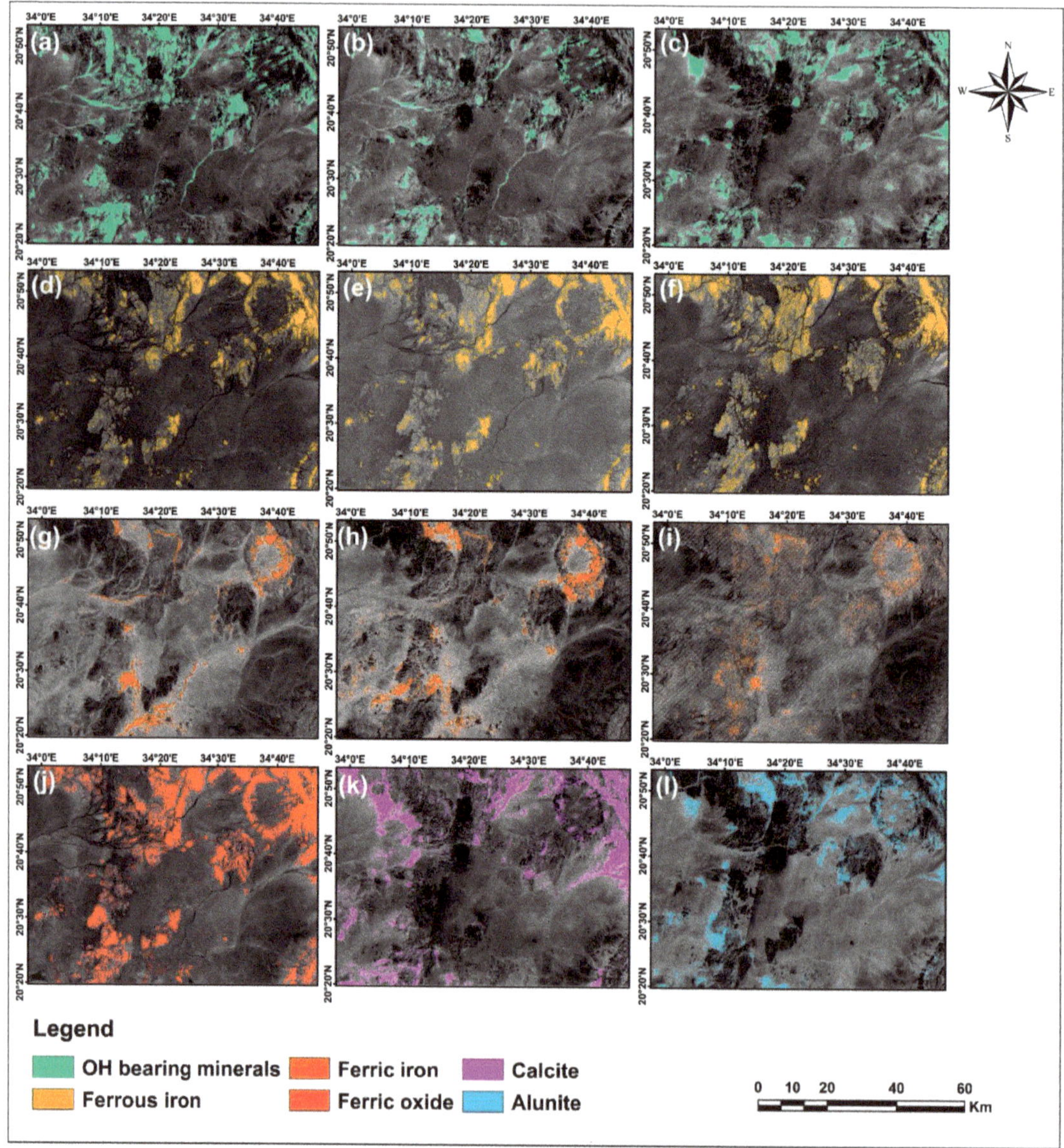

Figure 5. BR images derived from different sensors showing various targeted minerals in colored pixels (**a–c**) Hydroxyl-bearing minerals derived from Landsat-8 (6/7), Sentinel-2 (11/12), and ASTER (4/6), respectively; (**d–f**) Ferrous iron minerals derived from Landsat-8 (7/5 + 3/4), Sentinel-2 (12/8a + 3/4), and ASTER (5/3 + 1/2), respectively; (**g–i**) Ferric iron minerals derived from Landsat-8 (4/2), Sentinel-2 (4/3), and ASTER (2/1), respectively; (**j**) Ferric oxide minerals derived from Sentinel-2 (11/8a); (**k,l**) Calcite and Alunite minerals derived from ASTER BR 4/7 and 4/5, respectively.

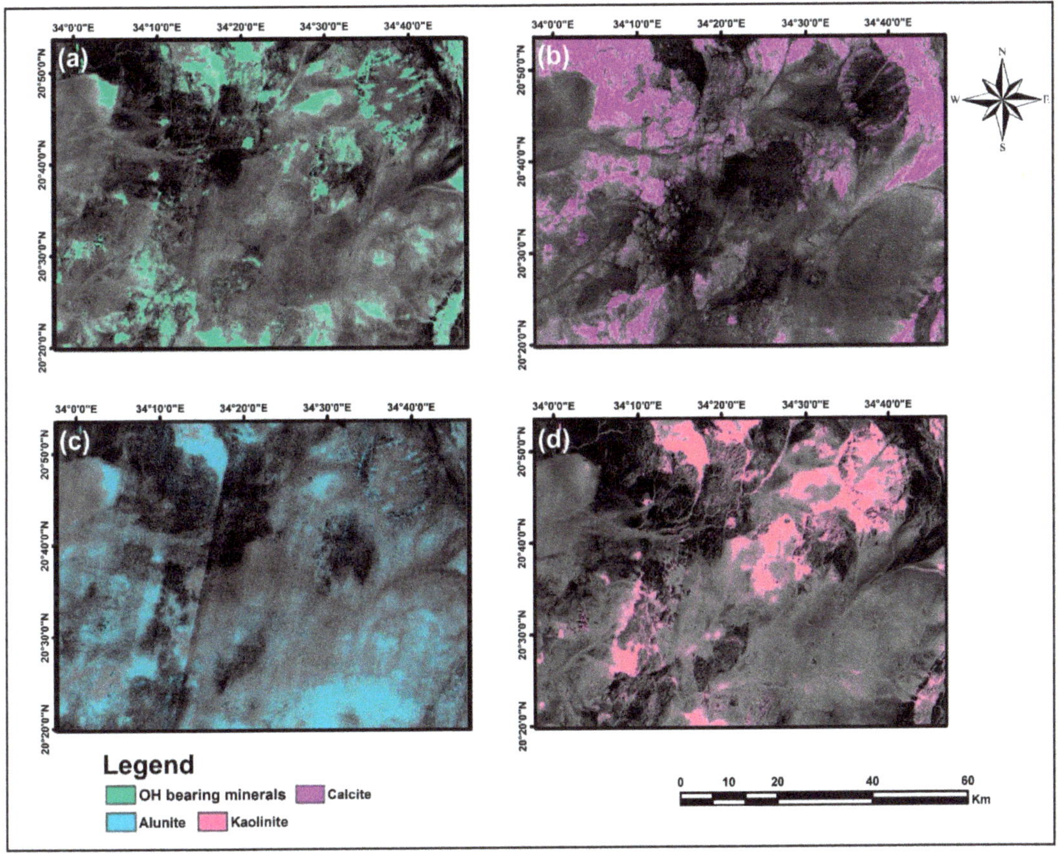

Figure 6. The mineralogical indices images derived from ASTER SWIR bands show the spatial distribution of targeted minerals in colored pixels: (**a**) OHI; (**b**) CLI; (**c**) ALI; (**d**) KLI.

In order to delineate minerals that indicate the existence of specific alteration zone, further BR and mineralogical indices were employed in the present study using ASTER data. Calcite, indicating propylitic alteration, was derived from BR 4/7 (Figure 5k) and cal-cite mineral index (CLI = 6/8 * 9/8) (Figure 6b). The prominent areas of calcite were marked as purple color in both images. The distribution of calcite is almost similar in both produced images, but it is more outspread in the northeastern part of the CLI image than the BR image. Argillic or advanced argillic alteration zone is characterized by the existence of kaolinite and alunite minerals. BR 4/5 and alunite mineral index (ALI = 7/5 * 7/8) were utilized to detect alunite altered mineral, while kaolinite mineral index (KLI = 4/5 * 8/6) was utilized to detect kaolinite. The identification of the alunite mineral in the BR image (Figure 5l) exhibits mineral distribution pattern different from the ALI image (Figure 6c). As shown in the images, the BR image mapped alunite similarly to the 4/6 ratio image (OH-bearing minerals), but it has a lower surface abundance. On the other hand, areas of alunite in the ALI image are highlighted in sky-blue tone in the drainage area and superficial deposits. Kaolinite mineral also coincides with drainage areas, but it is more concentrated in the northern part.

Three ASTER RBD images were specifically used for the detailed mapping of alteration zones (Figure 7). RBD1 (4+6/5), RBD2 (5+7/6), and RBD3 (6+9/7 + 8) were used to obtain argillic, phyllic, and propylitic alteration zone, respectively. The argillic alteration zone is illustrated by red pixels, which is more concentrated in the northern part around younger and older intrusions, meanwhile, it can also be seen in the southwestern part between younger intrusions and metavolcanics. The phyllic alteration zone is typically concentrated in the metavolcanic rock unit and partially scattered in the younger intrusion unit. The output of the third RBD, indicating the propylitic alteration zone, is similar to the image derived from CLI index (see Figure 6b).

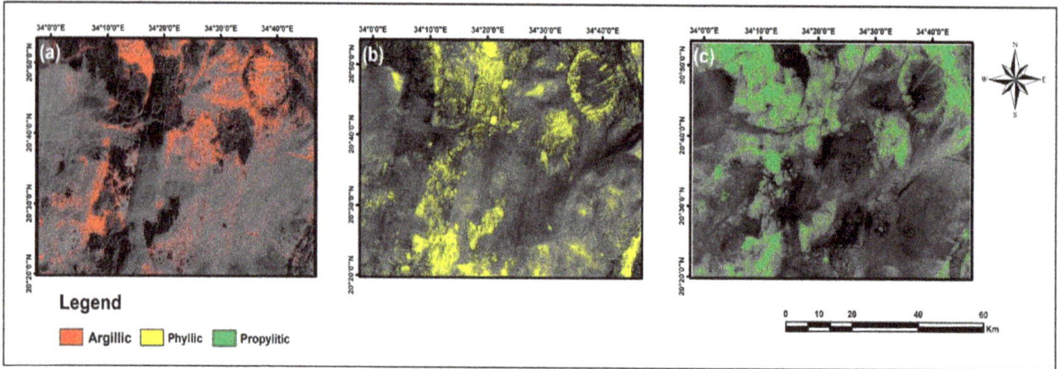

Figure 7. The RBD images derived from ASTER SWIR bands show the spatial distribution of different alteration zones in colored pixels: (**a**) RBD1; (**b**) RBD2; (**c**) RBD3.

Figure 8a–c show hydroxyl-bearing minerals derived from PCA using selective bands of Landsat-8, Sentinel-2, and ASTER, respectively. The eigenvector loadings corresponding to bands 2,5,6 and 7 of Landsat-8, bands 2,8a,11, and 12 of Sentinel-2, and bands 1,3,4, and 6 of ASTER, are listed in Table 5. After careful scanning of the eigenvectors, PCA derived from the selected bands of Landsat-8 shows a unique contribution of OH-bearing minerals, which corresponds to reflection in band 6 and absorption in band 7. This PCA has strong negative loading in the reflectance band (-0.7) followed by a strong positive loading in the absorption band (0.633). Hence, PCA 4 mapped OH-bearing minerals as dark pixels, due to the negative loading in the reflectance band. Thereafter, we inverted the dark pixels to bright ones by multiplying the image by -1. Similarly, PCA results using Sentinel-2 and ASTER data exhibit similar patterns for mapping OH-bearing minerals, since PCA4 in both datasets contains unique eigenvectors that correspond to the spectral feature. PCA4 of Sentinel-2 has strong loading in band 11 (-0.675) and band 12 (0.609), while the strong loadings in ASTER data correspond to band 4 (-0.595) and band 6 (0.692). both PCA 4 images of Sentinel-2 and ASTER were negated to display hydroxyl-bearing in bright pixels.

Moreover, PCA was applied on Landsat-8 bands 2,4,5, and 6, Sentinel-2 bands 2,4,8a, and 11, and ASTER bands 1,2,3, and 4, for mapping iron oxide minerals. Exploring eigenvector loadings displayed in Table 6 reveals that PCA2 in all three datasets has unique loadings corresponding to the spectral feature of iron oxide minerals. these PCAs showed moderate loadings with a positive sign in absorption bands (Landsat-8 B2 and B5, Sentinel-2 B2 and B8a, and ASTER B1 and B3), and strong loadings with a negative sign in the reflectance band (Landsat-8 B6, Sentinel-2 B11, and ASTER B4). The three PCA2 images were transferred to ArcGIS software and negated. Then, the pixels representing the iron oxide minerals were changed to orange color (Figure 8d–f). It is quite noticeable that the surface abundance of these minerals is lower compared to images derived from BR. Nevertheless, these minerals in PCA images are more distributed in northeast parts and have spatial agreement with hydroxyl-bearing minerals (see Figure 8a–f).

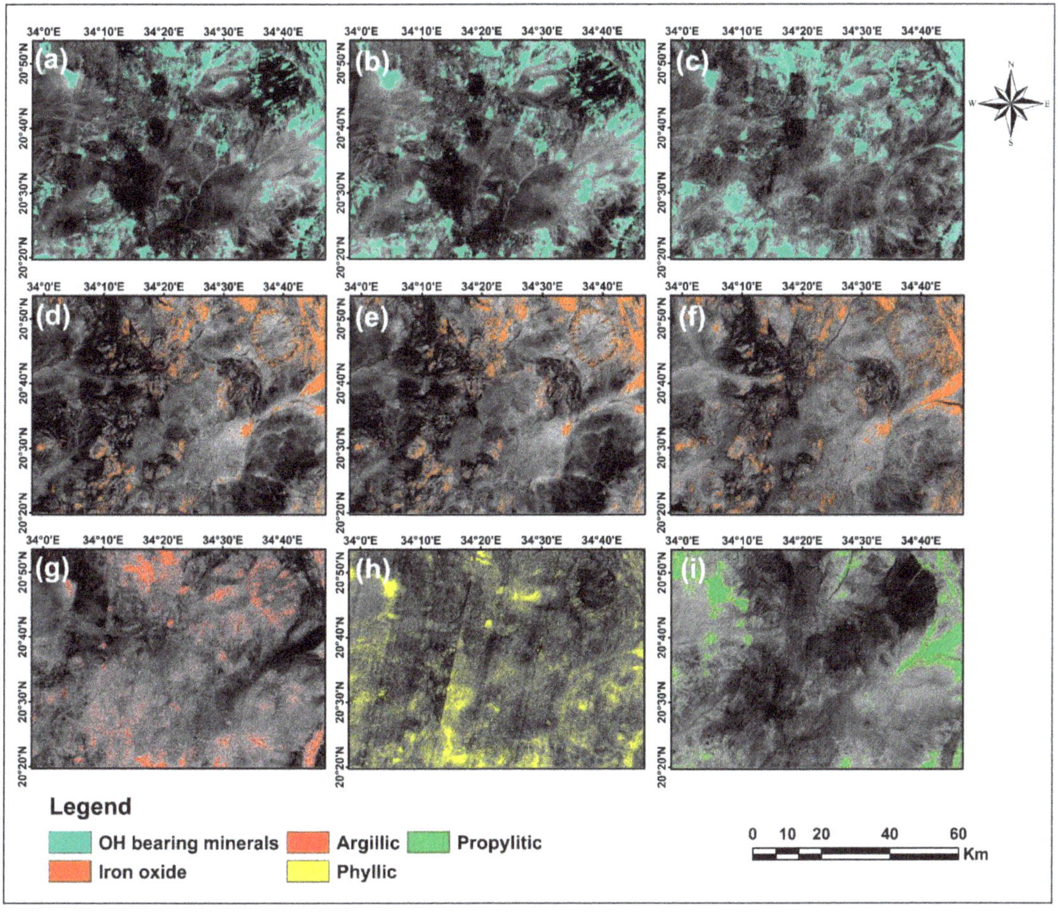

Figure 8. The PCA images derived from different sensors based on band selection in (Table 2). (**a**–**c**) PCA4 of Landsat-8, Sentinel-2, and ASTER, respectively, for mapping hydroxyl-bearing minerals. (**d**–**f**) PCA of Landsat-8, Sentinel-2 and ASTER, respectively, for mapping iron oxide minerals. (**g**–**i**) ASTER PCA for mapping argillic (PCA4), phyllic (PCA4), and propylitic (PCA3), respectively.

For more details about argillic, phyllic, and propylitic alteration zones, PCA method was also applied specifically to ASTER data. Table 7 shows the eigenvector loadings for argillic using bands 1, 4, 6, and 7, phyllic using bands 1, 3, 5, and 6, and propylitic using bands 1, 3, 5, and 8. The argillic zone has reflectance spectral features in bands 1 and 6, and absorption ones in bands 4 and 7 [29]. Subsequently, most loadings that correspond to this typical spectral feature are found in PCA 4, although the loadings are weaker in band 1 (−0.025) and band 4 (0.133) compared with band 6 (−0.767) and band 7 (0.631). PCA4 is selected to map phyllic alteration zone since it shows loadings with opposite signs in band 5 (−0.694) and band 6 (0.707). Thus, this loadings pattern corresponds to the assumption that band 5 could be considered as a reflection band since the absorption of muscovite mineral (typical mineral reveals phyllic alteration) is lower in band 5 than in band 6 [39]. Then PCA4 image of phyllic alteration was negated because band 5 has negative loading. PCA3 loadings in the eigenvector matrix of propylitic alteration, correspond to the calcite spectral properties. This PCA shows strong negative loading in band 5 (−0.722) and strong positive loading in band 8 (0.563). In this case band 5 was treated as a reflectance band because Fe, Mg-oh group has a lower absorption feature in band 5 compared with the strong absorption in band 8. Therefore, this PCA imagery was also negated. As seen in Figure 8g–i, PCA images are significantly different from those derived by RBD method. The surface abundance of the PCA images is much lower and the spatial distribution is almost different than RBD images.

Table 5. The eigenvector matrixes of PCA results for mapping hydroxyl-bearing minerals using different remote sensing data; Bold text represents the selected PCA and the unique eigenvalues.

Landsat-8	Eigenvector	Band 2	Band 5	Band 6	Band 7
	PCA1	0.248	0.552	0.591	0.534
	PCA2	0.468	0.646	−0.367	−0.479
	PCA3	0.816	−0.470	−0.166	0.291
	PCA4	**−0.230**	**0.239**	**−0.700**	**0.633**
Sentinel-2	Eigenvector	Band 2	Band 8a	Band 11	Band 12
	PCA1	0.215	0.557	0.606	0.526
	PCA2	0.402	0.688	−0.346	−0.495
	PCA3	0.834	−0.372	−0.240	0.329
	PCA4	**−0.310**	**0.280**	**−0.675**	**0.609**
ASTER	Eigenvector	Band 1	Band 3	Band 4	Band 6
	PCA1	0.399	0.576	0.536	0.470
	PCA2	0.497	0.517	−0.498	−0.488
	PCA3	0.695	−0.586	0.332	−0.251
	PCA4	**0.332**	**−0.240**	**−0.595**	**0.692**

Table 6. The eigenvector matrixes of PCA results for mapping iron oxide minerals using different remote sensing data; Bold text represents the selected PCA and the unique eigenvalues.

Landsat-8	Eigenvector	Band 2	Band 4	Band 5	Band 6
	PCA1	0.251	0.533	0.556	0.587
	PCA2	**0.328**	**0.460**	**0.243**	**−0.789**
	PCA3	0.829	−0.020	−0.532	0.169
	PCA4	−0.377	0.710	−0.590	0.075
Sentinel-2	Eigenvector	Band 2	Band 4	Band 8a	Band 11
	PCA1	0.221	0.512	0.566	0.607
	PCA2	**0.306**	**0.513**	**0.239**	**−0.766**
	PCA3	0.633	0.257	−0.700	0.206
	PCA4	0.676	−0.640	0.363	−0.045
ASTER	Eigenvector	Band 1	Band 2	Band 3	Band 4
	PCA1	0.388	0.530	0.557	0.508
	PCA2	**0.309**	**0.349**	**0.232**	**−0.854**
	PCA3	−0.778	0.020	0.618	−0.106
	PCA4	0.385	−0.773	0.503	−0.040

MNF method was employed to extract further information about alteration minerals and zones in the study area. After careful screening of the features presented in dark and bright colors in each MNF band. Three bands were selected and displayed in RGB colors. MNF 3, 4, and 5 of Landsat-8, and MNF 2, 3, and 4 of Sentinel-2 were selected. It can be seen in Figure 9a,b that altered rocks are presented as white to sky-blue tones. The white color demonstrates that there is important information in all three bands that were assigned to RGB colors. Combining negated MNF2, MNF3, and MNF4 clearly displays areas of alteration in white to yellow color (Figure 9).

Table 7. The eigenvector matrixes of applying PCA to ASTER selected bands for detailed mapping of argillic, phyllic, and propylitic alteration zones; Bold text represents the selected PCA and the unique eigenvalues.

Argillic	Eigenvector	Band 1	Band 4	Band 6	Band 7
	PCA1	0.409	0.565	0.496	0.518
	PCA2	0.910	−0.202	−0.263	−0.247
	PCA3	0.056	−0.792	0.311	0.522
	PCA4	−0.025	0.113	**−0.767**	**0.631**
Phyllic	Eigenvector	Band 1	Band 3	Band 5	Band 6
	PCA1	0.414	0.597	0.486	0.487
	PCA2	0.474	0.503	−0.510	−0.512
	PCA3	0.770	−0.619	0.148	−0.043
	PCA4	0.106	−0.084	**−0.694**	**0.707**
Propylitic	Eigenvector	Band 1	Band 3	Band 5	Band 8
	PCA1	0.403	0.584	0.474	0.522
	PCA2	0.525	0.483	−0.444	−0.542
	PCA3	**−0.287**	0.281	**−0.722**	**0.563**
	PCA4	0.693	−0.589	−0.239	0.340

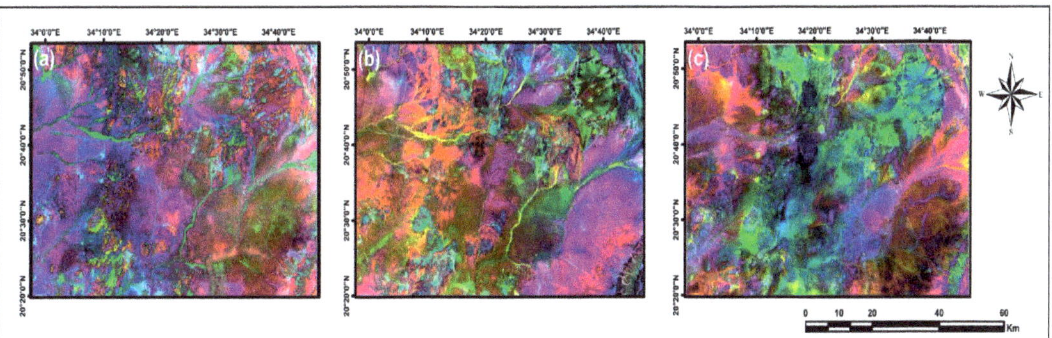

Figure 9. MNF results: (**a**) Landsat-8 MNF3, MNF4, and MNF5 in RGB.; (**b**) Sentinel-2 MNF2, MNF3, and MNF4 in RGB.; (**c**) ASTER MNF2 (negated), MNF3, and MNF 4 in RGB.

4.2. Generating Target Variable Using K-Means Clustering

All evidence layers of ASTER dataset that were mentioned earlier, including geological predictor maps, were utilized to classify the study area using k-means clustering. The purpose of this unsupervised method is to delineate non-prospective tracts, which then aids the process of selecting non-occurrence samples. Defining the proper number of clusters is the most critical step because these clusters will be assigned to different classes based on their spatial agreement with known gold occurrences. Each class prospectivity score is defined according to the percentage of captured deposits in the clusters. For instance, if each of the n clusters captured x deposits, then these n clusters will be classified as one class and the class prospectivity score is determined by the percentage of x deposits from the total known deposits. Therefore, increasing cluster numbers increases the number of clusters with no occurrence's association, which subsequently increases the area of non-prospective. Herein, we proposed that the number of clusters must be equal to or less than the number of known occurrences (k ≤ Au samples). In this case, the worst scenario will be if the frequency of occurrence in each cluster is one, which indicates that the k-means calculation process failed to find a connection between occurrences distribution and evidential layers.

Figure 10a shows the twenty clusters derived from applying k-means on ASTER dataset. As displayed in Figure 10b, the highest frequency is found to be 5 samples per cluster, which takes place in clusters 11 and 12. These two clusters were then classified as the very high prospective class. Clusters 14 and 8 captured four and three Au occurrence samples, respectively, so they were labeled as high and moderate classes. Each one of clusters 7,13, and 15 captured 2 occurrences, which were afterward combined to form the low prospective class. In the meantime, the pattern of one occurrence per cluster is found in clusters 16 and 20, which were defined as the very low prospective class. Finally, the rest of the clusters were classified as the non-prospective class, since none of the Au occurrences is spatially associated with these clusters. Figure 10c illustrates the selection of 25 non-occurrence points following the results of classifying k-means outputs, as well as the criteria described earlier in the methodology section.

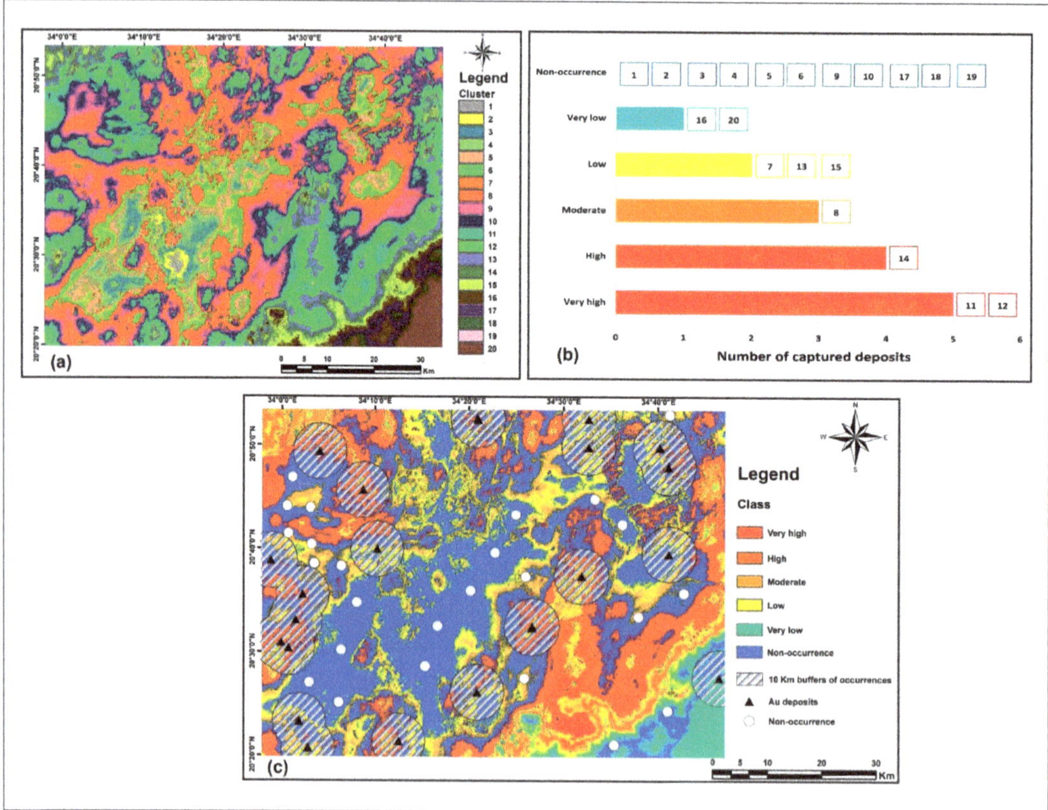

Figure 10. (**a**) K-means 20 clusters derived from ASTER and geological evidential layers.; (**b**) the number of Au occurrences spatially associated with different clusters.; (**c**) classified prospectivity clusters based on the frequency of the occurrences in each cluster obtained from (**b**).

4.3. Sensitivity of RF Predictive Model to Parameter Tuning

The success key for training data-driven models with higher accuracy prediction is the configuration of parameters (also called parametrization). Thus, due to its great impact on the robustness and generalization capacity of ML predictive models. The parameterization process was achieved using the GridSearchCV method based on 5-fold cross-validation. Figure 11 shows significant variations in MSE values of four RF models obtained from different parameter combinations and different datasets. Generally, RF is a very stable model since the higher MSE values are lower than 0.138 in all four datasets. Although there are complex variations of parameter selection using different datasets, the minimum score of MSE is very promising in the case of ASTER and data-integration datasets. The minimum score of MSE obtained by training ASTER and data-integration datasets were 0.096 and 0.093, respectively. Meanwhile, RF model had less accuracy in the case of using Sentinel-2 and Landsat-8, reaching minimum MSE values of 0.102 and 0.12, respectively. The results of MSE indicate that the complex architecture of RF does not lead to an accurate performance in different cases. For instance, the grid searching method selected only two features to be used in individual trees in both Landsat-8 and Sentinel-2 datasets. It is also suggested that the number of trees in the forest was set to 50 in training Sentinel-2 and data-integration datasets. The highest number of trees grown in the forest was 300 trees in the case of Landsat-8, while the highest number of features was 8 in the case of data-integration dataset.

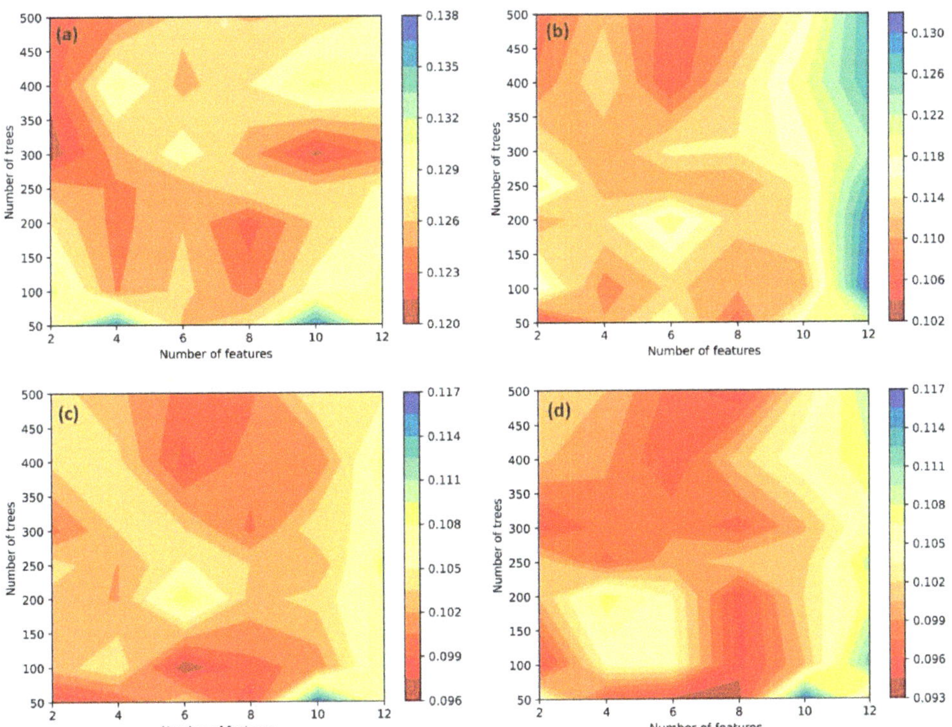

Figure 11. Contour maps showing the sensitivity of RF model based on MSE results obtained by training different datasets; the number of trees and number of features employed for training RF models based on: (**a**) Landsat-8; (**b**) Sentinel-2; (**c**) ASTER; (**d**) data integration.

4.4. Comparison of Various Datasets Performance

Different RF regression models were trained by the optimal parameter configurations to produce gold potential maps, where the prediction at each cell denotes the likelihood of gold occurrence by floating probability value (0–1) (Figure 12). The accuracy report of the classification performance is produced by labeling each cell into binary classes (prospective areas and barren areas), and thus by using a threshold value of 0.5 to define those areas. Table 8 lists all statistical metrics for measuring the classification performance of RF using four various datasets. In general, both ASTER and Data-integration datasets achieved an overall classification accuracy of 73.3%, which outperformed the classification of Sentinel-2 and Landsat-8 datasets. OA of Landsat-8 and Sentinel-2 were 60% and 66.7%, respectively. Although the OA of ASTER and data-integration datasets are the same, ASTER is more sensitive to correctly identifying 73.2 of the occurrence locations, while data-integration dataset has higher predictive values whether it is *PPV* or *NPV*. However, the highest predictive values (*PPV* and *NPV*) are found in Sentinel-2 dataset. RF models trained by Landsat-8 and Sentinel-2 have the worst specificity scores (28.6).

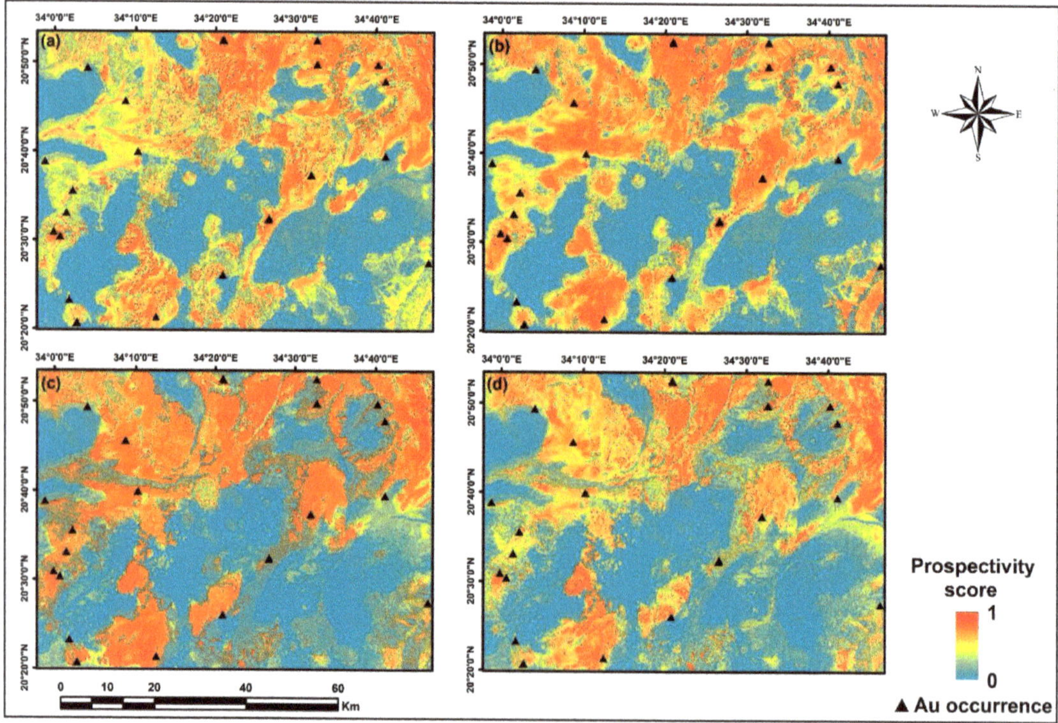

Figure 12. Predictive maps of likelihood score of gold propsectivity obtained from RF predictive modeling using: (**a**) Landsat-8; (**b**) Sentinel-2; (**c**) ASTER; and (**d**) Data-integration.

Table 8. Classification report of RF performance using different datasets.

Dataset	Sensitivity (%)	Specificity (%)	Positive Predictive Value (%)	Negative Predictive Value (%)	Accuracy (%)	Kappa
Landsat-8	58	28.6	62.5	66.6	60	0.167
Sentinel-2	64.3	28.6	80.8	100	66.7	0.299
ASTER	73.2	71.4	73.2	71.4	73.3	0.464
Data-Integration	72.3	57.1	75	80	73.3	0.454

Taking the cost of mineral exploration in the real world into counts, it is impractical to make a decision based on prospective tract delineation from the classification scenario (i.e., probability > 0.5) [3]. Therefore, it is essential to assess the predictive performance of high-probability zones using ROC curve. Figure 13 shows ROC curves and AUC values of various MLAs trained by four different input datasets. The closest ROC curve to the top left corner belongs to the data-integration dataset, whereby the AUC value is 0.938. Both ASTER and Sentinel-2 have AUC values of 0.875, which clarifies that both datasets have comparable prediction performance. Landsat-8 performs the weakest predictive capability with AUC value of 0.625.

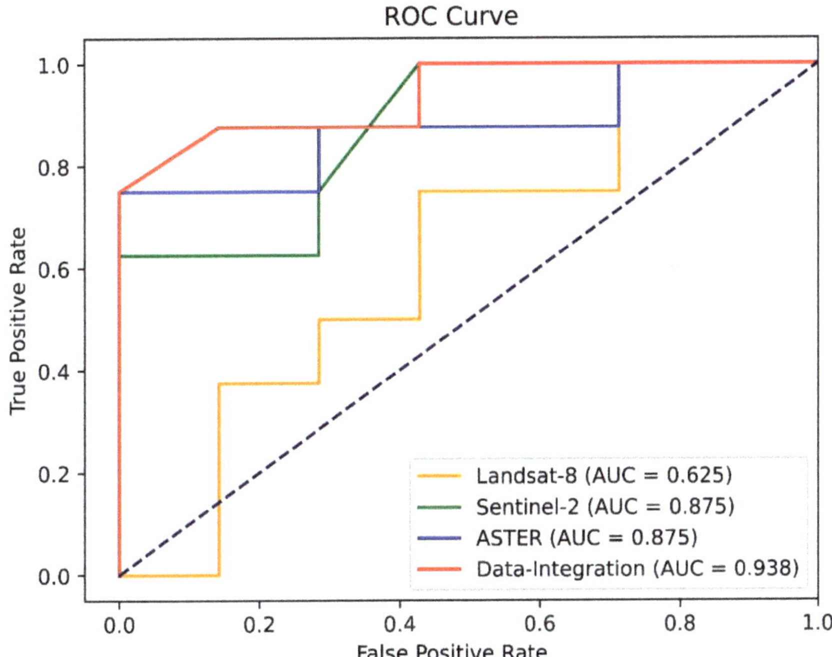

Figure 13. ROC curves showing AUC value of each RF model trained by Landsat-8, Sentinel-2, ASTER, and data integration datasets.

For a better understanding of the spatial distribution of Au deposits, and delineating exploration target areas, it is important to reclassify the MPM probability score into different levels (very high, high, moderate, and low). This can be achieved by classifying the success-rate curve based on the variations in the curve slope using four regression lines. The higher predictive region is defined by the steeper slope. Figure 14 shows the success-rate curves of four RF MPMs derived from different datasets, while the classified maps are displayed in Figure 15. The steepest curve is achieved by data-integration dataset, which indicates that this data predictive performance has the ability to define a smaller prospective area compared with datasets. The very high potential class of the data-integration identifies about 70.6% of the deposits in 11.5% of the total area. However, ASTER dataset was able to identify all of the occurrence locations in 33.3% of the total area, which is 4.5% lower than the total area that captured all occurrences in the MPM of data-integration. The total area of capturing all deposits is larger in the case of Landsat-8 and Sentinel-2, which are 47.2 and 42.1, respectively. As it is displayed in Figure 14a, the curves of ASTER and Landsat started similarly with a high angle, but they quickly become less steep by increasing the percentage of cumulative area. Hence, about 35% of the occurrence are captured in approximately 3.5% of the study area.

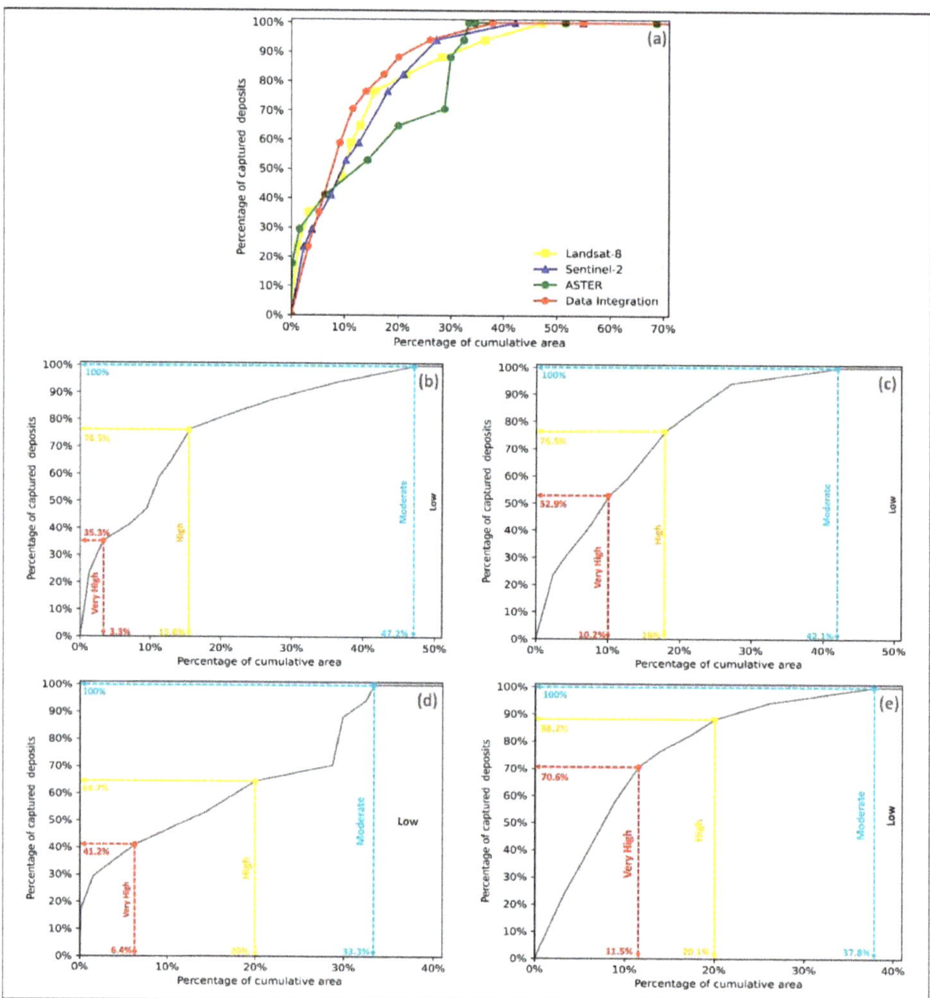

Figure 14. Success-rate curves of RF predictive maps trained using various datasets: (**a**) all success-rate curves; (**b**–**e**) curve of Landsat-8, Sentinel-2, ASTER, and data integration, respectively.

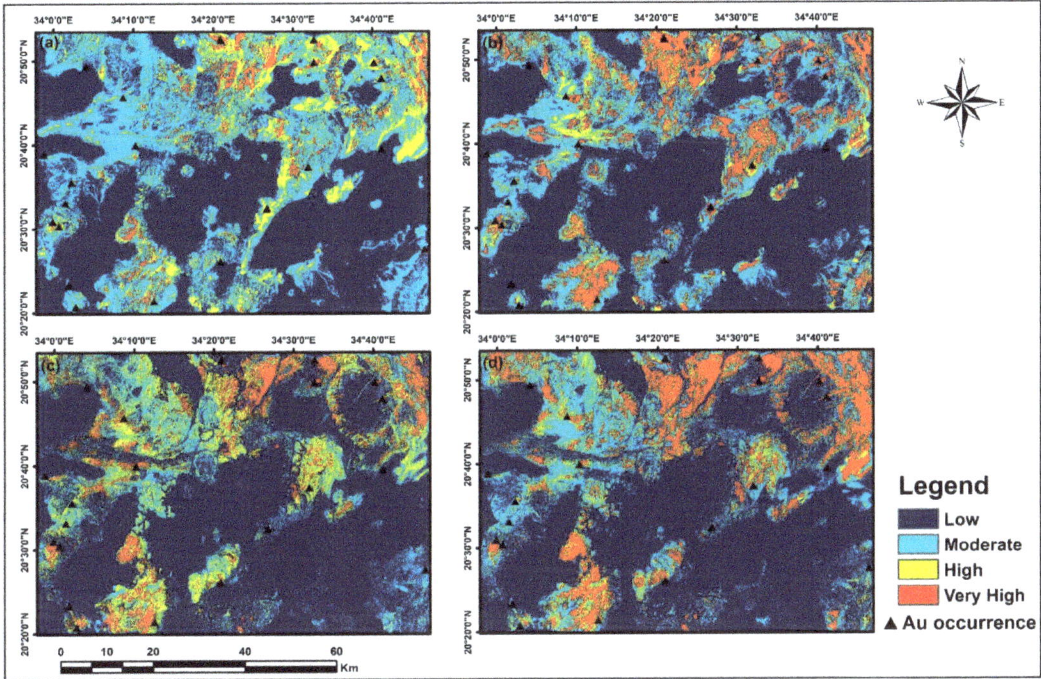

Figure 15. Reclassified gold prospectivity maps based on threshold values derived from success-rate curves: (**a**) Landsat-8; (**b**) Sentinel-2; (**c**) ASTER; (**d**) data-integration.

5. Discussion

The discovery of new prospective areas is deliberated as the most significant issue in mineral exploration. MPM has been successfully used to integrate geological features derived from multisource data to outline new undiscovered mineral deposits. Although remote sensing data represent a great source for recognizing surface alteration and other geological features (e.g., lineament and lithology), they have not yet been fully investigated as the main core of the input data for training mineral prospectivity predictive modeling. The comparison of Landsat-8, Sentinel-2, ASTER, as well as data fusion, for training 714 RF data-driven predictive model is successfully illustrated in the present study. The main findings are discussed below.

Several remote sensing enhancement techniques including BR, MNF, and PCA, have been employed in this study for generating predictor maps. MNF imagery is the only data used to produce color composite images, where the detection of altered rocks is specified by color tones. The selection of three MNF bands to composite images in RGB is the only subjective procedure in the study, which mainly depends on visual judgment and prior knowledge of hydrothermal alteration. Other methods are employed to produce grey-scale predictor maps, where the alteration zones or minerals are presented by the bright regions (higher value) of that image. These methods were extensively and successfully used in prior literature for mapping alteration zones associated with mineral deposits, which mainly depend on the spectral signatures of hydrothermal alteration minerals [29,31,39,68–70,82]. In this regard, all three multispectral sensors data have the capability to detect hydroxyl-bearing and iron-oxide minerals in general. Clay and carbonate minerals including kaolinite, alunite, muscovite, calcite, and dolomite, have high reflectance near 1.6 µm and absorption near 2.2 µm [15,61]. This reflectance signature relatively coincides with band 6 of Landsat-8, band 11 of Sentinel-2, and band 4 of ASTER, meanwhile, the absorption signature coincides with band 7 of Landsat-8, band 12 of Sentinel-2 and band 6 of ASTER. Therefore, these were employed to map OH-bearing minerals using BR method (Landsat-8: 6/7; Sentinel-2: 11/12; and ASTER: 4/6) and selective PCA method as well (Landsat-8: 2,5,6,7; Sentinel-2: 2,8a,11,12; and ASTER: 1,3,4,6). Likewise, iron (ferric and ferrous)/iron-oxide minerals, such as hematite, jarosite, and goethite, display significant absorption features in the VNIR region (from 0.4 μm to 1.3 μm) [61,83]. Specifically, iron-bearing minerals have two absorption features near 0.5 μm and 0.87 μm, which perfectly corresponds to bands 2 and 5 of Landsat-8, and bands 2 and 8a of Sentinel-2 [69,84]. Unfortunately, ASTER can only detect one diagnostic absorption feature near to 0.5 μm (band 2), due to its course spectral resolution in the VNIR region. Due to its higher spectral resolution in the VNIR than ASTER data, and its higher bandpass than Landsat-8, Sentinel-2 data have potential for MPM similar to ASTER data and greater than Landsat-8 data.

Unlike the limited capability of Landsat-8 and Sentinel-2 to map alteration minerals (mapping OH-bearing minerals in general), the higher resolution of ASTER data in the SWIR region allows it for detailed mapping of the hydrothermal alteration zones. Diagnosing Al-OH and Mg-OH groups of minerals helps define different alteration zones. The argillic alteration zone which is characterized by kaolinite and alunite minerals has a double absorption signature at 2.16 μm and 2.2 μm, which coincide with bands 5 and 6, respectively [15]. These bands, therefore, are used to enhance argillic to advance-argillic zone using 4/5 BR, (4 + 6)/5 RBD, and PCA (using bands 1, 4, 6, and 7). Identifying kaolinite and alunite minerals can be achieved using KLI and ALI mineral indices, respectively. The phyllic alteration can be recognized by the muscovite mineral, which shows double absorption features at 2.17 μm and 2.2 μm. The absorption at 2.2 μm (coinciding band 6) is stronger than that at 2.17 μm (coinciding band 5) [39]. This spectral feature is employed to map phyllic alteration using only two methods, which are (5 + 7)/6 RBD, and PCA using bands 1,3,5, and 6. For the optimum discovery of Mg-OH group minerals (e.g., chlorite, epidote, and calcite), band 8 is employed to detect such minerals. These minerals represent the propylitic alteration zone, which has a spectral absorption feature near 2.33 μm (coinciding with band 8). This high absorption property is used to detect propylitic zone using different methods, including propylitic RBD (6 + 9/7 + 8), calcite mineral index (CLI = (6/8) * (9/8)), and PCA (using bands 1,3,5, and 8). Although thermal bands of ASTER are not used in this study, they can be utilized to extend the number of predictor maps. TIR region helps identify minerals at the surface with specific emissivity and absorption features [37,65]. For example, silicate and carbonate can be mapped using BRs 13/12 and 13/14, respectively [24,66]. Moreover, Quartz Index (QI = 11 * 11/10 * 12) can be used as a predictor in the case of gold associated with Quartz dykes/veins [66]. It can be concluded that the possible number of predictor maps that are produced using ASTER data, is about 11 higher than those derived from other remote sensing data. Subsequently, this could be the main reason why ASTER dataset outperforms Landsat-8 and Sentinel-2 datasets in the classification performance of the MPM in the study area.

Since RF is trained using different input variable data, it is essential to assess the spatial association between these predictor variables and the gold occurrence (target variables). In the present study, predictor variables are produced from (i) different sources including geological and remote sensing data; (ii) different multispectral sensors including Landsat-8, Sentinel-, and ASTER; (iii) different processing methods including spatial analysis methods and remote sensing enhancement techniques. Hence, it is critical to measure the influence of each predictor variable on the prediction performance. As mentioned earlier, RF algorithm ranks the importance of the feature variables according to their marginal effect on the target variables [34]. Graphs in Figure 16 illustrate the importance of input feature variables in each dataset. Through all datasets, the most important geological-based predictor variable turns out to be lineaments. In both Landsat-8 and Sentinel-2 datasets, the lineaments density map yields the first rank of importance, while it comes second after propylitic RBD in ASTER and data-integration datasets. The second prominent pattern of the geological predictors through all datasets is that the distance from NW- faults is more important than the NE- faults, which indicates that the spatial association of known gold occurrences is much closer to NW-SE trending faults. RF did not vote for a specific enhancement technique method to be highly distinct from other methods. However, it can be noticed from data-integration dataset that four out of the first five important predictors are produced by the rationing technique. Predictor maps indicating iron-bearing minerals are much more important than those corresponding to hydroxyl-bearing minerals in Landsat-8 and Sentinel-2 datasets. In ASTER dataset, predictors of propylitic alteration zone are significantly more important than other alteration zones, since the propylitic RBD and calcite BR (4/7) are ranked as the first and the fourth important features. It can be noticed that mineralogical indices are relatively less important than other enhancement techniques. It is important to mention that predictors from different remote sensing sensors are highly representative in data-integration dataset. In other words, the rating of features' importance is roughly distributed between different remote sensing data.

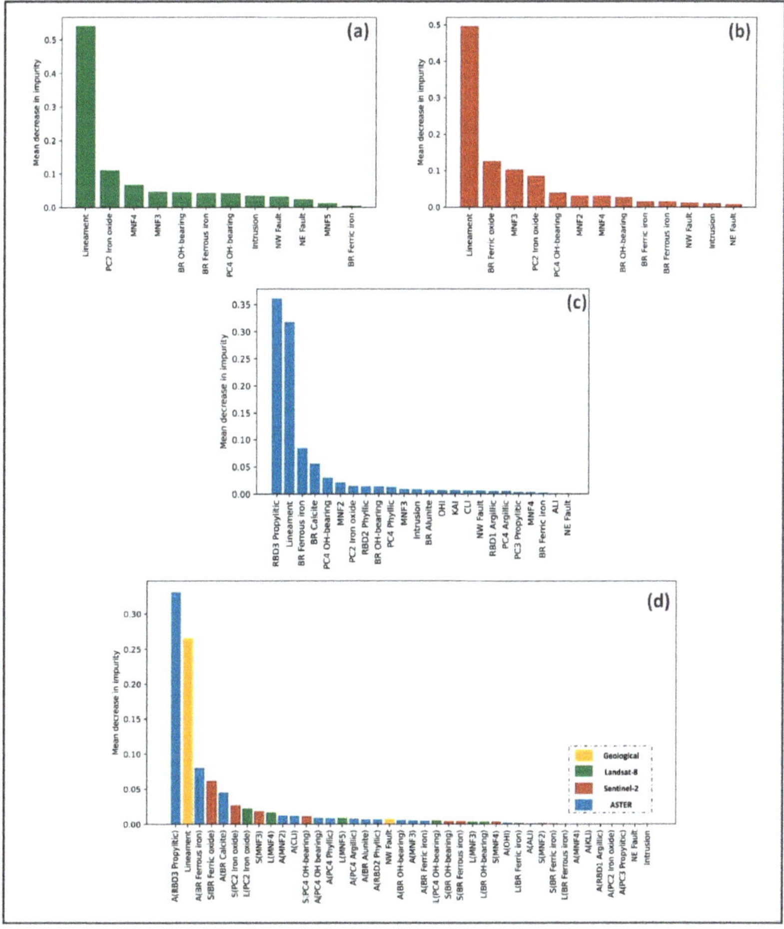

Figure 16. RF model important feature analysis results: (**a**) Landsat-8; (**b**) Sentinel-2; (**c**) ASTER; (**d**) data integration (Symbol 'L' represents Landsat-8; 'S' represents Sentinel-2; 'A' represents ASTER).

6. Conclusions

The investigation of various multispectral remote sensing data capabilities was carried out to produce mineral prospectivity map for gold mineralization in the Hamissana area, NE Sudan. Based on the combination of geological-based predictor maps (proximity to intrusion and faults, and density of lineaments) with remote sensing-based predictor maps (BR, PCA, and MNF), four input datasets including Landsat-8, Sentinel-2, ASTER, and data-integration datasets were prepared. The random forest algorithm was used as an objective tool for comparing the capabilities of various datasets.

As it is demonstrated by the comparison results and discussion, we conclude that Sentinel-2 and ASTER multispectral data have greater potential for mineral prospectivity modeling than Landsat-8. Both datasets achieved 0.875 AUC, while the overall classification accuracy of ASTER dataset (73.3%) is higher than Sentinel-2 (66.7%). Data-integration dataset boosts the prediction performance of RF up to (AUC: 0.938). The density of the lineaments plays a significant role in the prediction performance in all datasets.

Modeling results using different datasets suggest several prospecting regions. Nevertheless, considering the uncertainty of remote sensing data and MPM results, further geological investigation and exploration should be taken into account. Specifically, drilling, geophysical and geochemical surveys, and 3D modeling techniques are essential for future work and further accurate targeting.

In our future research, we plan to compare current multispectral remote sensing data with other data from multiple sources (e.g., comprehensive geochemical survey, gravity, and magnetic geophysical survey), which are not available at present. Moreover, we would like to conduct a comprehensive comparison using other machine learning algorithms such as a support vector machine and an artificial neural network. Finally, other deep learning techniques are preferable to be applied also in MPM, since deep learning is still a hot research topic in several geoscience fields.

Author Contributions: "Conceptualization, A.M.M.T. and Y.X.; methodology, A.M.M.T. and Q.H.; software, A.M.M.T., S.W. and X.L.; validation, A.M.M.T., Q.H. and A.H.; formal analysis, A.M.M.T. and Y.X.; data curation, A.M.M.T.; writing—original draft preparation, A.M.M.T.; writing—review and editing, A.M.M.T. and Y.X.; supervision, Y.X.; funding acquisition, Y.X. All authors have read and agreed to the published version of the manuscript.

Funding: This work was supported by the Priority Academic Program Development of Jiangsu Higher Education Institutions (PAPD).

Data Availability Statement: All remote sensing data used in this paper are freely available online on the websites mentioned in Section 3.1—Data. Python codes are available online on the first author's GitHub webpage (https://github.com/Abdallah-M-Ali).

Acknowledgments: We would like to express our respect and gratitude to the anonymous reviewers and editors for their professional comments and suggestions on improving the quality of this paper. The research was undertaken thanks to funding from the Priority Academic Program Development of Jiangsu Higher Education Institutions [PAPD].

Conflicts of Interest: The authors declare no conflict of interest.

References

1. Arndt, N.T.; Fontboté, L.; Hedenquist, J.W.; Kesler, S.E.; Thompson, J.F.; Wood, D.G. Future global mineral resources. *Geochem. Perspect.* **2017**, *6*, 1–171. [CrossRef]
2. Wang, K.; Zheng, X.; Wang, G.; Liu, D.; Cui, N. A Multi-Model Ensemble Approach for Gold Mineral Prospectivity Mapping: A Case Study on the Beishan Region, Western China. *Minerals* **2020**, *10*, 1126. [CrossRef]
3. Sun, T.; Li, H.; Wu, K.; Chen, F.; Zhu, Z.; Hu, Z. Data-Driven Predictive Modelling of Mineral Prospectivity Using Machine Learning and Deep Learning Methods: A Case Study from Southern Jiangxi Province, China. *Minerals* **2020**, *10*, 102. [CrossRef]
4. Carranza, E.J.M.; Laborte, A.G. Data-driven predictive mapping of gold prospectivity, Baguio district, Philippines: Application of Random Forests algorithm. *Ore Geol. Rev.* **2015**, *71*, 777–787. [CrossRef]
5. Porwal, A.; Carranza, E.J.M. Introduction to the Special Issue: GIS-based mineral potential modelling and geological data analyses for mineral exploration. *Ore Geol. Rev.* **2015**, *71*, 477–483. [CrossRef]

6. Carranza, E.J.M.; Hale, M.; Faassen, C. Selection of coherent deposit-type locations and their application in data-driven mineral prospectivity mapping. *Ore Geol. Rev.* **2008**, *33*, 536–558. [CrossRef]
7. Rodriguez-Galiano, V.; Sanchez-Castillo, M.; Chica-Olmo, M.; Chica-Rivas, M. Machine learning predictive models for mineral prospectivity: An evaluation of neural networks, random forest, regression trees and support vector machines. *Ore Geol. Rev.* **2015**, *71*, 804–818. [CrossRef]
8. Sun, T.; Chen, F.; Zhong, L.; Liu, W.; Wang, Y. GIS-based mineral prospectivity mapping using machine learning methods: A case study from Tongling ore district, eastern China. *Ore Geol. Rev.* **2019**, *109*, 26–49. [CrossRef]
9. Yousefi, M.; Nykänen, V. Introduction to the special issue: GIS-based mineral potential targeting. *J. Afr. Earth Sci.* **2017**, *128*, 1–4. [CrossRef]
10. Cheng, Q.M.; Agterberg, F.P. Fuzzy weights of evidence method and its application in mineral potential mapping. *Nat. Resour. Res.* **1999**, *8*, 27–35. [CrossRef]
11. Abedi, M.; Norouzi, G.-H.; Fathianpour, N. Fuzzy outranking approach: A knowledge-driven method for mineral prospectivity mapping. *Int. J. Appl. Earth Obs. Geoinf.* **2013**, *21*, 556–567. [CrossRef]
12. Yousefi, M.; Nykänen, V. Data-driven logistic-based weighting of geochemical and geological evidence layers in mineral prospectivity mapping. *J. Geochem. Explor.* **2016**, *164*, 94–106. [CrossRef]
13. Ghezelbash, R.; Maghsoudi, A.; Bigdeli, A.; Carranza, E.J.M. Regional-Scale Mineral Prospectivity Mapping: Support Vector Machines and an Improved Data-Driven Multi-criteria Decision-Making Technique. *Nat. Resour. Res.* **2021**, *30*, 1977–2005. [CrossRef]
14. Harris, J.R.; Grunsky, E.; Behnia, P.; Corrigan, D. Data- and knowledge-driven mineral prospectivity maps for Canada's North. *Ore Geol. Rev.* **2015**, *71*, 788–803. [CrossRef]
15. Abdelkareem, M.; Al-Arifi, N. Synergy of Remote Sensing Data for Exploring Hydrothermal Mineral Resources Using GIS-Based Fuzzy Logic Approach. *Remote Sens.* **2021**, *13*, 4492. [CrossRef]
16. Rowan, L.C.; Mars, J.C. Lithologic mapping in the Mountain Pass, California area using advanced spaceborne thermal emission and reflection radiometer (ASTER) data. *Remote Sens. Environ.* **2003**, *84*, 350–366. [CrossRef]
17. Pazand, K.; Pazand, K. Identification of hydrothermal alteration minerals for exploring porphyry copper deposit using ASTER data: A case study of Varzaghan area, NW Iran. *Geol. Ecol. Landsc.* **2020**, *6*, 217–223. [CrossRef]
18. Pour, A.B.; Hashim, M. Identification of hydrothermal alteration minerals for exploring of porphyry copper deposit using ASTER data, SE Iran. *J. Asian Earth Sci.* **2011**, *42*, 1309–1323. [CrossRef]
19. Guha, A.; Mondal, S.; Chatterjee, S.; Kumar, K.V. Airborne imaging spectroscopy of igneous layered complex and their mapping using different spectral enhancement conjugated support vector machine models. *Geocarto Int.* **2020**, *37*, 349–365. [CrossRef]
20. Rao, D.A.; Guha, A. Potential Utility of Spectral Angle Mapper and Spectral Information Divergence Methods for mapping lower Vindhyan Rocks and Their Accuracy Assessment with Respect to Conventional Lithological Map in Jharkhand, India. *J. Indian Soc. Remote* **2018**, *46*, 737–747. [CrossRef]
21. Rani, N.; Mandla, V.R.; Singh, T. Spatial distribution of altered minerals in the Gadag Schist Belt (GSB) of Karnataka, Southern India using hyperspectral remote sensing data. *Geocarto Int.* **2016**, *32*, 225–237. [CrossRef]
22. Rani, N.; Mandla, V.R.; Singh, T. Evaluation of atmospheric corrections on hyperspectral data with special reference to mineral mapping. *Geosci. Front.* **2017**, *8*, 797–808. [CrossRef]
23. Noori, L.; Pour, A.; Askari, G.; Taghipour, N.; Pradhan, B.; Lee, C.-W.; Honarmand, M. Comparison of Different Algorithms to Map Hydrothermal Alteration Zones Using ASTER Remote Sensing Data for Polymetallic Vein-Type Ore Exploration: Toroud–Chahshirin Magmatic Belt (TCMB), North Iran. *Remote Sens.* **2019**, *11*, 495. [CrossRef]
24. Pour, A.B.; Hashim, M. Identifying areas of high economic-potential copper mineralization using ASTER data in the Urumieh–Dokhtar Volcanic Belt, Iran. *Adv. Space Res.* **2012**, *49*, 753–769. [CrossRef]
25. Pour, A.B.; Hashim, M.; Makoundi, C.; Zaw, K. Structural Mapping of the Bentong-Raub Suture Zone Using PALSAR Remote Sensing Data, Peninsular Malaysia: Implications for Sediment-hosted/Orogenic Gold Mineral Systems Exploration. *Resour. Geol.* **2016**, *66*, 368–385. [CrossRef]
26. Pour, A.B.; Hashim, M.; Park, Y. Application of ASTER SWIR bands in mapping anomaly pixels for Antarctic geological mapping. *J. Phys. Conf. Ser.* **2017**, *852*, 012025. [CrossRef]
27. Pour, A.B.; Park, Y.; Park, T.-Y.S.; Hong, J.K.; Hashim, M.; Woo, J.; Ayoobi, I. Regional geology mapping using satellite-based remote sensing approach in Northern Victoria Land, Antarctica. *Polar Sci.* **2018**, *16*, 23–46. [CrossRef]
28. Son, Y.-S.; Lee, G.; Lee, B.H.; Kim, N.; Koh, S.-M.; Kim, K.-E.; Cho, S.-J. Application of ASTER Data for Differentiating Carbonate Minerals and Evaluating MgO Content of Magnesite in the Jiao-Liao-Ji Belt, North China Craton. *Remote Sens.* **2022**, *14*, 181. [CrossRef]
29. Bahrami, Y.; Hassani, H.; Maghsoudi, A. Investigating the capabilities of multispectral remote sensors data to map alteration zones in the Abhar area, NW Iran. *Geosyst. Eng.* **2018**, *24*, 18–30. [CrossRef]
30. Fereydooni, H.; Mojeddifar, S. A directed matched filtering algorithm (DMF) for discriminating hydrothermal alteration zones using the ASTER remote sensing data. *Int. J. Appl. Earth Obs. Geoinf.* **2017**, *61*, 1–13. [CrossRef]
31. Chen, Q.; Zhao, Z.-F.; Xia, J.-S.; Zhao, X.; Yang, H.-Y.; Zhang, X.-L. Improving the accuracy of hydrothermal alteration mapping based on image fusion of ASTER and Sentinel-2A data: A case study of Pulang Cu deposit, Southwest China. *Geocarto Int.* **2022**, 1–26. [CrossRef]

32. Joly, A.; Porwal, A.; McCuaig, T.C.; Chudasama, B.; Dentith, M.C.; Aitken, A.R.A. Mineral systems approach applied to GIS-based 2D-prospectivity modelling of geological regions: Insights from Western Australia. *Ore Geol. Rev.* **2015**, *71*, 673–702. [CrossRef]

33. Carranza, E.J.M. Natural Resources Research Publications on Geochemical Anomaly and Mineral Potential Mapping, and Introduction to the Special Issue of Papers in These Fields. *Nat. Resour. Res.* **2017**, *26*, 379–410. [CrossRef]

34. Carranza, E.J.M.; Laborte, A.G. Random forest predictive modeling of mineral prospectivity with small number of prospects and data with missing values in Abra (Philippines). *Comput. Geosci.* **2015**, *74*, 60–70. [CrossRef]

35. Zuo, R.; Carranza, E.J.M. Support vector machine: A tool for mapping mineral prospectivity. *Comput. Geosci.* **2011**, *37*, 1967–1975. [CrossRef]

36. Brown, W.M.; Gedeon, T.D.; Groves, D.I.; Barnes, R.G. Artifcial neural network: A new method for mineral prospectivity mapping. *Aust. J. Earth Sci.* **2000**, *47*, 757–770. [CrossRef]

37. Xi, Y.; Mohamed Taha, A.M.; Hu, A.; Liu, X. Accuracy comparison of various remote sensing data in lithological classification based on random forest algorithm. *Geocarto Int.* **2022**, 1–29. [CrossRef]

38. Mansouri, E.; Feizi, F.; Jafari Rad, A.; Arian, M. Remote-sensing data processing with the multivariate regression analysis method for iron mineral resource potential mapping: A case study in the Sarvian area, central Iran. *Solid Earth* **2018**, *9*, 373–384. [CrossRef]

39. Bolouki, S.M.; Ramazi, H.R.; Maghsoudi, A.; Beiranvand Pour, A.; Sohrabi, G. A Remote Sensing-Based Application of Bayesian Networks for Epithermal Gold Potential Mapping in Ahar-Arasbaran Area, NW Iran. *Remote Sens.* **2019**, *12*, 105. [CrossRef]

40. Rodriguez-Galiano, V.F.; Chica-Olmo, M.; Chica-Rivas, M. Predictive modelling of gold potential with the integration of multisource information based on random forest: A case study on the Rodalquilar area, Southern Spain. *Int. J. Geogr. Inf. Sci.* **2014**, *28*, 1336–1354. [CrossRef]

41. Gaboury, D.; Nabil, H.; Ennaciri, A.; Maacha, L. Structural setting and fluid composition of gold mineralization along the central segment of the Keraf suture, Neoproterozoic Nubian Shield, Sudan: Implications for the source of gold. *Int. Geol. Rev.* **2020**, *64*, 45–71. [CrossRef]

42. Mohamed, M.T.A.; Al-Naimi, L.S.; Mgbeojedo, T.I.; Agoha, C.C. Geological mapping and mineral prospectivity using remote sensing and GIS in parts of Hamissana, Northeast Sudan. *J. Pet. Explor. Prod.* **2021**, *11*, 1123–1138. [CrossRef]

43. Bierlein, F.; Reynolds, N.; Arne, D.; Bargmann, C.; McKeag, S.; Bullen, W.; Al-Athbah, H.; McKnight, S.; Maas, R. Petrogenesis of a Neoproterozoic magmatic arc hosting porphyry Cu-Au mineralization at Jebel Ohier in the Gebeit Terrane, NE Sudan. *Ore Geol. Rev.* **2016**, *79*, 133–154. [CrossRef]

44. Zeinelabdein, K.A.E.; Nadi, A.H.H.E. The use of Landsat 8 OLI image for the delineation of gossanic ridges in the Red Sea Hills of NE Sudan. *Am. J. Earth Sci.* **2014**, *1*, 62–67.

45. Sasmaz, A. The Atbara porphyry gold–copper systems in the Red Sea Hills, Neoproterozoic Arabian–Nubian Shield, NE Sudan. *J. Geochem. Explor.* **2020**, *214*, 106539. [CrossRef]

46. El Khidir, S.O.; Babikir, I.A. Digital image processing and geospatial analysis of landsat 7 ETM+ for mineral exploration, Abidiya area, North Sudan. *Int. J. Geomat. Geosci.* **2013**, *3*, 645–658.

47. Ali, A.; Pour, A. Lithological mapping and hydrothermal alteration using Landsat 8 data: A case study in ariab mining district, red sea hills, Sudan. *Int. J. Basic Appl. Sci.* **2014**, *3*, 199–208. [CrossRef]

48. Breiman, L. Random Forests. *Mach. Learn.* **2001**, *45*, 5–32. [CrossRef]

49. Breiman, L. bagging predictors. *Mach. Learn.* **1996**, *24*, 123–140. [CrossRef]

50. Breiman, L.; Friedman, J.; Olshen, R.; Stone, C. *Classification and Regression Trees*; Wadsworth & Brooks; Cole Statistics/Probability Series; Chapman & Hall: London, UK, 1984.

51. Belgiu, M.; Drăguţ, L. Random forest in remote sensing: A review of applications and future directions. *ISPRS J. Photogramm. Remote Sens.* **2016**, *114*, 24–31. [CrossRef]

52. Shah, S.H.; Angel, Y.; Houborg, R.; Ali, S.; McCabe, M.F. A Random Forest Machine Learning Approach for the Retrieval of Leaf Chlorophyll Content in Wheat. *Remote Sens.* **2019**, *11*, 920. [CrossRef]

53. Kulkarni, A.D.; Lowe, B. Random forest algorithm for land cover classification. *Int. J. Recent Innov. Trends Comput. Commun.* **2016**, *4*, 58–63.

54. Pal, M. Random forest classifier for remote sensing classification. *Int. J. Remote Sens.* **2005**, *26*, 217–222. [CrossRef]

55. Maepa, F.; Smith, R.S.; Tessema, A. Support vector machine and artificial neural network modelling of orogenic gold prospectivity mapping in the Swayze greenstone belt, Ontario, Canada. *Ore Geol. Rev.* **2021**, *130*, 103968. [CrossRef]

56. Zeinelabdein, K.E.; Albiely, A. Ratio image processing techniques: A prospecting tool for mineral deposits, Red Sea Hills, NE Sudan. *Int. Arch. Photogramm. Remote Sens. Spat. Inf. Sci.* **2008**, *37*, 1295–1298.

57. Abdullah, A.; Nassr, S.; Ghaleeb, A. Remote Sensing and Geographic Information System for Fault Segments Mapping a Study from Taiz Area, Yemen. *J. Geol. Res.* **2013**, *2013*, 201757. [CrossRef]

58. Adiria, Z.; Hartia, A.E.; Jelloulia, A.; Lhissoua, R.; Maachab, L.; Azmib, M.; Zouhairb, M.; Bachaouia, E.M. Comparison of Landsat-8, ASTER and Sentinel 1 satellite remote sensing data in automatic lineaments extraction: A case study of Sidi Flah Bouskour inlier, Moroccan Anti Atlas. *Adv. Space Res.* **2017**, *60*, 2355–2367. [CrossRef]

59. Pour, A.B.; Hashim, M. ASTER, ALI and Hyperion sensors data for lithological mapping and ore minerals exploration. *SpringerPlus* **2014**, *3*, 130. [CrossRef]

60. Li, N. Textural and Rule-Based Lithological Classification of Remote Sensing Data, and Geological Mapping in Southwestern Prieska Sub-Basin, Transvaal Supergroup, South Africa. Ph.D. Thesis, LMU, München, Germany, 2010.

61. Zhang, T.; Yi, G.; Li, H.; Wang, Z.; Tang, J.; Zhong, K.; Li, Y.; Wang, Q.; Bie, X. Integrating Data of ASTER and Landsat-8 OLI (AO) for Hydrothermal Alteration Mineral Mapping in Duolong Porphyry Cu-Au Deposit, Tibetan Plateau, China. *Remote Sens.* **2016**, *8*, 890. [CrossRef]

62. Sabins, F.F. Remote sensing for mineral exploration. *Ore Geol. Rev.* **1999**, *14*, 157–183. [CrossRef]

63. Inzana, J.; Kusky, T.; Higgs, G.; Tucker, R. Supervised classifications of Landsat TM band ratio images and Landsat TM band ratio image with radar for geological interpretations of central Madagascar. *J. Afr. Earth Sci.* **2003**, *37*, 59–72. [CrossRef]

64. Ninomiya, Y. *Lithologic Mapping with Multispectral ASTER TIR and SWIR Data*; SPIE: Bellingham, WA, USA, 2004.

65. Ninomiya, Y.; Fu, B.; Cudahy, T.J. Detecting lithology with Advanced Spaceborne Thermal Emission and Reflection Radiometer (ASTER) multispectral thermal infrared "radiance-at-sensor" data. *Remote Sens. Environ.* **2005**, *99*, 127–139. [CrossRef]

66. Rajan Girija, R.; Mayappan, S. Mapping of mineral resources and lithological units: A review of remote sensing techniques. *Int. J. Image Data Fusion* **2019**, *10*, 79–106. [CrossRef]

67. Ninomiya, Y. A stabilized vegetation index and several mineralogic indices defined for ASTER VNIR and SWIR data. In Proceedings of the IGARSS 2003, 2003 IEEE International Geoscience and Remote Sensing Symposium, Proceedings (IEEE Cat. No.03CH37477), Toulouse, France, 21–25 July 2003; pp. 1552–1554.

68. van der Meer, F.D.; van der Werff, H.M.A.; van Ruitenbeek, F.J.A. Potential of ESA's Sentinel-2 for geological applications. *Remote Sens. Environ.* **2014**, *148*, 124–133. [CrossRef]

69. Ge, W.; Cheng, Q.; Jing, L.; Wang, F.; Zhao, M.; Ding, H. Assessment of the Capability of Sentinel-2 Imagery for Iron-Bearing Minerals Mapping: A Case Study in the Cuprite Area, Nevada. *Remote Sens.* **2020**, *12*, 3028. [CrossRef]

70. Timkin, T.; Abedini, M.; Ziaii, M.; Ghasemi, M.R. Geochemical and Hydrothermal Alteration Patterns of the Abrisham-Rud Porphyry Copper District, Semnan Province, Iran. *Minerals* **2022**, *12*, 103. [CrossRef]

71. Crosta, A.; De Souza Filho, C.; Azevedo, F.; Brodie, C. Targeting key alteration minerals in epithermal deposits in Patagonia, Argentina, using ASTER imagery and principal component analysis. *Int. J. Remote Sens.* **2003**, *24*, 4233–4240. [CrossRef]

72. Ourhzif, Z.; Algouti, A.; Algouti, A.; Hadach, F. Lithological Mapping Using Landsat 8 Oli and Aster Multispectral Data in Imini-Ounilla District South High Atlas of Marrakech. *Int. Arch. Photogramm. Remote Sens. Spat. Inf. Sci.* **2019**, *XLII-2/W13*, 1255–1262. [CrossRef]

73. Saljoughi, B.S.; Hezarkhani, A. A comparative analysis of artificial neural network (ANN), wavelet neural network (WNN), and support vector machine (SVM) data-driven models to mineral potential mapping for copper mineralizations in the Shahr-e-Babak region, Kerman, Iran. *Appl. Geomat.* **2018**, *10*, 229–256. [CrossRef]

74. Torppa, J.; Nykänen, V.; Molnár, F. Unsupervised clustering and empirical fuzzy memberships for mineral prospectivity modelling. *Ore Geol. Rev.* **2019**, *107*, 58–71. [CrossRef]

75. Bachri, I.; Hakdaoui, M.; Raji, M.; Teodoro, A.C.; Benbouziane, A. Machine Learning Algorithms for Automatic Lithological Mapping Using Remote Sensing Data: A Case Study from Souk Arbaa Sahel, Sidi Ifni Inlier, Western Anti-Atlas, Morocco. *ISPRS Int. J. Geo-Inf.* **2019**, *8*, 248. [CrossRef]

76. Cracknell, M.J.; Reading, A.M. Geological mapping using remote sensing data: A comparison of five machine learning algorithms, their response to variations in the spatial distribution of training data and the use of explicit spatial information. *Comput. Geosci.* **2014**, *63*, 22–33. [CrossRef]

77. Heydari, S.S.; Mountrakis, G. Effect of classifier selection, reference sample size, reference class distribution and scene heterogeneity in per-pixel classification accuracy using 26 Landsat sites. *Remote Sens. Environ.* **2018**, *204*, 648–658. [CrossRef]

78. Barsi, Á.; Kugler, Z.; László, I.; Szabó, G.; Abdulmutalib, H.M. Accuracy Dimensions in Remote Sensing. *Int. Arch. Photogramm. Remote Sens. Spat. Inf. Sci.* **2018**, *XLII-3*, 61–67. [CrossRef]

79. Agterberg, F.P.; Bonham-Carter, G.F. Measuring the Performance of Mineral-Potential Maps. *Nat. Resour. Res.* **2005**, *14*, 1–17. [CrossRef]

80. Landgrebe, T.C.W.; Paclik, P. The ROC skeleton for multiclass ROC estimation. *Pattern Recognit. Lett.* **2010**, *31*, 949–958. [CrossRef]

81. Bradley, A.P. The use of the area under the ROC curve in the evaluation of machine learning algorithms. *Pattern Recognit.* **1997**, *30*, 1145–1159. [CrossRef]

82. Hua, B.; Xua, Y.; Wana, B.; Wua, X.; Yi, G. Hydrothermally altered mineral mapping using synthetic application of Sentinel-2A MSI, ASTER and Hyperion data in the Duolong area, Tibetan Plateau, China. *Ore Geol. Rev.* **2018**, *101*, 384–397. [CrossRef]

83. Ge, W.; Cheng, Q.; Jing, L.; Armenakis, C.; Ding, H. Lithological discrimination using ASTER and Sentinel-2A in the Shibanjing ophiolite complex of Beishan orogenic in Inner Mongolia, China. *Adv. Space Res.* **2018**, *62*, 1702–1716. [CrossRef]

84. Ge, W.; Cheng, Q.; Tang, Y.; Jing, L.; Gao, C. Lithological Classification Using Sentinel-2A Data in the Shibanjing Ophiolite Complex in Inner Mongolia, China. *Remote Sens.* **2018**, *10*, 638. [CrossRef]

Article

Fusion of Multispectral Remote-Sensing Data through GIS-Based Overlay Method for Revealing Potential Areas of Hydrothermal Mineral Resources

Saad S. Alarifi [1,*], Mohamed Abdelkareem [2,*], Fathy Abdalla [2], Ismail S. Abdelsadek [2], Hisham Gahlan [1], Ahmad. M. Al-Saleh [1] and Mislat Alotaibi [3]

[1] Department of Geology and Geophysics, College of Science, King Saud University, P.O. Box 2455, Riyadh 11451, Saudi Arabia
[2] Geology Department, South Valley University, Qena 83523, Egypt
[3] Department of Geosciences, University of Arkansas, Fayetteville, AR 72704, USA
* Correspondence: ssalarifi@ksu.edu.sa (S.S.A.); mohamed.abdelkareem@sci.svu.edu.eg (M.A.)

Abstract: Revealing prospective locations of hydrothermal alteration zones (HAZs) is an important technique for mineral prospecting. In this study, we used multiple criteria inferred from Landsat-8 OLI, Sentinel-2, and ASTER data using a GIS-based weighted overlay multi-criteria decision analysis approach to build a model for the delineating of hydrothermal mineral deposits in the Khnaiguiyah district, Saudi Arabia. The utilized algorithms revealed argillic, phyllic, and propylitic alteration characteristics. The HAZs map resulted in the identification of six zones based on their mineralization potential, providing a basis for potential hydrothermal mineral deposit assessment exploration, which was created by the fusion of mineral bands indicators designated very low, low, moderate, good, very good, and excellent and covers 31.36, 28.22, 20.49, 10.99, 6.35, and 2.59%. Based on their potential for hydrothermal mineral potentiality, the discovered zones match gossans related to sulfide mineral alteration zones, as demonstrated by previous studies.

Keywords: mineral exploration; ASTER; OLI; Sentinel-2; GIS; Khnaiguiyah; Saudi Arabia

Citation: Alarifi, S.S.; Abdelkareem, M.; Abdalla, F.; Abdelsadek, I.S.; Gahlan, H.; Al-Saleh, A.M.; Alotaibi, M. Fusion of Multispectral Remote-Sensing Data through GIS-Based Overlay Method for Revealing Potential Areas of Hydrothermal Mineral Resources. *Minerals* **2022**, *12*, 1577. https://doi.org/10.3390/min12121577

Academic Editors: Amin Beiranvand Pour, Omeid Rahmani and Mohammad Parsa

Received: 23 October 2022
Accepted: 6 December 2022
Published: 9 December 2022

Publisher's Note: MDPI stays neutral with regard to jurisdictional claims in published maps and institutional affiliations.

1. Introduction

Remote sensing techniques have provided valuable tools for characterizing and delineating geological, structural, and lithological features that have aided in the identification of mineralization regions [1,2]. Because of its fine geospatial, radiometric, and spectral resolution, remotely sensed data provides significant information for mineral exploration. One of the main aims of remote sensing investigations is the delineation of hydrothermal alteration zones and the identification of the mineralogical signature [1,3,4]. Hydrothermal alteration zones (HAZs) and their grade must be characterized in order to identify possible mineral resource locations [1,2,5,6]. This is due to the fact that such a process is frequently linked to the economic concentration of base metals like Au, Cu, and Ag. Several studies were conducted using multispectral remotely sensed data to characterize the extent of the hydrothermally altered areas and to identify the minerals forming zones [3,7–15].

Data from satellites can be used to detect new prospects prior to detailed and expensive ground research [4,15]. Landsat Operational Land Imager (OLI) and Advanced Spaceborne Thermal Emission and Reflection Radiometer (ASTER) images were used to process and analyze remote sensing multispectral datasets. Electromagnetic (EM) radiation reflected, transmitted, or backscattered from the Earth's surface is sensitive to remote sensing devices such as OLI and ASTER. With passive or active systems, remote sensing sensors can monitor wavelengths of EM radiation in the visible near-infrared and short-wave infrared (VIS/NIR/SWIR) to microwave. Landsat satellite image data have been utilized for lithologic mapping using image transformation techniques [7,8,16–18].

Although Landsat data had been widely used in characterizing hydrothermal alteration zones for decades [8,9,19,20], the introduction of the Advanced Spaceborne Thermal Emission and Reflection Radiometer (ASTER) data in 1999 added meaningful data to the research area of mineral deposits [5,11,12,15,21]. This is due to the fact that, as compared to Landsat data, such data have better spectral, spatial, and radiometric resolutions, allowing for greater information regarding mineral properties.

The capacity of ASTER (e.g., SWIR) spectral bands to distinguish the alteration zones was tested using a variety of methodologies, including band ratios, principal component analysis (PCA), and spectral analysis [5,13,14,21,22]. To improve the spectral disparities across bands and eliminate topographic effects, band ratios were adopted [7,8,13,23]. Mineral indices [11,12,22] and relative absorption band depth (RBD; [24]) were also used. Despite the fact that band ratios, PCA, and RBD have been successful in delineating hydrothermal alteration zones. Many studies have used the band ratios technique to distinguish between different rock units or minerals [4,6].

Because earlier studies did not have such data to use, little emphasis was made on delineating alteration zones and extracting certain important hydrothermal minerals linked with the above-mentioned deposits utilizing remote sensing data in the study area. ASTER spectral bands are thus used in this study to identify the alteration zones associated with Zn-Cu deposits and extract the major hydrothermal alteration zones. This is performed in order to identify prospective mineralization sites in the study area.

Using a GIS-based process to develop mineral development capabilities based on remote data has thus become a rapid and accurate tool for identifying target areas for mineral exploration [4,25], particularly during the reconnaissance stage. Developments in revealing promising areas of hydrothermal mineral resources have been made with the emergence of GIS-based spatial analytic tools [26–29]. This is because employing a GIS method to integrate spatially distributed remote-sensing data is a key approach to mineral exploration since it allows for the combination of different data utilizing digital overlay methods to optimize mineral prospection maps [30]. The GIS-based knowledge-driven technique, for example, is effective in producing predicted maps based on expert opinion [25] since each GIS predictive layer is given a weight that reflects its value in the process.

Prior to the advent of high spectral resolution, multi-spectral sensors, it was challenging to detect alteration zones linked with hydrothermal deposits like those associated with the Khnaiguiyah Zn mineralization. Three sensors' data, e.g., ASTER, Sentinel-2, and OLI spectral bands, are thus used in this study to identify the alteration zones associated with sulfide deposits and extract the major hydrothermal alteration zones. This is performed in order to identify prospective mineralization regions in the Khunayqiyah region.

2. Study Area

The present study is a part of Arabian Shield, Khnaiguiyah, Saudi Arabia. It extends between latitudes 24°13′25.48″ and 24°17′43.72″ and longitudes 45°2′47.90″ and 45°6′58.37″, covering an area of about 57 sq km.

The district of Khnaiguiyah is located at the eastern periphery of the Arabian Shield (Figure 1), which is the exposed Precambrian basement of the Arabian Plate. The Arabian Shield (ANS) is the northernmost extension of the East African Orogen [31,32] and consists of a collage of tectonostratigraphic terranes with ensialic and ensimatic arc affinities [33,34]. The convergence of East and West Gondwana caused the terrane amalgamation/accretion during the Pan-African event (780–600 Ma) [34]. The final suturing (680–610 Ma) of the ANS coincided with the development of the Nabitah fault zone, gneiss domes, and massive Molasse basins [35,36]. The NW–SE trending Najd fault system/Najd Orogeny (620–540 Ma) subsequently formed due to escape tectonics concomitant with the assembly of the Gondwana supercontinent [37,38]. By the Cambrian (541 Ma), the ANS was established as a stable juvenile continental block forming the northeastern margin of Gondwana [39].

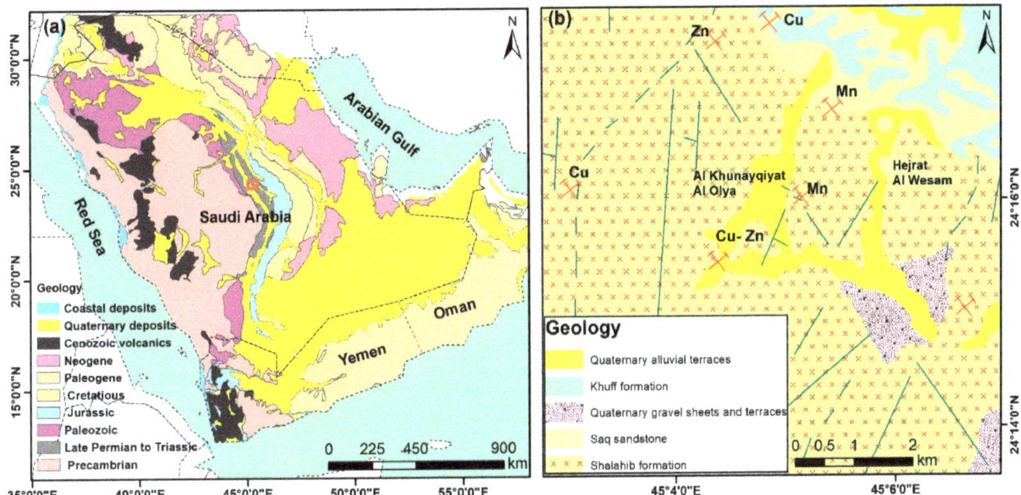

Figure 1. (**a**) General geologic map of the Arabian Peninsula showing the location of the Khnaiguiyah mineralized district; (**b**) simplified geologic map of the Khnaiguiyah area denoting the main mineral occurrences.

Khnaiguiyah Zn–Cu deposits represent a substantial zinc resource and attracted considerable exploration efforts in the last three decades [40]. The four ore bodies of the Khnaiguiyah district comprise mineable reserves of up to 11 Mt averaging 7.41% Zn and 0.82% Cu [41]. Khnaiguiyah-type deposit features are consistent with metamorphism and deformation of volcanogenic massive sulfide (VMS) mineralization; the related stratiform Mn-rich units are mainly suggestive of a seafloor hydrothermal setting.

The Khnaiguiyah ores are hosted by the Shalahib Formation (1500 m thick), which is made up of felsic volcano-sedimentary rocks interlayered with carbonates [42]. The Shalahib formation predominantly comprises andesite and rhyolite volcanic rocks and associated pyroclastics and ignimbrites, and the whole sequence is affected by low-temperature greenschist-facies regional metamorphism. The Khnaiguiyah deposit lies within an area of 3 × 3 km. Four mineralized orebodies are interpreted hydrothermal mineral deposits containing Zinc and Copper that are hosted by strongly sheared and folded late Proterozoic medium to felsic volcanics/volcaniclastics. The shear zones, which are tens of meters thick, are oriented NS and dip 10 to 70 to the west. The hydrothermally altered rocks occur within discontinuous anastomosed bands 50 to 100 m wide and several Kilometers long and are regionally oriented along with the north–south regional foliation. Detailed analysis of surface and drill-core samples shows that the hydrothermal alteration zones and associated Zn-Cu-Fe-Mn mineralization are controlled by a shearing deformation phase that post-dated the first phase of regional folding. The hydrothermal alteration zone contains illite, kaolinite, quartz, albite, hematite, and calcite.

At this locality, the Precambrian basement is overlain by the basal conglomerates and cross-bedded red sandstone of the middle Cambrian Saq formation, which is in turn overlain by the Permo-Triassic shallow marine carbonates of the Khuff formation.

3. Data Used and Methods

This study used visible/infrared satellite-derived imagery to characterize mineralization associated with hydrothermal alteration zones. Landsat 8-OLI, Sentinel-2, and ASTER data (Figure 2) were all employed to detect altered features and structural patterns. A comparison of these sensors is shown in Figure 2 [19].

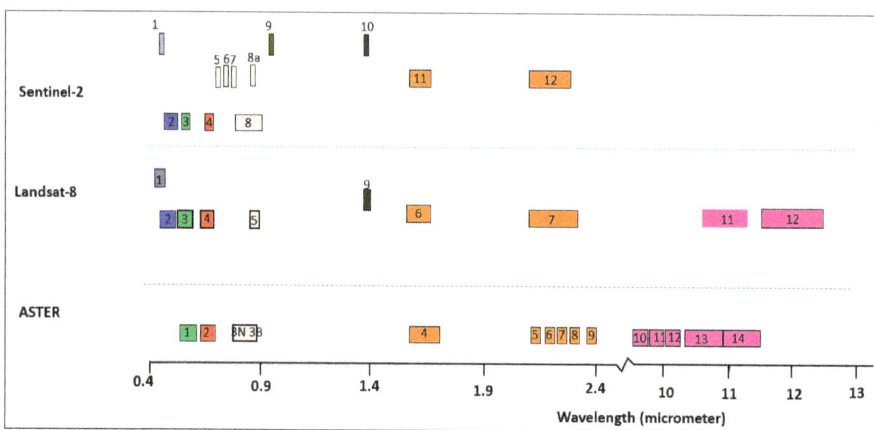

Figure 2. Comparison between ASTER, Landsat 8, and Sentinel-2.

On 11 February 2013, the Landsat-8 (OLI) satellite was launched. Landsat-8 scene dimension is 85-km-cross-track-by-180-km-along-track. There are nine VIS/NIR and SWIR ranges reported, as well as two longwave thermal ranges. The pixel size of OLI channels was stated to be 30 m; however, TIRS has a spatial resolution of 100 m. The quantization level is 12-bit data that permits additional bits to be used to acquire optimal data, enabling the assessment of minor surface disturbances. Landsat-OLI scene (path/row 166/43; ID: LC08_L1TP_166043_20211209_20211215_01_T1) that was acquired on 9 December 2021.

NASA and METI (Japan's Ministry of Economic Trade and Industry) deployed ASTER, an advanced multispectral satellite imaging system, onboard the TERRA spacecraft in December 1999. NASA's Land Processes Distributed Active Archive Centre provided the ASTER data (LP DAAC). ASTER data includes spectral ranges in the visible and near-infrared (VNIR), shortwave infrared (SWIR), and thermal infrared (TIR): three bands (with 15 m spatial resolution) in the VNIR, six bands (with 30 m spatial resolution) in the SWIR, and five bands (with 90 m spatial resolution) in the TIR (TIR). In this investigation, the ASTER SWIR spectral bands (30 m spatial resolution) are used to measure between ~1.60 and 2.45 µm to allow discriminating between Al-OH, Fe, Mg-OH, H-O-H, and CO_3 absorption features [43].

Preprocessing of the obtained ASTER scene (ASTER data ID: ASTB061106074035) included cross-talk correction and data orthorectification using the ENVI software application. Using ENVI v.5 software, the Log-residual (LR) technique was used to calibrate, normalize, and decrease noise from sensors and solar illumination in SWIR bands [2,14]. On ASTER data, this approach was used to remove atmospheric and topography impacts. As a result, the data became more reflective of the target area's composition and lithology. It was also possible to compare the retrieved endmembers of SWIR bands to reference spectra from the spectral library of the United States Geological Survey (USGS). This approach was used to reveal minerals using SWIR data [14,44].

On 23 June 2015, the Sentinel-2A satellite was launched, and the first data was taken a few days later. Sentinel-2 sensors gather data in the VIS/NIR, and SWIR, TIR wavelength ranges. These bands have a spatial resolution of 10–60 m. Sentinel-2 captures 13 bands in the VIS/NIR and SWIR spectrum. The VIS/NIR bands: blue B2 (490 nm), green B3 (560 nm), red B4 (665 nm), and infrared B8 (842 nm) have a 10 m pixel size, whilst the coastal band B1 (443 nm) has a 60 m pixel geometry. The pixel sizes of the SWIR bands (B11: 1610 nm, B12: 2190 nm) are both 20 m. Sentinel-2 scene (S2A_MSIL1C_20211221T073331_N0301_R049_T38QNM_20211221T084355) is delivered as zip-compressed files in Sentinel's own SAFE format. The spectral bands are stored as jpg files in this SAFE file in three different geometric resolutions (10 m, 20 m, and 60 m. The jpg files of bands B2, B3, B4, and B8 with a spatial resolution of 10 m, and B11 and B12 with 20 m are stacked into a single GeoTIFF file of a uniform pixel size of 10 m. A subset of these data was conducted during preprocessing using SNAP software in order to minimize the computational time and the data.

To analyze multispectral data, several methods have been used, including PCA, the utilization of band ratios, relative absorption band depth (RBD; 24), and mineral indices [11,12,22], as well as spectral analysis. Band ratios have been employed to investigate spectral differences between bands and to reduce topographic impacts [7,8,23,45]. The intensity of the hydrothermal activity can be used to reveal hydrothermal mineral assemblages [13,28]. Sub-pixel spectral classifications can thus be attributed to specific important hydrothermal minerals associated with propylitic (epidote, chlorite, calcite), phyllic (muscovite, sericite, illite), argillic (montmorillonite, kaolinite, dickite), and advanced argillic (alunite–pyrophyllite) alteration zones.

The band ratio is a transformation procedure for enhancing spectral differences in remote sensing data. It works by dividing pixels from one band by pixels from another band [46] and sometimes dividing bands of the numerator or/and denominator after mathematical calculation. The goal of this technique is to reveal the spectral characteristics of material so that variables on Earth's surface can be distinguished better [47]. Band ratios can be used to distinguish between soils, rock types, and land use effects [48–51]. The ENVI software and ArcGIS software packages v. 10.8 are utilized in the present study. The PCA process has been used to transform a large number of correlated spectral bands into a smaller number of uncorrelated spectral bands, which is a statistical approach used in image transformation. In the mapping of hydrothermal potential alteration zones, the selective "principal components" (PCs) technique has been frequently used [5,15]. Based on the eigenvectors of the selected bands, statistical parameters were examined to determine which PC image could be utilized to emphasize the particular minerals.

Spectral mapping was used to differentiate mixed pixels from unwanted pixels during the mineral extraction process. This enabled the mapping and identification of possible minerals based on the end member's spectral signature in comparison to those in the spectral library [5,14]. The MNF transformation [52] was used to derive a PPI that represented the input image's most spectrally pure pixels. This was utilized to detect endmembers using n-D visualization for mineral identifications based on spectral classifications. Using MNF and PPI, the n-D visualize viewer can locate, characterize, cluster (group), and pick the purest pixels (endmembers) in n-spaces. Each class indicated a mineral with a high absorption capacity.

Digital overlay approaches have been utilized to create integrative maps using Geographic Information System (GIS) technologies [28,29,53]. Predictive maps have also been created using knowledge-driven systems using weighted overlay analysis of ArcGIS that integrate multi-criteria decision-making based on expert judgment [30,54]. Each evidential image was reclassified into five classes using the Natural Breaks method; the class of high intensity of hydrothermal alteration is given "5" and the opposite given "1. As a result, the final prospective map can be created by combining several evidential maps [54].

4. Results

4.1. Lithologic and Structural Characteristics

Goethite, hematite, and jarosite are examples of iron minerals that have diagnostic spectral characteristics near 0.43 m, 0.65 m, 0.85 m, and 0.93 m, which are close to Sentinel-2 band 1, band 4, band 8/8A, and band 9 [55]. Moreover, both hematite and jarosite exhibit reflectance characteristics near 0.72 m and 0.74 m, which are both close to Sentinel-2 band 6. Hematite also displays a distinguishing absorbance pattern at a wavelength of 0.88 m, which corresponds to Sentinel-2A band 8A. Thus, Sentinel-2 band ratios of 6/1, 6/8A, and (6 + 7)/8A were utilized to distinguish hematite + goethite, hematite + jarosite, and a mixture of iron-bearing minerals (see more information on iron mineral spectra in Ge et al. [55]) from felsic or sedimentary deposits in red from basement mafic to intermediate variations in cyan (See Figure 3a) as they contain ferromagnesian minerals [2]. Sentinel-2 (Figure 3b) uses 11/8A, 3/4, and (6 + 7)/8A of ferric, ferrous, and a combination of iron-containing minerals to indicate likely areas rich in hydrothermal alteration in purple, mafic varieties in yellow, and sedimentary deposits in the northeast in red-orange, but vegetation in green. Images from the Sentinel-2 satellite (Figure 3c) 11/12, 11/8A, and (6 + 7)/8A displayed white-toned patches in HAZs [2]. The study's heights (Figure 3d) vary from 789 to 933 m above sea level, and the majority of the structural patterns are visible at these elevations.

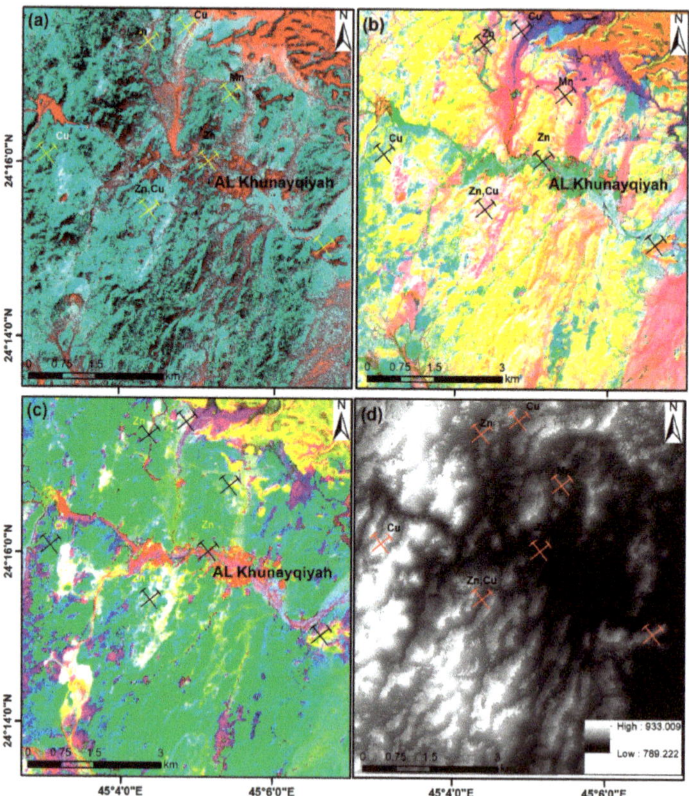

Figure 3. (**a**) band ratio 6/1, 6/8A, and (6 + 7)/8A; (**b**) 11/8A, 3/4, and (6 + 7)/8A; (**c**) 11/12, 11/8A, and (6 + 7)/8A; (**d**) DEM derived from Geo-Eye data.

4.2. Hydrothermal Alteration Zones

4.2.1. Landsat-8

Band ratios are used to enhance hydrothermally altered zones and the oxidation zone that reflects the abundance of certain minerals. Band ratios 6/7, 6/2, and 6/5 * 4/5 [16] were employed to improve the identification of rocks and minerals based on content mineralogy (Figure 4a,b). Band ratio 6/7 is susceptible to OH-bearing minerals, and band ratio 6/2 highlights rocks rich in FeO composition, so mafic igneous rocks have lower reflectance than other igneous rocks; and band 6/5 * 4/5 is useful to distinguish between mafic and non-mafic rocks based on their sensitivity to high Fe-bearing aluminosilicate concentration. Iron-bearing minerals and OH-bearing minerals are abundant in these areas. Sultan et al. [16] demonstrated that rationing in the 6/5 * 4/5 band is possible. Sultan et al. [16] use the sensitivity of the ratio to Fe-bearing aluminosilicates to distinguish mafic rocks (bluish color) from other rocks. Ramadan et al. [56] revealed the potential locations of hydrothermal mineral deposits in 6/7, 6/5, and 5 (Figure 4c,d). Using the band ratios 6/7, 4/2, and 5/6 in R, G, and B [7], felsic rocks are colored green, mafic rocks are colored blue, and areas of extensive hydrothermal alteration are colored light pink, yellow, and light red (Figure 4a,c,e, respectively). The 6/7 ratio emphasizes hydrothermal alteration and surface weathering oxides and hydroxides [23,57,58]. Clay minerals have a high Band 6 reflectance and a strong Band 7 absorption [10,58]. The 4/2 ratio is important for detecting iron oxide-bearing rocks (Figure 4e) due to considerable absorption in Band 2 and reflectance characteristics in Band 4 for iron oxides [7]. The felsic rocks in 4/2 appear in green color.

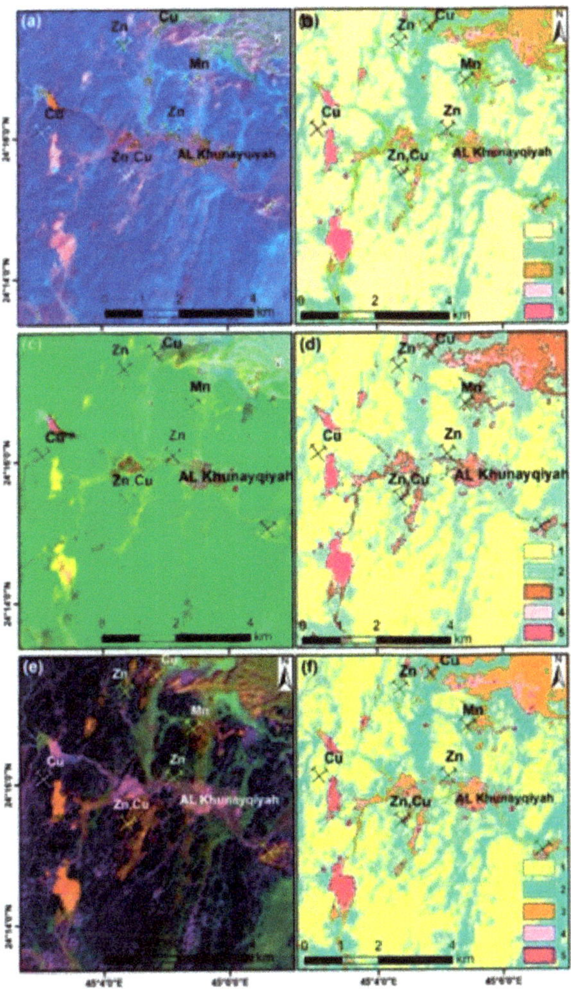

Figure 4. (**a**) Band ratios 6/7, 6/2, and 6/5 * 4/5, (**b**) reclassify of 6/7, 6/2, and 6/5 * 4/5, (**c**) 6/7, 6/5, 5 of Ramadan; (**d**) reclassify of 6/7, 6/5, 5; (**e**) Band ratios composite (6/7, 4/2, 5/6) Abrams; (**f**) reclassify of (6/7, 4/2, 5/6).

4.2.2. Sentinel-2

As shown in (Figure 5a), [59] offered three band ratios in R, G, and B: 11/12, 11/8, and 4/2. These band ratios were utilized in this research to distinguish the alteration zones. These ratios are proportional to the occurrence of OH-bearing minerals (11/12), iron oxides (4/2), and the band ratio 6/5, which is utilized to enhance the presence of ferrous oxides. Metavolcanics are colored green in this ratio, indicating high ferrous oxide content. Some wadi deposits have a purple tint due to the occurrence of clay minerals and iron-bearing minerals in high concentrations. The locations of probable HAZs are revealed in a yellow color due to their high presence of clay and ferrous oxides.

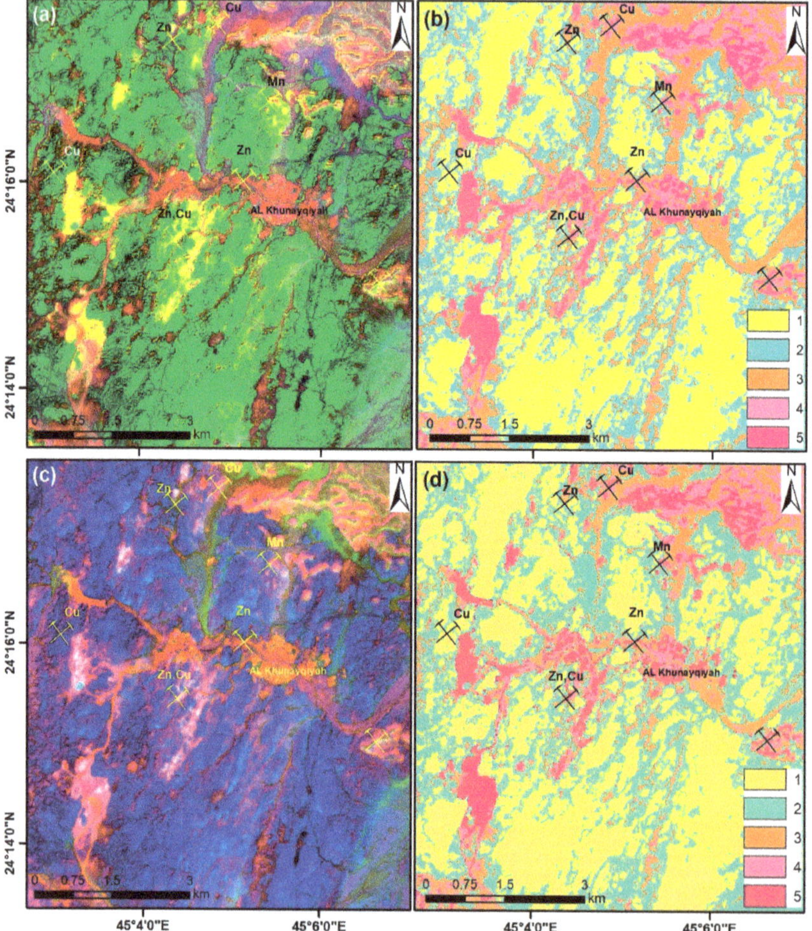

Figure 5. Sentinel-2 (**a**) Band ratios 11/12, 11/8, and 4/2; (**b**) reclassify of 11/12, 11/8, and 4/2; (**c**) 11/12, 11/2, and 11/8 * 4/8; (**d**) reclassify of 11/12, 11/2, and 11/8 * 4/8.

To improve the identification of rock units based on mineral content, ratio bands of 11/12, 11/2, and 11/8 * 4/8 [16] were used (Figure 5b,c). Band ratio 11/12 is amenable to OH-bearing minerals; band ratio 11/2 is associated with the content of opaque minerals (e.g., FeO) in rocks, so mafic rock types have lesser reflectance than some other igneous rocks; and band 11/8 * 4/8 can be used to distinguish between mafic and non-mafic rocks based on their sensitivity to high Fe-bearing aluminosilicate concentration. Iron-bearing minerals and OH-bearing minerals are widespread in such regions.

4.2.3. ASTER

The OH-bearing minerals have reflectance at 1.656 µm (band 4). The argillic minerals (montmorillonite and kaolinite) contain absorption features at 2.205 (band 6), and kaolinite display double-shaped absorption features around 2.165 (band 5) and 2.205 (band 6), in contrast to phyllic minerals (muscovite and illite), which have a single deep absorption feature at 2.205 (band 6) (Figure 6; from Mars and Rowan [3]). Propylitic minerals with a 2.335 m absorption characteristic. This most likely matches minerals like calcite and chlorite that contain CaCO3 or Mg-OH [3].

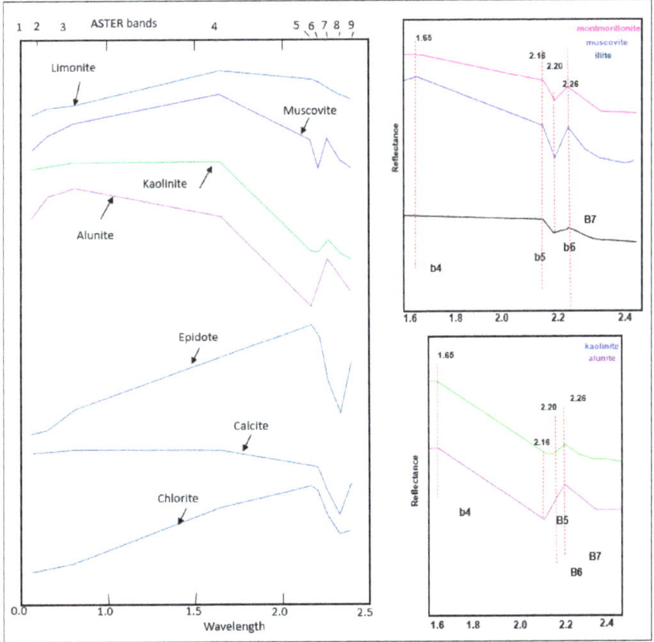

Figure 6. Mineral spectra of minerals and ASTER bands.

The ASTER band ratios 4/6, 4/5, and 4/7 boost argillic and sericitic alteration zones, respectively [6]. Furthermore, in these images, the ASTER– 4/5 band ratio defines the advanced argillic alteration (e.g., alunite and dickite). As a result, the combination of band ratios 4/6, 4/5, and 4/7 in R, G, and B of ASTER is utilized to depict HAZs. Figure 7 depicts the classification of these combined ratios into five ranks of hydrothermal alteration, with the highest rank (0.60–0.64) in red, denoting locations with both argillic and sericitic regions of alteration.

Figure 7c,d shows the results of merging band ratio images 4/6, 7/6, and (5 + 7)/6. Regions richer in white mica were identified using the band ratio of 7/6. Using such ratios, locations rich in Al-OH minerals are depicted in a white tone. The largest band ratios of 4/6 and 7/6 refer to the phyllic zone [60], and the values of (5 + 7)/6 are also true. The green-colored area has less hydrothermal alteration than the first, but pink-colored portions have the least amount of HAZs. Band ratios 4/6 in this composite highlight the white tone's alteration zones (Figure 7). Furthermore, band ratios of 7/6 were applied to determine the areas of an abundance of Al-OH minerals as white mica (muscovite) as revealed in white tone, which occupies the middle part of the present study area (Figure 7).

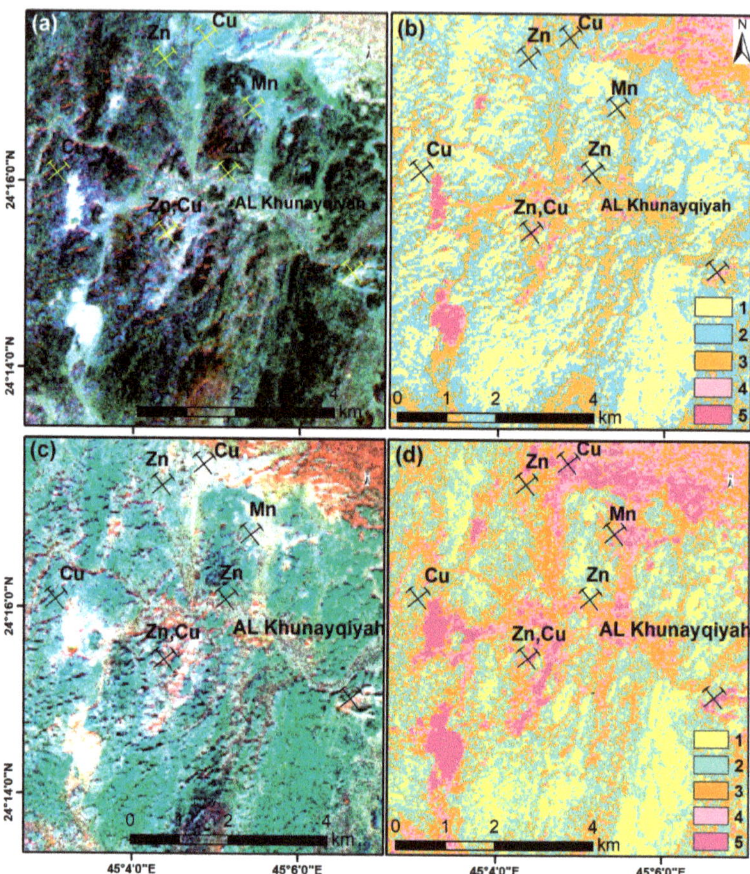

Figure 7. ASTER (**a**) band ratios 4/6, 4/5, and 4/7; (**b**) reclassify band ratios 4/6, 4/5, and 4/7; (**c**) 4/6, 7/6, and (5 + 7)/6; (**d**) reclassify 4/6, 7/6, and (5 + 7)/6.

The contrast between hydrothermally altered zones was emphasized by the combination of band ratios (5 + 7)/6, (4 + 6)/5, and (7 + 9)/8 in R, G, and B, respectively (Figure 8a). This allowed areas richer in phyllic, argillic [6,61], and propylitic minerals to be detected, respectively. In this combination, yellow areas revealed argillic and phyllic hydrothermal alterations. PCA was used to map areas of argillic hydrothermal alteration using ASTER bands 4, 5, and 6b. Table 1 shows the eigenvector values obtained using the specific bands (B4, B5, and B6) for the PCA method (Table 1). PC2 indicates a negative loading of band 4 (−0.780) and a positive loading of bands 5 (0.605) and 6 (= 0.157) according to eigenvector loadings. The locations of hydrothermal alteration are revealed in a white tone when negated (multiplied by −1) PC2 is displayed in greyscale (Figure 8b).

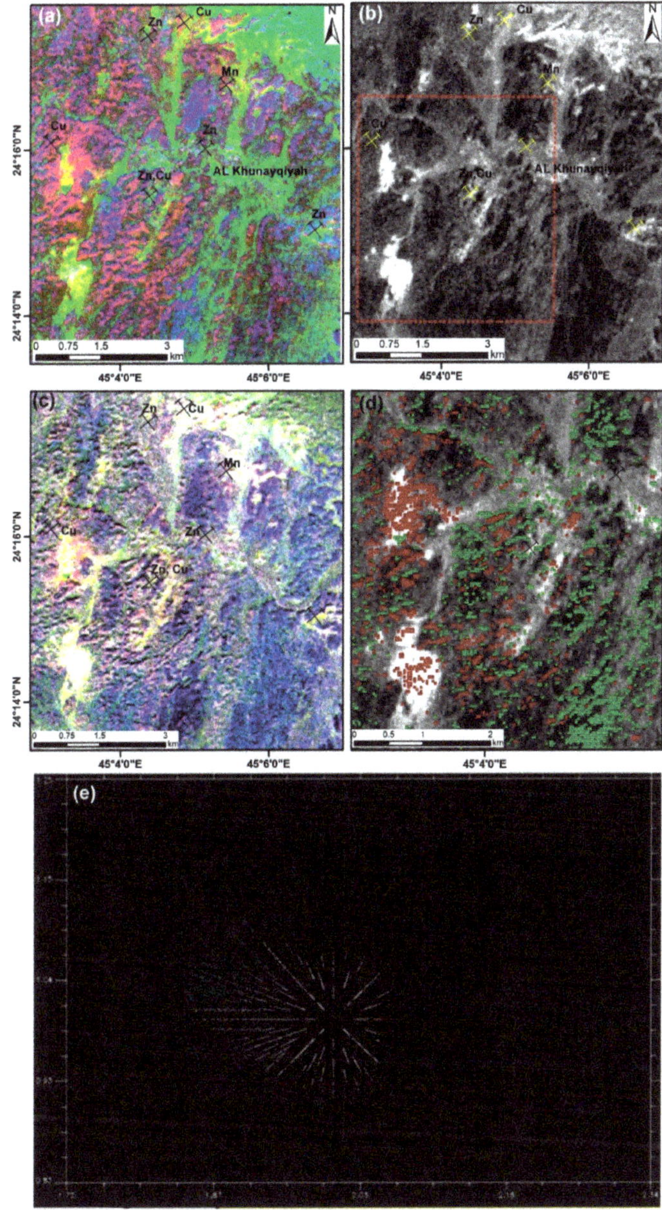

Figure 8. ASTER (**a**) band ratios composite $(5 + 7)/6$, $(4 + 6)/5$, and $(7 + 9)/8$ in R, G, and B; (**b**) OHI, KAI, and ALI in R, G, and B; (**c**) Negated PC2 of selected bands 4, 5, and 6; (**d**) subset of "c" image overlain by extracted interest pixels of Scattergram of the ASTER derived bands $5 + 7/6$ 'Al-OH content' vs. bands $5/7$ (outside absorption) 'Al-OH composition' (Cudahy et al. 2008) in "(**e**)". The AL-OH area rich in minerals marked in red are consistent with the areas of high hydrothermal alteration.

Table 1. PCA of selected bands 4, 5, and 6.

Eigenvector	Band 4	Band 5	Band 6	Eigenvalue
PC1	−0.5646	−0.57235	−0.59468	99.923
PC2	−0.7798	0.605985	0.157136	0.057
PC3	−0.27043	−0.55245	0.788457	0.021

The band ratios OHI, KAI, and ALI were integrated into R, G, and B to distinguish between potential sites of argillic and phyllic alteration. The areas of increased hydrothermal alteration are congruent with structural features associated with granitic rocks, according to the classifications of these combined values. Each fraction's greatest value is denoted by a white tone. The minerals indices OHI, KLI, and ALI were displayed in R, G, and B, respectively (Figure 8c), and locations with abundances of the three indices of OHI, KAI, and ALI are emphasized in white tone.

OHI bearing altered minerals Index (OHI) = [band 7/band 6] × [band 4/band 6]

Kaolinite Index (KLI) = [band 4/band 5] × [band 8/band 6],

Alunite Index (ALI) = [band 7/band 5] × [band 7/band 8]

A subset image of PC2 (Figure 6d) is overlain by extracted interest pixels derived from the scattergram (Figure 6e). A two-dimensional (2D) scatter plot of band ratios (5 + 7)/6 (Al-OH content) vs. 5/7 (Al-OH composition) was used to explore ASTER Al-OH composition [62]. The diagram's extreme far bottom right side indicated regions with no Al-OH minerals, whereas the extreme top left side of the diagram revealed places with no Al-OH minerals. Cudahy et al. [62] found that plotting band ratios (5 + 7)/6 (Al-OH content) vs. 5/7 (Al-OH composition) clearly separated areas rich in Al-OH from those with no Al-OH, confirming prior findings. Higher values, indicating higher Al-OH concentrations, are highlighted in red and correspond to locations of significant hydrothermal alteration. As illustrated by the green in Figure 8e, this area was clearly delimited by lower Al-OH concentration and higher Mg-OH content.

5. XRD Analysis of Hydrothermal Alteration Zones

Samples were taken from the two main alteration zones (Figure 9), and 21 representative specimens were selected for XRD analysis at the National Research Center (Egypt). The results of the analyses revealed the presence of silica minerals (mostly quartz), gypsum, anhydrite, kaolinite, illite, clinochlore, and hematite, with a small percentage of microcline, calcite, and halite (Tables 2 and 3). Clay minerals (kaolinite, illite, sericite) in these zones are mainly the products of the decomposition of plagioclase feldspar, and the presence of ferric iron oxides and hydroxides (hematite, goethite) is related to the weathering of ferromagnesian minerals (mostly hornblende and clinopyroxene). The abundance of sulfates in the form of gypsum and anhydrite is highly indicative of the former presence of disseminated sulfide phases, which is compatible with these zones being either mature gossans or conduits within the feeder zones beneath massive sulfide mounds.

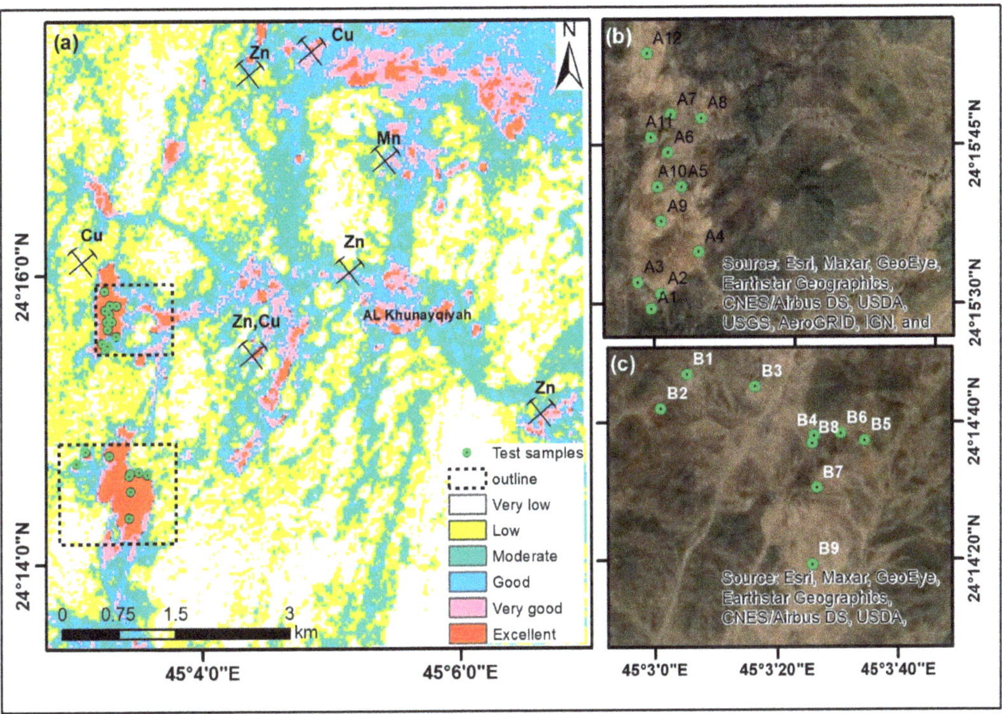

Figure 9. (**a**) Mineral perspective map of the study area; (**b**,**c**) Maps show sample locations that were selected for XRD analysis. A shows sample location for Table 2. B shows the samples location for Table 3.

Table 2. The results of XRD analysis; see Figure 9b for the location map.

Sample Name	Compound Name	Chemical Formula	Vol %
A1	Quartz	SiO_2	25.8
	Gypsum	$CaSO_4 \cdot 2H_2O$	61.9
	Illite	$K_{0.5}(Al,Fe,Mg)3Si,Al)4O_{10}(OH)_2$	9.7
	Anhydrite	$CaSO_4$	2.6
A2	Quartz	SiO_2	5.7
	Gypsum	$CaSO_4 \cdot 2H_2O$	90.2
	Kaolinite	$Al_2Si_2O_5(OH)_4 / Al_2O_3 \cdot 2SiO_2 \cdot 2H_2O$	3.0
	Anhydrite	$CaSO_4$	7.1
A3	Quartz	SiO_2	43.1
	Kaolinite	$Al_2Si_2O_5(OH)_4 / Al_2O_3 \cdot 2SiO_2 \cdot 2H_2O$	18.9
	Illite	$KAl_2Si_3AlO_{10}(OH)2$	31.5
	Anhydrite	$CaSO_4$	6.6
A4	Quartz	SiO_2	62.50
	Gypsum	$CaSO_4 \cdot 2H_2O$	12.30
	Illite	$KAl_2Si_3AlO_{10}(OH)_2$	18.8
	Hematite	Fe_2O_3	6.4

Table 2. *Cont.*

Sample Name	Compound Name	Chemical Formula	Vol %
A5	Quartz	SiO_2	41.5
	Gypsum	$CaSO_4 \cdot 2H_2O$	24.9
	Illite	$KAl_2Si_3AlO_{10}(OH)_2$	3.1
	Kaolinite	$Al_2Si_2O_5(OH)_4 / Al_2O_3 \cdot 2SiO_2 \cdot 2H_2O$	17.1
	Hematite	Fe_2O_3	2.0
	Clinochlore	$Mg_5Fe_0 \cdot 2Al_2Si_3O_{10}(OH)_8$	11.3
A6	Quartz	SiO_2	62.8
	Hematite	Fe_2O_3	3.1
	Illite	$KAl_2Si_3AlO_{10}(OH)_2$	24.5
	Anhydrite	$CaSO_4$	9.6
A7	Quartz	SiO_2	47.2
	Gypsum	$CaSO_4 \cdot 2H_2O$	21.2
	Illite	$KAl_2Si_3AlO_{10}(OH)_2$	10.6
	Halite	$NaCl$	3.2
	Bassanite	$CaSO_4 \cdot 0.5H_2O$	17.7
A8	Quartz	SiO_2	20.0
	Gypsum	$CaSO_4 \cdot 2H_2O$	72.5
	Illite	$KAl_2Si_3AlO_{10}(OH)_2$	7.6
A9	Quartz	SiO_2	70.8
	Gypsum	$CaSO_4 \cdot 2H_2O$	21.2
	Calcite	$CaCO_3$	8.0
A10	Quartz	SiO_2	38.3
	Gypsum	$CaSO_4 \cdot 2H_2O$	35.8
	Illite	$KAl_2Si_3AlO_{10}(OH)_2$	20.1
	Anhydrite	$CaSO_4$	5.9
A11	Quartz	SiO_2	48.4
	Gypsum	$CaSO_4 \cdot 2H_2O$	18.1
	Illite	$KAl_2Si_3AlO_{10}(OH)_2$	15.9
	Anhydrite	$CaSO_4$	4.9
	Hematite	Fe_2O_3	4.1
	Kaolinite	$Al_2Si_2O_5(OH)_4 / Al_2O_3 \cdot 2SiO_2 \cdot 2H_2O$	8.6
A12	Quartz	SiO_2	57.3
	Albite	$NaAlSi_3O_8$	10.2
	Illite	$KAl_2Si_3AlO_{10}(OH)_2$	6.4
	Microcline	$KAlSi_3O_8$	13.6
	Calcite	$CaCO_3$	12.4

Table 3. The results of XRD analysis; see Figure 8b for the location map.

Sample Name	Compound Name	Chemical Formula	vol %
B1	Quartz	SiO_2	37.5
	Gypsum	$CaSO_4 \cdot 2H_2O$	28.1
	Kaolinite	$Al2Si2O5(OH)4 / Al2O3.2SiO_2 \cdot 2H_2O$	28.1
	Illite	$KAl_2Si_3AlO_{10}(OH)_2$	3.4
	Bassanite	$CaSO4.0.5H2O$	2.8
B2	Quartz	SiO_2	11.4
	Gypsum	$CaSO_4 \cdot 2H_2O$	73.0
	Albite	$NaAlSi_3O_8$	8.7
	Calcite	$CaCO_3$	7.0

Table 3. *Cont.*

Sample Name	Compound Name	Chemical Formula	vol %
B3	Quartz	SiO_2	65.3
	Kaolinite	$Al_2Si_2O_5(OH)4/Al_2O_3 \cdot 2SiO_2 \cdot 2H_2O$	9.8
	Calcite	$CaCO_3$	8.2
	Hematite	Fe_2O_3	0.8
	Microcline	$KAlSi_3O_8$	15.9
B4	Quartz	SiO_2	4.3
	Gypsum	$CaSO_4 \cdot 2H_2O$	91.8
	Kaolinite	$Al_2Si_2O_5(OH)_4/Al_2O_3 \cdot 2SiO_2 \cdot 2H_2O$	3.8
B5	Quartz	SiO_2	29.9
	Albite	$NaAlSi_3O_8$	70.1
B6	Quartz	SiO_2	3.1
	Gypsum	$CaSO_4 \cdot 2H_2O$	77.0
	Kaolinite	$Al_2Si_2O_5(OH)_4/Al_2O_3 \cdot 2SiO_2 \cdot 2H_2O$	19.2
	Anatase	TiO_2	0.6
B7	Quartz	SiO_2	75.3
	Gypsum	$CaSO_4 \cdot 2H_2O$	4.5
	Illite	$KAl_2Si_3AlO_{10}(OH)_2$	12.8
	Kaolinite	$Al_2Si_2O_5(OH)_4/Al_2O_3.2SiO_2 \cdot 2H_2O$	5.9
	Anhydrite	$CaSO_4$	1.5
B8	Quartz	SiO_2	47.3
	Gypsum	$CaSO_4 \cdot 2H_2O$	21.3
	Kaolinite	$Al_2Si_2O_5(OH)4/Al_2O_3.2SiO_2 \cdot 2H_2O$	2
	Calcite	$CaCO_3$	3.2
	Albite	$NaAlSi_3O_8$	4.6
	Minamite	$(Na,Ca)1-xAl_3(SO_4)2(OH)_6$	2.4
	Halite	$NaCl$	0.8
B9	Quartz	SiO_2	31.3
	Gypsum	$CaSO_4 \cdot 2H_2O$	49.2
	Kaolinite	$Al_2Si_2O_5(OH)4/Al_2O_3 \cdot 2SiO_2 \cdot 2H_2O$	6.4
	Illite	$KAl_2Si_3AlO_{10}(OH)_2$	4.1
	Albite	$NaAlSi_3O_8$	8.9

6. Mineral Potential Map

The final potential location of mineralization was created by combining multi-criteria data. This allows for revealing the prospective areas of hydrothermal mineralization associated with hydrothermal alteration zones; thus, we use a variety of ways to emphasize alteration zones. The results of remote sensing analysis data were integrated using GIS approaches (band ratio, PC, mineral indices). These images were quantified and divided into various zones (different probability values). To obtain the prospective or promising map of mineral exploration, a succession of evidential maps is used. The approach of merging data in a GIS enabled the promotion and identification of the best exploration and mining locations. This aided exploration and the prediction of new mineralized zones. The recent innovative procedures that use recent digital technologies and creative geo-information approaches have allowed for the detection of the optimal mineral resource area. Multiple datasets have been aggregated and integrated since the birth of the GIS to locate new mineralized zones [63] reliably.

Using a GIS-based overlay method, the likely locations of hydrothermal mineral deposits were revealed by combining several evidence hydrothermal alteration maps (Figure 9). The mineral prospective map is divided into six groups (Figure 9a) based on their prospective for hydrothermal alteration amplitude: very low, low, moderate, good, very good, and excellent, and covers 31.36, 28.22, 20.49, 10.99, 6.35, and 2.59 % percent of the research region, respectively. The red color indicates the most promising mineral deposit zone. The resulting map (Figure 9) demonstrates a pattern of coherence in the hydrothermal-ore deposits found in the area's mines. Many sections of the potentially high zone, meanwhile, were restricted to wadi deposits and areas of sedimentary cover in the northeast of the research area.

7. Discussion

The use of various methodologies for three different sensors, including OLI, Sentinel-2, and ASTER data, clearly shows that hydrothermal alteration processes dominate the examined area. This is because identifying HAZs through fracture/fault zones is required when exploring mineral deposits that originated from hydrothermal processes [2,5,8]. As a result, the severity of the alteration can help determine where the ore body is located.

The areas of HAZs (Figure 4) were identified using band ratios generated from OLI sensors such as 6/7, 6/2, and 6/5 * 4/5 [16], 6/7, 6/5, 5 [56], and 6/7, 4/2, 5/6 [7]. Because hydrothermal activities alter the physical and chemical characteristics of country rocks, they change. Band ratio 6/7 emphasized OH-bearing minerals such as kaolinite–smectite, micas, and amphiboles [23]. Iron-bearing minerals, on the other hand, are delineated utilizing band ratios such as 4/2, 6/5, and 6/5 * 4/5 [16,17]. Furthermore, applying Sentinel-2 band ratio 3/4 characterizes the ferrous iron, the ferric oxides (Fe^{3+}) represented by 11/8A, and ferrous iron (Fe^{2+}) represented by (3/4) [55]. In addition to Sentinel-2 band ratios, 11/12 marks the OH-bearing minerals [2]. This is because the integration of iron-bearing minerals mixed with OH-bearing minerals from different sensors (OLI and Sentinel-2) characterized the gossans and iron-rich zones [28,64], as displayed in Figures 3 and 4.

Following that, the SWIR ASTER data was analyzed using various band ratios to look for areas of hydrothermal alteration, comprising 4/6, 4/5, and 4/7; 4/6, 7/6, and (5 + 7)/6; (5 + 7)/6, PC2, MNF3; OHI, KAI, and ALI, and PC2 of PCA OHI, KAI, and ALI and Calcite. Delineation of OH–bearing minerals was possible because of the use of a band ratio of 4/6 (λ = 1.656/2.209 m) (Figure 7). The 4/6 ratio is excellent for accentuating hydrous minerals like kaolinite, illite, and montmorillonite because they have a high absorption signature in band 6 and a high reflectance in band 4. Furthermore, ASTER band ratios 4/5 and 4/7 boost argillic and sericitic alteration zones, respectively [2,6]. The white tone in Figures 7 and 8 highlights areas of hydrothermal alteration, which for the most part, aligns with structural connections.

The relative band depth (5 + 7)/6 was efficiently adopted (Figures 7c and 8a) for excellent detection of Al–smectite, muscovite, sericite, and illite [2,62,65], and Al/Fe-OH minerals, such as muscovite, kaolinite, and jarosite [66]. Moreover, ASTER bands 5 + 7/6 'Al-OH content' vs. bands 5/7 (outside absorption) 'Al-OH composition' [62]. This diagram (Figure 8c) shows that the selected red pixels are rich in AL-OH, but the lowest ones are in green. The AL-OH area rich in minerals that are marked in red is consistent with the areas of high hydrothermal alteration.

Using GIS-based weighted overlay analysis to confirm the findings of band ratios and mineral indices acquired from Landsat-OLI, Sentinel-2, and ASTER data that revealed iron-containing and Al-OH-carrying minerals revealed useful information regarding places rich in gossans. Such gossans that consist of limonite, goethite, hematite, malachite, and azurite reveal the existence of massive sulfide; porphyry and skarn deposits [64] are consistent with areas of high hydrothermal alteration intensity.

8. Conclusions

The Khnaiguiyah area, Saudi Arabia, is tested to delineate the area of probable mineral resources. The ability of multispectral remote sensing data to detect and characterize the hydrothermal alteration minerals is significant for mineral exploration. The present study used ASTER, Sentinel-2, and Landsat-OLI to identify potential areas of HAZs. The HAZs generated from these various multispectral sensors were combined through GIS to highlight the potential areas of HAZs. The highest grade of HAZs, which covers about 2.59 %, is compatible with areas of significant hydrothermal changes and has been verified with areas of gossans that revealed the presence of sulfide minerals.

Author Contributions: Conceptualization, M.A. (Mohamed Abdelkareem); methodology, M.A. (Mohamed Abdelkareem) and S.S.A.; software, I.S.A.; validation, M.A. (Mohamed Abdelkareem), S.S.A., and F.A.; investigation, M.A. (Mohamed Abdelkareem); resources, S.S.A.; writing—original draft preparation, M.A. (Mohamed Abdelkareem), F.A., A.M.A.-S., I.S.A., H.G. and M.A. (Mislat Alotaibi); writing—review and editing, M.A. (Mohamed Abdelkareem); visualization, M.A. (Mohamed Abdelkareem); funding acquisition, S.S.A. All authors have read and agreed to the published version of the manuscript.

Funding: This research was funded by King Saud University, grant number RSP2022R496. The APC was funded by King Saud University.

Acknowledgments: This research was supported by Researchers Supporting Project number (RSP2022R496), King Saud University, Riyadh, Saudi Arabia.

Conflicts of Interest: The authors declare no conflict of interest.

References

1. Shi, K.; Chang, Z.; Chen, Z.; Wu, J.; Yu, B. Identifying and evaluating poverty using multisource remote sensing and point of interest (POI) data: A case study of Chongqing, China. *J. Clean. Prod.* **2020**, *255*, 120245. [CrossRef]
2. Abdelkareem, M.; Al-Arifi, N. Synergy of Remote Sensing Data for Exploring Hydrothermal Mineral Resources Using GIS-Based Fuzzy Logic Approach. *Remote Sens.* **2021**, *13*, 4492. [CrossRef]
3. Mars, J.C.; Rowan, L.C. Regional mapping of phyllic- and argillic-altered rocks in the Zagros magmatic arc, Iran, using Advanced Spaceborne Thermal Emission and Refl ection Radiometer (ASTER) data and logical operator algorithms. *Geosphere* **2006**, *2*, 161–186. [CrossRef]
4. Abdelkareem, M.; El-Baz, F. Characterizing hydrothermal alteration zones in Hamama area in the central Eastern Desert of Egypt by remotely sensed data. *Geocarto Int.* **2018**, *33*, 1307–1325. [CrossRef]
5. Crosta, A.P.; De Souza Filho, C.R.; Azevedo, F.; Brodie, C. Targeting key alteration minerals in epithermal deposits in Patagonia, Argentina, using ASTER imagery and principal component analysis. *Int. J. Remote Sens.* **2003**, *24*, 4233–4240. [CrossRef]
6. Testa, F.J.; Villanueva, C.; Cooke, D.R.; Zhang, L. Lithological and Hydrothermal Alteration Mapping of Epithermal, Porphyry and Tourmaline Breccia Districts in the Argentine Andes Using ASTER Imagery. *Remote. Sens.* **2018**, *10*, 203. [CrossRef]
7. Abrams, M.J.; Brown, D.; Lepley, L.; Sadowski, R. Remote sensing for porphyry copper deposits in southern Arizona. *Econ. Geol.* **1983**, *78*, 591–604. [CrossRef]
8. Sabins, F. *Remote Sensing Principles and Interpretation*, 3rd ed.; W.H. Freeman Company: New York, NY, USA, 1997; p. 494.
9. Sabins, F.F. Remote sensing for mineral exploration. *Ore Geol. Rev.* **1999**, *14*, 157–183. [CrossRef]
10. Shi, X.; Al-Arifi, N.; Abdelkareem, M.; Abdalla, F. Application of remote sensing and GIS techniques for exploring potential areas of hydrothermal mineralization in the central Eastern Desert of Egypt. *J. Taibah Univ. Sci.* **2020**, *14*, 1421–1432. [CrossRef]
11. Ninomiya, Y. A stabilized vegetation index and several mineralogic indices defined for ASTER VNIR and SWIR data. In Proceedings of the 2003 IEEE International Geoscience and Remote Sensing Symposium, Toulouse, France, 21–25 July 2003; pp. 1552–1554.
12. Ninomiya, Y. Advanced remote lithologic mapping in ophiolite zone with ASTER multispectral thermal infrared data. In Proceedings of the 2003 IEEE International Geoscience and Remote Sensing Symposium, Toulouse, France, 21–25 July 2003; pp. 1561–1563.
13. Rowan, L.C.; Mars, J.C. Lithologic mapping in the Mountain Pass, California area using advanced spaceborne ther-mal emission and reflection radiometer (ASTER) data. *Remote Sens. Environ.* **2003**, *84*, 350–366. [CrossRef]
14. Azizi, H.; Tarverdi, M.; Akbarpour, A. Extraction of hydrothermal alterations from ASTER SWIR data from east Zanjan, northern Iran. *Adv. Space Res.* **2010**, *46*, 99–109. [CrossRef]
15. Zhang, X.; Pazner, M.; Duke, N. Lithologic and mineral information extraction for gold exploration using ASTER data in the south Chocolate Mountains (California). *ISPRS J. Photogramm. Remote. Sens.* **2007**, *62*, 271–282. [CrossRef]
16. Sultan, M.; Arvidson, R.E.; Sturchio, N.C. Mapping of serpentinites in the Eastern Desert of Egypt by using Landsat thematic mapper data. *Geology* **1986**, *14*, 995–999. [CrossRef]
17. Madani, A.M. Utilization of Landsat ETM+ Data for Mapping Gossans and Iron Rich Zones Exposed at Bahrah Area, Western Arabian Shield, Saudi Arabia. *J. King Abdulaziz Univ. Sci.* **2009**, *20*, 35–49. [CrossRef]
18. Ramadan, T.M.; Kontny, A. Mineralogical and structural characterization of alteration zones detected by orbital re-mote sensing at Shalatein District, SE Desert, Egypt. *J. Afr. Earth Sci.* **2004**, *40*, 89–99. [CrossRef]
19. Cardoso-Fernandes, J.; Teodoro, A.C.; Lima, A.; Perrotta, M.; Roda-Robles, E. Detecting Lithium (Li) Mineralizations from Space: Current Research and Future Perspectives. *Appl. Sci.* **2020**, *10*, 1785. [CrossRef]
20. El-Din, G.M.K.; El-Noby, M.E.; Abdelkareem, Z.M.; Hamimi, Z. Using multispectral and radar remote sensing data for geological investigation, Qena-Safaga Shear Zone, Eastern Desert, Egypt. *Arab. J. Geosci.* **2021**, *14*, 997. [CrossRef]
21. Galvão, L.S.; Almeida-Filho, R.; Vitorello, I. Spectral discrimination of hydrothermally altered materials using ASTER short-wave infrared bands: Evaluation in a tropical savannah environment. *Int. J. Appl. Earth Obs. Geoinf.* **2005**, *7*, 107–114. [CrossRef]

22. Yamaguchi, Y.; Naito, C. Spectral indices for lithologic discrimination and mapping by using the ASTER SWIR bands. *Int. J. Remote Sens.* **2003**, *24*, 4311–4323. [CrossRef]
23. Gupta, R.P. *Remote Sensing Geology*, 2nd ed.; Springer-Verlag: Berlin/Heidelberg, Germany, 2003.
24. Crowley, J.K.; Brickey, D.W.; Rowan, L.C. Airborne imaging spectrometer data of the Ruby Mountains, Montana: Mineral discrimination using relative absorption band-depth images. *Remote Sens. Environ.* **1989**, *29*, 121–134. [CrossRef]
25. Zhang, N.; Zhou, K. Mineral prospectivity mapping with weights of evidence and fuzzy logic methods. *J. Intell. Fuzzy Syst.* **2015**, *29*, 2639–2651. [CrossRef]
26. Harris, J.R.; Wilkinson, L.; Heather, K.; Fumerton, S.; Bernier, M.A.; Ayer, J.; Dahn, R. Application of GIS Processing Techniques for Producing Mineral Prospectivity Maps—A Case Study: Mesothermal Au in the Swayze Greenstone Belt, Ontario, Canada. *Nonrenewable Resour.* **2001**, *10*, 91–124. [CrossRef]
27. Abdelkareem, M.; Othman, I.; Kamal El Din, G. Lithologic mapping using remote sensing data in Abu Marawat Area, Eastern Desert of Egypt. *Int. J. Adv. Remote Sens. GIS.* **2017**, *6*, 2171–2177. [CrossRef]
28. Abdelkareem, M.K.; El-Din, G.; Osman, I. An integrated approach for mapping mineral resources in the Eastern Desert of Egypt. *Int. J. Appl. Earth Obs. Geoinf.* **2018**, *73*, 682–696. [CrossRef]
29. Sun, T.; Chen, F.; Zhong, L.; Liu, W.; Wang, Y. GIS-based mineral prospectivity mapping using machine learning methods: A case study from Tongling ore district, eastern China. *Ore Geol. Rev.* **2019**, *109*, 26–49. [CrossRef]
30. Campos, L.D.; de Souza, S.M.; de Sordi, D.A.; Tavares, F.M.; Klein, E.L.; Lopes, E.C.D.S. Predictive mapping of prospectivity in the Gurupi orogenic gold belt, north–northeast Brazil: An example of districtscale mineral system approach to exploration targeting. *Nat. Resour. Res.* **2017**, *26*, 509–534. [CrossRef]
31. Stern, R.J.; Johnson, P.R.; Kröner, A.; Yibas, B. Neoproterozoic ophiolites of the Arabian-Nubian shield. *Dev. Precambrian Geol.* **2004**, *13*, 95–128.
32. Kröner, A.; Stern, R.J. Pan-African Orogeny. *Encycl. Geol.* **2004**, *1*, 1–12.
33. Stoeser, D.B.; Camp, V.E. Pan-African microplate accretion of the Arabian Shield. *Geol. Soc. Am. Bull.* **1985**, *96*, 817–826. [CrossRef]
34. Johnson, P.R.; Woldehaimanot, B. *Development of the Arabian-Nubian Shield: Perspectives on Accretion and Defor-Mation in the Northern East African Orogen and the Assembly of Gondwana*; Geological Society, London, Special Publications: London, UK, 2003; Volume 206, pp. 289–325.
35. Genna, A.; Nehlig, P.; Le Goff, E.; Guerrot, C.; Shanti, M. Proterozoic tectonism of the Arabian Shield. *Precambrian Res.* **2002**, *117*, 21–40. [CrossRef]
36. Nehlig, P.; Genna, A.; Asfirane, F.; BRGM France; Guerrot, C.; Eberlé, J.; Kluyver, H.; Lasserre, J.; Le Goff, E.; Nicol, N.; et al. A review of the Pan-African evolution of the Arabian Shield. *GeoArabia* **2002**, *7*, 103–124. [CrossRef]
37. Hamimi, Z.; Fowler, A.R. Najd Shear System in the Arabian-Nubian Shield. In *The Geology of the Arabian-Nubian Shield*; Springer: Cham, Switzerland, 2021; pp. 359–392.
38. Hamimi, Z.; Abdelkareem, M.; Fowler, A.R.; Younis, M.H.; Matsah, M.; Abdalla, F. Remote sensing and structural studies of the Central Asir Shear Zone, Western Arabian Shield: Implications for the late Neoproterozoic EW Gondwana assembly. *J. Asian Earth Sci.* **2021**, *215*, 104782. [CrossRef]
39. Johnson, P.R.; Andresen, A.; Collins, A.S.; Fowler, A.R.; Fritz, H.; Ghebreab, W.; Kusky, T.; Stern, R.J. Late Cryoge-nian–Ediacaran history of the Arabian–Nubian Shield: A review of depositional, plutonic, structural, and tectonic events in the closing stages of the northern East African Orogen. *J. Afr. Earth Sci.* **2011**, *61*, 167–232. [CrossRef]
40. Doebrich, J.L.; Al-Jehani, A.M.; Siddiqui, A.A.; Hayes, T.S.; Wooden, J.L.; Johnson, P.R. Geology and metallogeny of the Ar Rayn terrane, eastern Arabian shield: Evolution of a Neoproterozoic continental-margin arc during assembly of Gondwana within the East African orogen. *Precambrian Res.* **2007**, *158*, 17–50. [CrossRef]
41. BRGM Geoscientists. *Khnaiguiyah Zinc-Copper Deposit Prefeasibility Study: Saudi Arabian Directorate General of Mineral Resources*; Technical Report BRGM-TR-13-4; French Geological Survey: Orléans, France, 1994.
42. Vaslet, D.; Delfour, J.; Manivit, J.; Le Nindre, Y.M.; Brosse, J.M.; Fourniquet, J. *Geologic Map of the Wadi ar Rayn Quadrangle, Sheet 23H, Kingdom of Saudi Arabia*; Saudi Arabian Deputy Ministry for Mineral Resources: Jeddah, Saudi Arabia, 1983.
43. Mahdevar, M.R.; Ketabi, P.; Saadatkhah, N.; Rahnamarad, J.; Mohammadi, S.S. Application of ASTER SWIR data on detection of alteration zone in the Sheikhabad area, eastern Iran. *Arab. J. Geosci.* **2015**, *8*, 5909–5919. [CrossRef]
44. Hosseinjani Zadeh, M.; Tangestani, M.H.; Roldan, F.V.; Yusta, I. Mineral exploration and alteration mapping using mixture tuned matched filtering approach on ASTER data at the central part of Dehaj-Sarduiyeh copper belt, SE Iran. *IEEE J. Sel. Topics Appl. Earth Obs. Remote Sens.* **2014**, *7*, 284–289. [CrossRef]
45. Rowan, L.C.; Goetz, A.F.H.; Ashley, R.P. Discrimination of Hydrothermally Altered and Unaltered Rocks in Visible and Near Infrared Multispectral Images. *Geophysics* **1977**, *42*, 522–535. [CrossRef]
46. Mather, P.M.; Koch, M. *Computer Processing of Remotely-Sensed Images: An Introduction*; John Wiley & Sons: Hoboken, NJ, USA, 2011.
47. Goetz, A.F.H.; Rock, B.N.; Rowan, L.C. Remote sensing for exploration; an overview. *Econ. Geol.* **1983**, *78*, 573–590. [CrossRef]
48. Inzana, J.; Kusky, T.; Tucker, R. Comparison of TM Band Ratio Images, Supervised Classifications, and Merged TM and Radar Imagery for Structural Geology Interpretations of the Central Madagascar Highlands. In Proceedings of the American Society of Photogrammetry and Remote Sensing, Annual Meeting, Abstracts with Programs, London, UK, 12–14 September 2001; p. 34.

49. Kusky, T.; Ramadan, T.M. Structural controls on Neoproterozoic mineralization in the South Eastern Desert, Egypt: An integrated field, Landsat TM, and SIR-C/X SAR approach. *J. Afr. Earth Sci.* **2002**, *35*, 107–121. [CrossRef]

50. Volesky, J.C.; Stern, R.J.; Johnson, P.R. Geological control of massive sulfide mineralization in the Neoproterozoic Wadi Bidah shear zone, southwestern Saudi Arabia, inferences from orbital remote sensing and field studies. *Precambrian Res.* **2003**, *123*, 235–247. [CrossRef]

51. Gad, S.; Kusky, T. Lithological mapping in the Eastern Desert of Egypt, the Barramiya area, using Landsat thematic mapper (TM). *J. Afr. Earth Sci.* **2006**, *44*, 196–202. [CrossRef]

52. Boardman, W.; Kruse, F.A. *Automated Spectra Analysis: A Geologic Example Using AVIRIS Data, North Grapevine on Geologic Remote Sensing*; Environmental Research Institute of Michigan: Ann Arbor, MI, USA, 1994; pp. 407–418.

53. Joly, A.; Porwal, A.; McCuaig, T.C.; Chudasama, B.; Dentith, M.C.; Aitken, A.R.A. Mineral systems approach applied to GIS-based 2D-prospectivity modelling of geological regions: Insights from Western Australia. *Ore Geol. Rev.* **2015**, *71*, 673–702. [CrossRef]

54. Carranza, E.; Hale, M.; Faassen, C. Selection of coherent deposit-type locations and their application in data-driven mineral prospectivity mapping. *Ore Geol. Rev.* **2008**, *33*, 536–558. [CrossRef]

55. Carranza, E.J.M.; Laborte, A.G. Data-driven predictive mapping of gold prospectivity, Baguio district, Philippines: Application of random forests algorithm. *Ore Geol. Rev.* **2015**, *71*, 777–787. [CrossRef]

56. Ge, W.; Cheng, Q.; Jing, L.; Wang, F.; Zhao, M.; Ding, H. Assessment of the Capability of Sentinel-2 Imagery for Iron-Bearing Minerals Mapping: A Case Study in the Cuprite Area, Nevada. *Remote. Sens.* **2020**, *12*, 3028. [CrossRef]

57. Ramadan, E.; Feng, X.-Z.; Cheng, Z. Satellite remote sensing for urban growth assessment in Shaoxing City, Zhejiang Province. *J. Zhejiang Univ. A* **2004**, *5*, 1095–1101. [CrossRef] [PubMed]

58. Ahmed, A.; Abdelkareem, M.; Asran, A.M.; Mahran, T.M. Geomorphic and lithologic characteristics of Wadi Feiran basin, southern Sinai, Egypt, using remote sensing and field investigations. *J. Earth Syst. Sci.* **2017**, *126*, 1–25. [CrossRef]

59. El-Din, G.K.; Abdelkareem, M. Integration of remote sensing, geochemical and field data in the Qena-Safaga shear zone: Implications for structural evolution of the Eastern Desert, Egypt. *J. Afr. Earth Sci.* **2018**, *141*, 179–193. [CrossRef]

60. Olmo, M.C.; Abarca, F.; Rigol, J.P.; Rigol-Sanchez, J.P. Development of a Decision Support System based on remote sensing and GIS techniques for gold-rich area identification in SE Spain. *Int. J. Remote Sens.* **2002**, *23*, 4801–4814. [CrossRef]

61. Brandmeier, M. Remote sensing of Carhuarazo volcanic complex using ASTER imagery in Southern Peru to detect alteration zones and volcanic structures–A combined approach of image processing in ENVI and ArcGIS/ArcScene. *Geocarto Int.* **2010**, *25*, 629–648. [CrossRef]

62. Bolouki, S.M.; Ramazi, H.R.; Maghsoudi, A.; Pour, A.; Sohrabi, G. A Remote Sensing-Based Application of Bayesian Networks for Epithermal Gold Potential Mapping in Ahar-Arasbaran Area, NW Iran. *Remote Sens.* **2020**, *12*, 105. [CrossRef]

63. Cudahy, T.J.; Jones, M.; Thomas, M.; Laukamp, C.; Caccetta, M.; Hewson, R.D.; Rodger, A.D.; Verrall, M. *Next Generation Mineral Mapping: Queensland Airborne HyMap and Satellite ASTER Surveys, 2006–2008*; CSIRO Exploration and Mining: Canberra, Australia, 2008.

64. Zeghouane, H.; Allek, K.; Kesraoui, M. GIS-based weights of evidence modeling applied to mineral prospectivity mapping of Sn-W and rare metals in Laouni area, Central Hoggar, Algeria. *Arab. J. Geosci.* **2016**, *9*, 373. [CrossRef]

65. Gahlan, H.; Ghrefat, H. Detection of Gossan Zones in Arid Regions Using Landsat 8 OLI Data: Implication for Mineral Exploration in the Eastern Arabian Shield, Saudi Arabia. *Nat. Resour. Res.* **2018**, *27*, 109–124. [CrossRef]

66. Sekandari, M.; Masoumi, I.; Pour, A.B.; Muslim, A.M.; Rahmani, O.; Hashim, M.; Zoheir, B.; Pradhan, B.; Misra, A.; Aminpour, S.M. Application of Landsat-8, Sentinel-2, ASTER and WorldView-3 Spectral Imagery for Exploration of Carbonate-Hosted Pb-Zn Deposits in the Central Iranian Terrane (CIT). *Remote Sens.* **2020**, *12*, 1239. [CrossRef]

minerals

MDPI

Article

Depositional Environment and Hydrocarbon Distribution in the Silurian–Devonian Black Shales of Western Peninsular Malaysia Using Spectroscopic Characterization

Monera Adam Shoieb [1], Haylay Tsegab Gebretsadik [1,*], Syed Muhammad Ibad [1] and Omeid Rahmani [2]

[1] Southeast Asia Clastic and Carbonate Research Laboratory, Geoscience Department, Universiti Teknologi PETRONAS (UTP), Seri Iskandar 32610, Perak, Malaysia
[2] Department of Natural Resources Engineering and Management, School of Science and Engineering, University of Kurdistan Hewlêr (UKH), Erbil 44001, Kurdistan Region, Iraq
* Correspondence: haylay.tsegab@utp.edu.my; Tel.: +60-53687347

Abstract: The present study aimed to evaluate the hydrocarbon functional groups, aromaticity degree, and depositional environment in the Silurian–Devonian Kroh black shales of western peninsular Malaysia. Fourier transform infrared spectroscopy (FTIR) was applied to measure the hydrocarbon functional groups in the sedimentary succession and associated organic matter of the black shale samples. The results showed that aromatic C=C stretching, aromatic C-H out-of-plane, aromatic C-H in-plane, and aliphatic =C–H bending are the major hydrocarbon functional groups in the Kroh shales. Also, ultraviolet-visible spectroscopy (UV-Vis) was used to evaluate the type of humic substance and analyze the sample extract ratios of E4/E6. It was revealed that the methanol-treated Kroh shale samples ranged from 0.00048 to 0.12 for E4 and 0.0040 to 0.99 for E6. The lower E4/E6 ratio (>5) indicates the dominance of humic acid over fulvic acid in the Kroh shales. The Kroh shale samples' total organic carbon content (TOC) ranges from 0.33 to 8.5 wt.%, analyzed by a multi-N/C 3100 TOC/TNb analyzer. The comparison study revealed that the TOC content of the Kroh shale has close obtainable values for the Montney shales of Canada. Furthermore, both hydrocarbon functional groups from FTIR, and the E4/E6 ratio from UV-Vis show no correlation with TOC content. It is revealed that humic acid, aromatic, and aliphatic hydrocarbons are not the controlling factors of the enrichment of organic matter in the Kroh shales. Conversely, a positive correlation between aliphatic and aromatic hydrocarbons in the Kroh shales indicated that organic matter is thermally overmatured. The presence of humic acid and enrichment of aromatic hydrocarbons in the Kroh shales demonstrated that the organic matter in these shales contains plant-derived hydrophilic minerals, i.e., terrestrial in origin. These findings may provide clues on the depositional and thermal maturation of organic matter for the exploration efforts into the pre-Tertiary sedimentary successions of the peninsular.

Keywords: Kroh shale; aromatic hydrocarbons; spectroscopy; humic acid; organic carbon

Citation: Shoieb, M.A.; Gebretsadik, H.T.; Ibad, S.M.; Rahmani, O. Depositional Environment and Hydrocarbon Distribution in the Silurian–Devonian Black Shales of Western Peninsular Malaysia Using Spectroscopic Characterization. *Minerals* **2022**, *12*, 1501. https://doi.org/10.3390/min12121501

Academic Editor: Behnam Sadeghi

Received: 28 September 2022
Accepted: 22 November 2022
Published: 24 November 2022

Publisher's Note: MDPI stays neutral with regard to jurisdictional claims in published maps and institutional affiliations.

1. Introduction

Many studies have shown that the characterization of organic-rich sedimentary successions is a must-have step in looking into the fundamental properties affecting the concentration of aromatic hydrocarbons and their correlation with the total organic carbon (TOC) content of the sediments [1–3]. Fourier transform infrared spectroscopy (FTIR) is a frequently used technique to differentiate the hydrocarbon functional groups in shale and coal [4]. The functional groups of aromatic and aliphatic hydrocarbons are potent tools for evaluating the origin, richness, and for interpretation of depositional environments. The aromatic and aliphatic hydrocarbons are determined through the vibrational characteristics of their structural and chemical bonds. The use of Attenuated Total Reflection (ATR) accessories to enhance the surface sensitivity by using tough crystals (such as germanium, silica, zinc selenide, and diamond) characterized by their range of hardness values and optical properties has further advanced the use of FTIR in soils, shale, and coal materials [5]. Dilution with KBr is no longer necessary, reproducibility is increased, and the non-destructive nature of this analysis allows the sample to be re-used for other analyses. Like FTIR, UV-Vis is increasingly employed for in-field applications [6], for laboratory studies of crude oils, and in determining the type of humic substance [7]. For the organic chemist, UV-Vis is mainly concerned with conjugated systems with electronic transitions; the intensities and positions of the absorption band largely depend on the specific system under consideration [8].

Alkyl naphthalenes are widespread and constitute geological and geochemical materials. They are commonly found in oil and several types of sedimentary rocks, including shales and coals [9]. It has been suggested that alkyl naphthalenes are derived mainly from the de-functionalization of terpenoids; hence, they have the potential to provide information about their precursor, as well as the depositional environment [10].

Spectroscopic methods such as ATR-FTIR and UV-Vis have been used to evaluate the liquid petroleum yield of hydrocarbon source rocks for correlation of source and tracing migration paths [11,12]. Therefore, in this study, FTIR and UV-Vis are used to identify aromatic hydrocarbons such as alkyl naphthalenes, aliphatic hydrocarbons, and humic substances of Silurian–Devonian Kroh shales from peninsular Malaysia. The spectroscopic analysis is used to evaluate the origin of organic matter in the source or reservoir rocks, which have received little attention to date. Therefore, the objective of the present study is to use spectroscopic analysis to obtain the hydrocarbon distribution and humic substance type, which will determine the shale's source and depositional environment of organic matter. Furthermore, the effect of TOC on hydrocarbon functional groups and humic acid is also being investigated to determine the controlling factor for the enrichment of the organic matter in the Kroh shales from western peninsular Malaysia.

2. Study Area

Peninsular Malaysia is situated at the southernmost tip of the Asian mainland, and it shares borders with Thailand in the north, Singapore in the south, the South China Sea in the east, and the Straits of Malacca in the west (Figure 2). The peninsula covers a total area of 130, 268 km^2, and forms part of Sundaland and the shallow seas from which several smaller islands emerge. It is elongated in an NNW-SSE direction and characterized by a dense network of streams and rivers that expose Paleozoic rocks [13,14].

Most of Malaysia's Paleozoic rocks are in peninsular Malaysia and account for about 25% of the land-based portion [15]. The Kroh Formation is situated in the Western Belt of peninsular Malaysia, about 34 kilometers north of Gerik along the Malaysia-Thailand frontier (Figure 1). The locality has three main formations: the Kroh, the Kati, and the Nenering Tertiary. The Kroh Formation is widespread in north Perak and extensively accessible in Pengkalan Hulu, Kelian Intan, and Kerunai. It is comprised of black shale, sub-mature arenite, chert, limestone, and calcareous shale (Figure 2). This study is mainly focused on shale samples of the Kroh Formation. They date to the upper Silurian lower Devonian period. The samples were obtained from seven outcrops in the Gerik area (Figure 1).

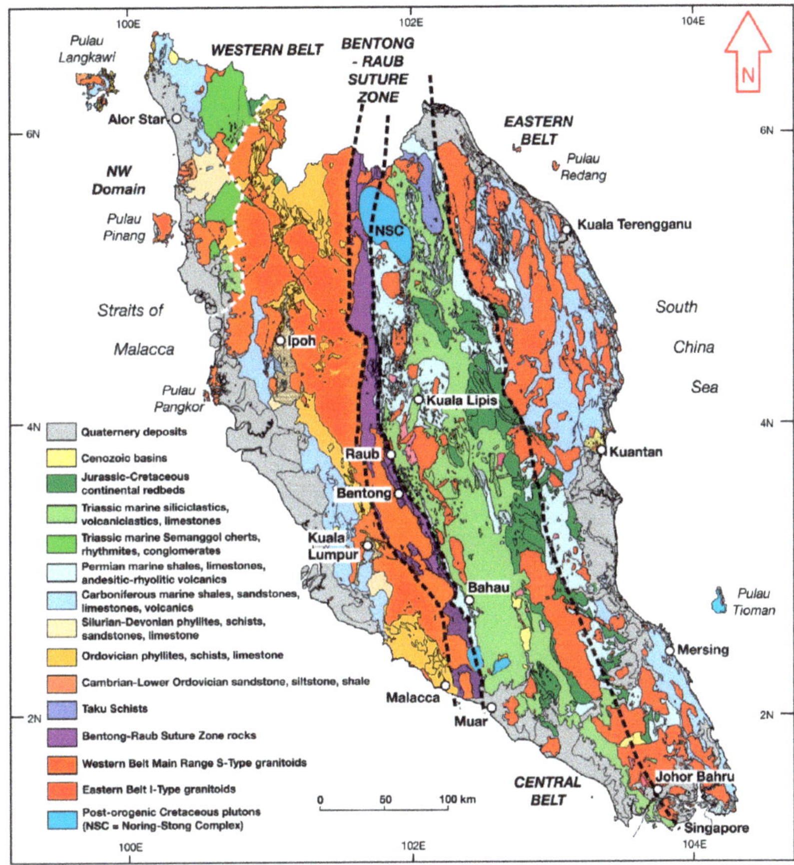

Figure 1. A locational map of the study area, changed from [17], with the prominent outcrops (in purple circles) where the samples were taken from.

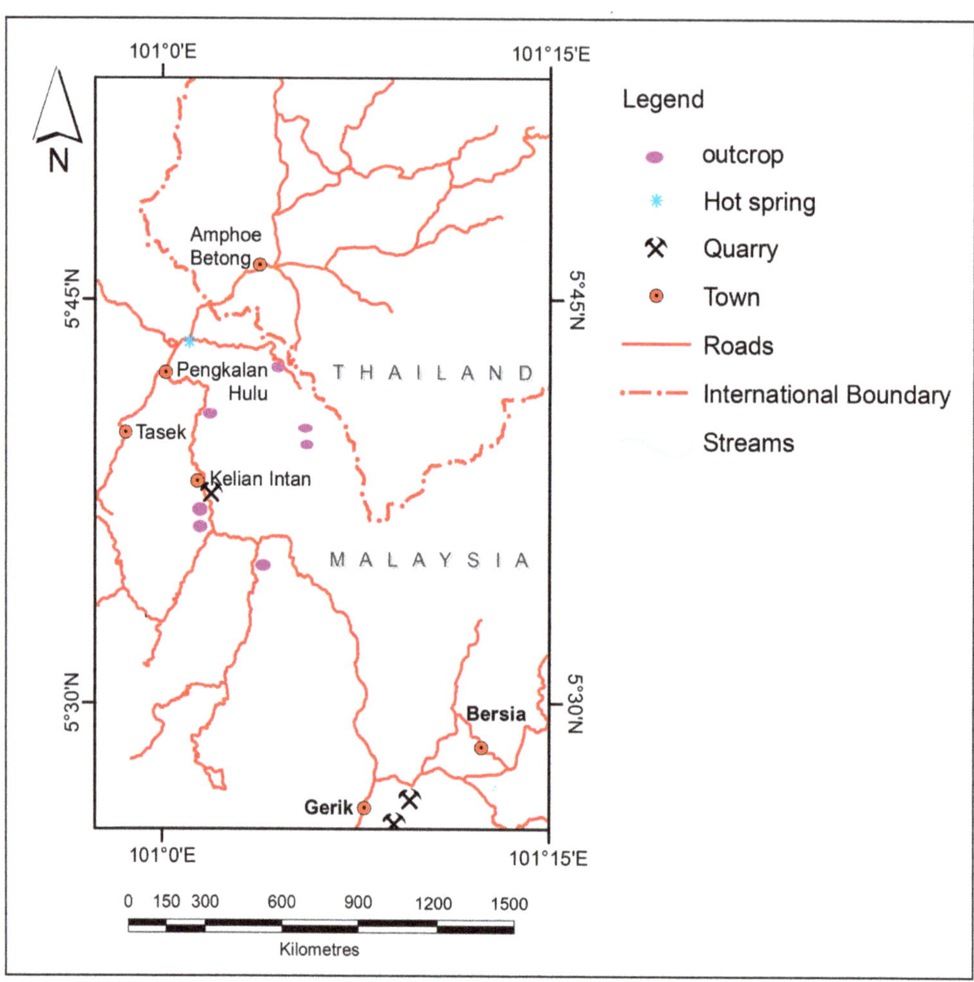

Figure 2. Simplified geological map of peninsular Malaysia; adopted from [16] with Elsevier's License Number 5397511182486.

3. Materials and Methods

3.1. Sampling

Seventy-three representative samples were taken from the Gerik area, upper Perak (Figure 1). The outcrop samples were obtained using a channel-profile sampling strategy, and freshly exposed faces were chosen to prevent inclusion of weathered and oxidized materials. All shale samples were analyzed for TOC, while fifty-six were studied for hydrocarbon functional groups. Thirty-three samples were analyzed for humic and fulvic acid content.

3.2. Fourier-Transform Infrared Spectroscopy (FTIR)

Infrared (IR) is part of the electromagnetic radiation between the visible region's high-frequency end and the microwave region's low-frequency end. FTIR is based on the principle that covalent bonds have resonating frequencies in the mid-infrared region ($4000\ \mathrm{cm}^{-1}$ to $400\ \mathrm{cm}^{-1}$) at which they vibrate, and these frequencies depend on the bond type and the bonded atoms [18]. For the past few decades, FTIR has been extensively used to assess hydrocarbon bonds in geological samples such as shale, coal, silicate glass, and microfossils to identify and characterize clays and other minerals [19,20]. The spectra generated from the FTIR analysis in this study were interpreted based on studies by Coates [21] and Stuart [22].

FTIR was used to determine the distribution of hydrocarbon functional groups and the differences in the composition of the studied shale samples. The Kroh shale samples were analyzed using Shimadzu 8400S Fourier-transform infrared (FTIR) spectroscopy (Shimadzu, Kyoto, Japan). An Attenuated Total Reflection (ATR) was attached to the machine, allowing the sample to be analyzed quickly and directly. About 2 mg of the samples were scanned over a wavelength of 400–4000 cm^{-1}, collecting 32 scans at a resolution of 8 cm^{-1}. The limit of detection of the instrument was 0.08%. Background scans were collected using the same settings as the sample analyses. Replicate spectra collected on selected samples showed consistent peak positions and absorbance intensities. The data collected was further analyzed using Essential FTIR software (Monona, WI, USA). The area percentage of hydrocarbon functional groups was calculated by summing the absorbance intensities between the respective wavelengths [23]. The absorbance of hydrocarbon functional groups in all Kroh shale samples has been calculated by using Essential FTIR software.

3.3. Ultraviolet-Visible Spectroscopy (UV-Vis)

The Ultraviolet-Visible Spectrum (UV-Vis) is obtained using a diluted sample solution in a glass tube (cuvette). The sides of the cuvette are 1 cm, and the overall volume is 2–3 cm^3. UV or visible light should pass through the sample, and the transmitted light intensity is recorded across the wavelength spectrum of the instrument. The UV-Vis analysis for this study focused primarily on the E4/E6 ratio. The ratio of optical densities or absorbance of dilute aqueous humic acid and fulvic acid solutions using the UV-Vis techniques at 465 nm and 665 nm is commonly used in the characterization of organic matter in soil science [24]. UV-Vis spectroscopy (Chongqing Gold Mechanical & Electrical Equipment Co. Ltd., Chongqing, China) was used to analyze the sample extracts to identify their E4/E6 ratios. Two fundamental wavelengths widely used to describe the humic matter for one-dimensional UV-Vis are 465 nm and 665 nm [7,24,25].

Each black shale sample weighed two grams and was put into a glass flask with a cap. The samples were then subjected to three consecutive extractions using 8 mL of methanol, 3 min of ultrasonic stirring by Thornton Unique 1450USC ultrasonic cleaner (Santa Cruz County, CA, USA), and 5 min of centrifugation at 2500 rpm by Janetzk T23 centrifuge (Hein Janetzk KG, Engelsdorf Leipzig, Germany). The methanol extract solution was analyzed using a Lambda 750 UV Vis Spectrophotometer (Perkin Elmer, NJ, USA) with liquid samples placed in quartz cells. This spectrophotometer is equipped with a tungsten lamp and a D2 lamp to provide the radiation source. The scanning wavelength ranged from 200–800 nm.

3.4. Total Organic Carbon (TOC)

Total organic carbon (TOC in wt. %) content is a significant parameter that has been used to assess the amount of organic matter and hydrocarbon generation potential of the source rock [26]. Representative black shale samples were analyzed using a Multi N/C 3100 TOC/TNb analyzer (Analytik Jena GmbH, Jena, Germany) at Core Laboratories Malaysia Sdn Bhd (Perak, Malaysia), using the direct approach suggested by Dow and Pearson [27]. The TOC measurement was performed on fresh near-surface samples, implying a loss of organic content due to thermal degradation; weathering; and biodegradation. The samples have been pulverized using an automated grinding machine. About 3 g of each pulverized sample was pre-treated with a concentration of 37% hydrochloric acid by 10% to remove the inorganic carbon fraction from the samples, which might have come from carbonate minerals. It was then left for 12 h in the fume chamber before being rinsed three times with reverse osmosis water and then dried for 24 h in the oven at 60 °C. 60 mg of the sample was weighed after drying and placed on a ceramic boat. Measurements were run in duplicate, and the results were averaged. The residual material was heated to temperatures exceeding 850 °C to determine the TOC by combustion analysis. Analysis of the organic matter content in this study has been used as an implication for petroleum exploration.

4. Results

4.1. Hydrocarbon Functional Groups

The infrared spectral analysis of the samples was obtained using the method described in Section 3.1. Figure 3 shows the absorption spectra from the representative shale samples, and each absorption band represents the presence of a functional group. As shown in Figure 3, six aromatic hydrocarbon peaks are observed in the 500–2000 cm^{-1} absorption band. There are four additional peaks in the range of 670–900 cm^{-1}, which are representative of aromatic C-H out-of-plane bend, and the other two peaks in the absorption band of 950–1225 cm^{-1} are assigned as aromatic C-H in-plane bend. Alkyl naphthalenes (pentylnaphthalenes) are also confirmed by the presence of two strong and two weak bands in the C-H out-of-plane bend (700–900 cm^{-1}) vibration region [28]. These are 694 cm^{-1} (w), 779 cm^{-1} (s), 797 cm^{-1} (s), and 827 cm^{-1} (w). One aliphatic hydrocarbon peak is detected in the 600–700 cm^{-1}, region, which is representative of alkyne =C–H bending. Two ATR diamond peaks are observed at 2300–2400 cm^{-1}, while two peaks of OH compounds are found at 3600–3800 cm^{-1}.

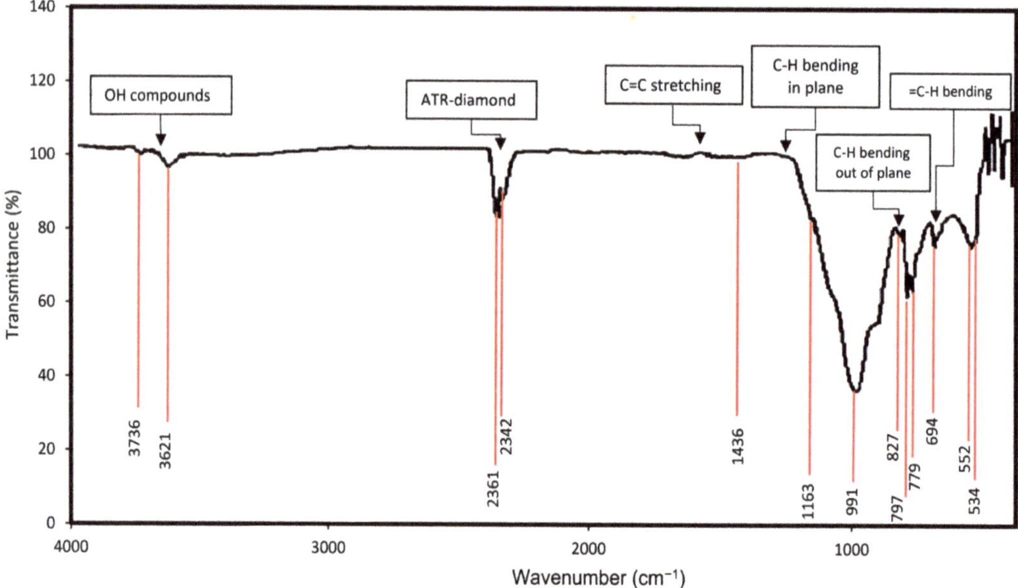

Figure 3. FTIR spectra of the represented shale sample showing hydrocarbon functional groups in the wavenumber range from 500 to 4000 cm^{-1}. ATR: attenuated total reflectance.

Table 1 and Figure 4 represent hydrocarbon functional groups (avg.) from an FTIR analysis for each shale sample.

Table 1. Hydrocarbon functional groups in the Kroh shale samples.

Samples	Aromatic Hydrocarbon			Aliphatic Hydrocarbon
	1600–1430 C=C Stretching Absorbance	900–690 Out of Plane C-H Bending Absorbance	1275–1000 In-Plane C–H Bending Absorbance	700–600 C–H Bending Absorbance
KR1-2	0.903	1.761	2.18	1.2
KR1-3	N.A	1.105	1.361	0.72
KR1-4	0.806	1.57	2.602	1.064
KR1-5	0.708	1.416	2.443	0.943
KR1-6	0.734	1.408	1.804	0.932
KR1-7	0.745	1.279	1.838	0.929
KR1-8	0.712	1.167	1.494	0.89
KR1-9	0.531	0.749	1.224	0.627
KR1-10	0.97	1.503	2.267	1.14
KR1-11	0.845	1.381	2.04	1.342
KR1-12	0.63	1.173	1.928	0.815
KR2-2	0.534	0.73	0.887	0.604
KR2-3	1.019	1.505	2.148	1.224
KR2-4	0.737	1.038	1.518	0.846
KR2-6	0.981	1.344	2.113	1.151
KR2-8	0.599	0.809	0.974	0.685
KR2-9	0.823	1.077	1.602	0.92
KR2-12	0.63	0.868	1.257	0.698
KR2-14	0.615	0.724	1.049	0.648
KR3-1	N.A	0.436	0.529	0.44
KR3-3	0.453	0.617	0.793	0.485
KR3-4	0.619	0.927	1.129	0.661
KR3-5	N.A	0.413	0.573	0.381
KR3-6	0.757	1.163	1.284	0.827
KR3-7	0.613	0.849	1.086	0.668
KR3-8	N.A	0.895	1.03	0.767
KR3-10	N.A	0.646	0.806	0.535
KR3-13	N.A	1.328	1.328	0.94
KR4-1	0.834	1.661	3	1.008
KR4-2	0.63	1.244	1.552	0.748
KR4-3	N.A	0.591	0.967	0.44
KR4-4	1.023	1.974	2.698	1.27
KR4-5	0.633	1.036	1.636	0.72
KR4-6	0.689	1.29	1.578	0.816
KR4-7	0.61	0.98	1.651	0.689
KR4-8	0.707	1.346	1.673	0.84
KR5-1	N.A	0.395	0.643	0.318
KR5-2	0.676	1.29	1.829	0.877

Table 1. *Cont.*

Samples	Aromatic Hydrocarbon			Aliphatic Hydrocarbon 700–600 C–H Bending Absorbance
	1600–1430 C=C Stretching Absorbance	900–690 Out of Plane C-H Bending Absorbance	1275–1000 In-Plane C–H Bending Absorbance	
KR6-1	0.927	1.251	2.92	1.232
KR6-2	N.A	1.021	2.031	1.086
KR6-3	0.602	1.014	1.74	0.784
KR6-4	0.579	1.019	1.744	0.745
KR6-5	0.667	1.211	2.113	0.831
KR6-6	0.431	0.703	2.055	0.568
KR6-8	0.488	0.769	1.524	0.611
KR6-9	0.683	1.181	1.982	0.925
KR6-10	0.97	1.503	2.267	1.14
KR6-11	1.059	1.521	2.585	1.317
KR6-12	0.732	1.175	2.455	1.395
KR6-13	N.A	0.341	1.511	0.617
KR6-14	0.591	0.698	1.61	0.856
KR6-15	0.556	0.665	1.69	0.801
KR6-18	1.094	1.709	2.148	1.384
KR6-19	N.A	1.267	2.187	1.385
KR6-20	N.A	1.038	1.928	1.174
KR7-1	1.098	1.95	2.455	1.395

N.A: not applicable.

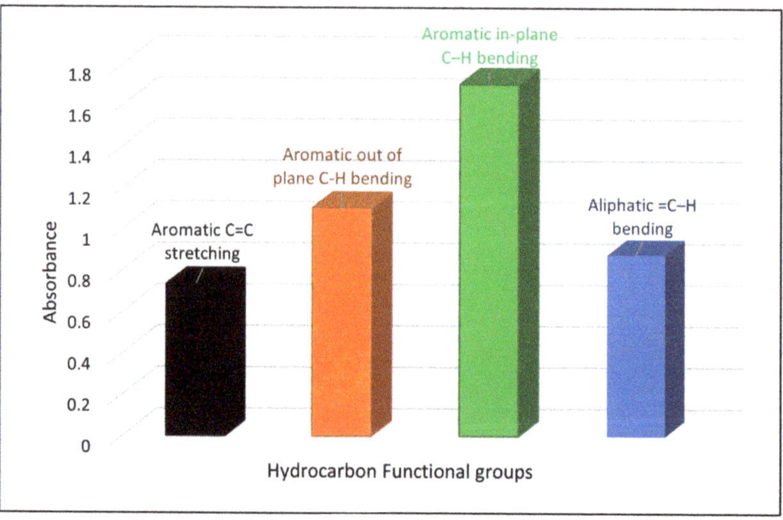

Figure 4. Hydrocarbon functional groups in the Kroh shales.

4.2. Distribution of E4 and E6

The ratio of aliphatic to aromatic compounds in the rocks has been determined by calculating the E4/E6 ratio of extracts from the samples. E4 was determined at an absorption frequency of 465 nm and E6 at 665 nm. Figure 5 represents the spectroscopic UV-visible ratio (E4/E6) results of Kroh black shale samples. The E4 treated with methanol ranges from 0.00048 to 0.12, while the value of E6 ranges from 0.0040 to 0.99 in the Kroh black shale samples. The E4/E6 values range from 0.008 to 8.1, while 3.4 is the average for the studied samples.

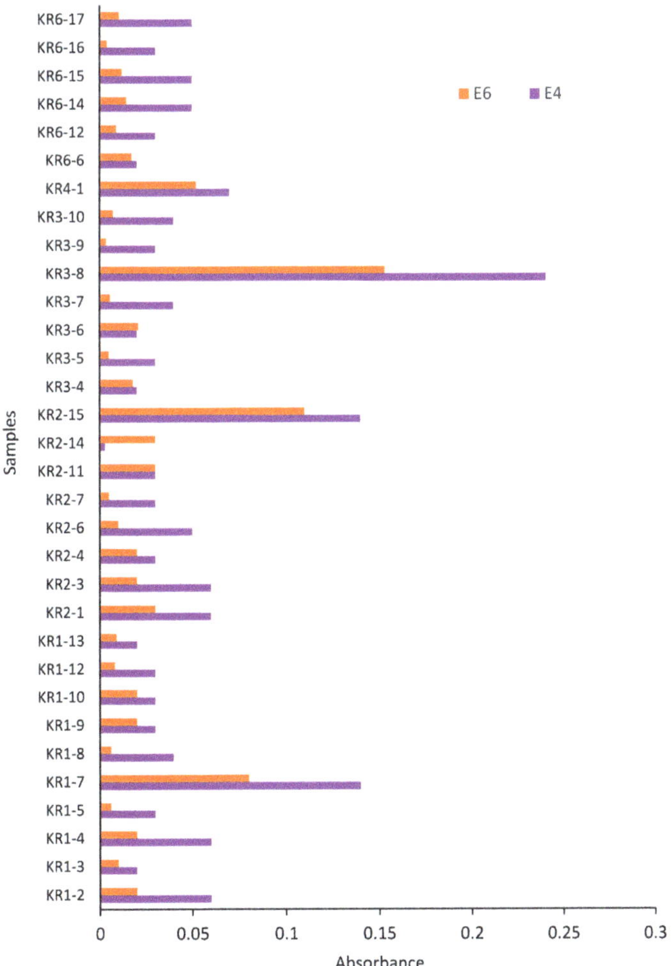

Figure 5. Distribution of E4 and E6 in the Kroh shales.

4.3. Total Organic Carbon (TOC)

Representative samples from the seven different outcrops in the Kroh Formation were analyzed using a total organic carbon analyzer. Figure 6 presents the measured total organic carbon (TOC) values of the analyzed samples of black shale. The TOC present in black shale samples is high, ranging from 0.33 wt.% to 8.58 wt.%, with an average of 1.71 wt.%.

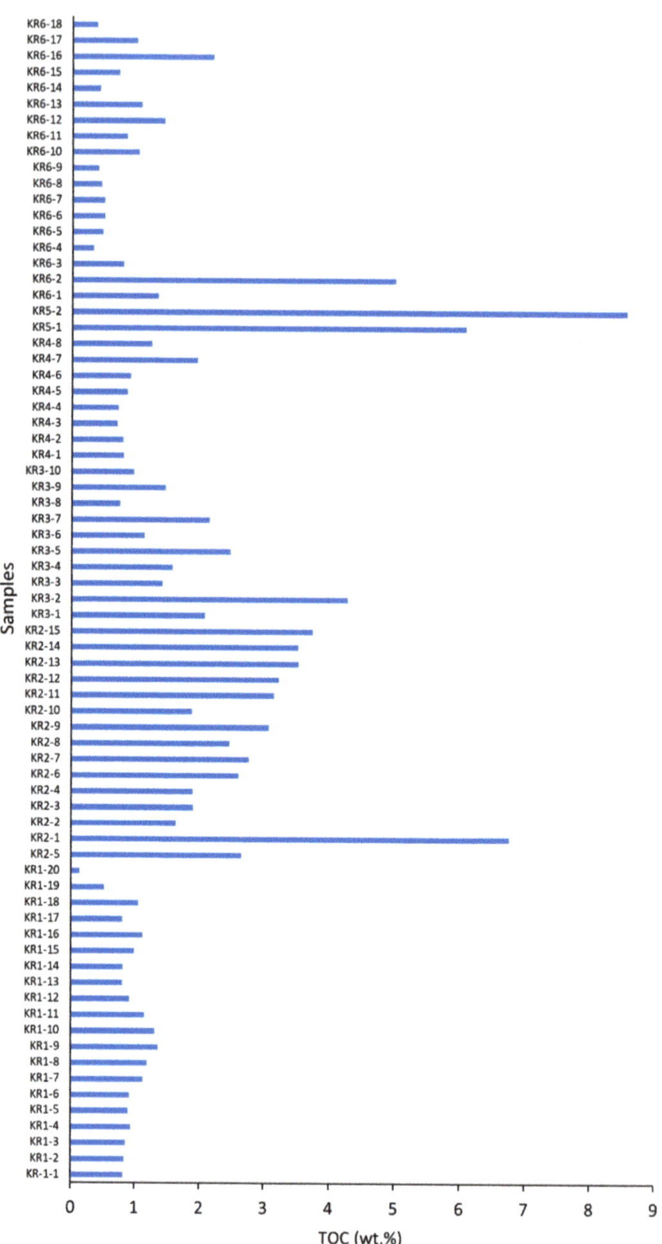

Figure 6. TOC (wt.%) of shale samples from the Kroh Formation.

5. Discussion

5.1. Hydrocarbon Functional Groups Distribution

The FTIR analysis was conducted to determine the compositional variations in hydrocarbon functional groups in the black shale. Under IR radiation, the absorbency of molecular vibrations was proportional to the abundance of the functional groups. The maximum height frequently determines the integrated area between the baseline and a peak or the absorbance for each band of molecular vibration. In geology, FTIR deals with the MIR (mid-infrared region) of light between 4000 cm^{-1} to 500 cm^{-1}. Under the radiation of IR, the molecular vibration absorbances were proportional to the functional group's abundance. The maximum height is determined by the integrated area between the baseline and a peak or the absorbance for each band of molecular vibration.

The FTIR spectra of different shale exhibited similar absorption bands and characteristic absorption peaks based on the vibration of the atoms in a molecule. The spectrum obtained depends on the fraction of the incident radiation absorbed in particular energy. The C-H functional group for aromatic compounds appears to be different in its absorption, with different members suggesting the different concentrations of these compounds in the samples. According to Stuart [22], the differences in peak absorbance were a sign of the variation in the available group quantity, and higher peak intensities display a higher amount in the samples. There was no dominant peak in the 2800–3200 cm^{-1} range in the Kroh shale FTIR spectra (Figure 3), which is associated with an aliphatic hydrocarbon functional group [21]. It was observed that all Kroh shale samples had a high concentration of aromatic C-H out-of-plane and C-H in-plane compared to aromatic C=C stretching and aliphatic =C–H bending (see Table 1 and Figure 4).

The processes controlling the level of aromatic and aliphatic hydrocarbons in shales are complex. Key factors that may influence it are: (1) the sediment's composition, i.e., clay and TOC content; (2) patterns of the sedimentary depositional environment; and (3) the chemical properties of the compounds, particularly their water solubility [29,30].

We have investigated the effect of the aliphatic hydrocarbon functional group on aromatic hydrocarbon functional groups. As shown in Figure 7, the aliphatic hydrocarbons of shale samples from the Kroh Formation display a strong positive relationship (i.e., $R^2 = 0.82$) with all three aromatic hydrocarbon functional groups. The relative increase in the proportion of naphthenic and aliphatic hydrocarbons to aromatic hydrocarbons might have happened because of organic matter thermal maturity. Similar findings were reported by [31,32]. Our recent study also supports this interpretation of thermal over-maturity by employing Rock-Eval pyrolysis and vitrinite reflectance of the Kroh shale samples [15].

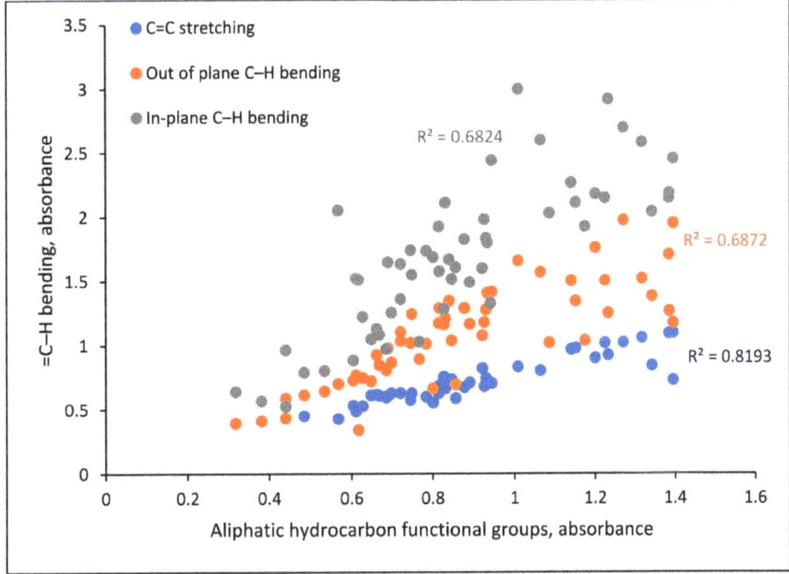

Figure 7. Relationship of aliphatic and aromatic (=C–H bending) hydrocarbon functional groups in the Kroh shales.

5.2. Relationship of TOC with Functional Groups

Total organic carbon (wt.%) content is a significant parameter used to assess the amount of organic matter and hydrocarbon generation potential of source rocks [26]. Analysis of the organic matter content in this study has been used as an implication for petroleum exploration. The TOC values of Silurian–Devonian Kroh shales (TOC = 1.71 wt.%) are remarkable and similar to hot shales from China (Longmaxi shale, TOC = 2.32 wt.%) and Canada (Montney shales, TOC = 2.64 wt.%), and close to Muskwa, Besa and Fort Simpson Canada (Devonian–Mississippian shales, TOC = 1.39 wt.%) and El Sebaiya Egypt (Duwi Formation, TOC = 1.4 wt.%) (Table 2).

Table 2. Comparing the TOC content of Paleozoic Kroh black shales with other worldwide shale gas rocks.

Formation	Age	TOC (wt.%)	Average TOC (wt.%)	Reference
Longmaxi, China	Lower Silurian	0.44–4	2.32	[33]
Niutitang, China	Lower Cambrian	0.39–10.2	5.26	[33]
Muskwa, Besa & Fort Simpson, Canada	Devonian–Mississippian	0.18–4.72	1.39	[34]
Barnett shale, USA	Mississippian	2.62–11.47	4.66	[35]
Gufeng, China	Lower Permian	0.04–22.1	3.4	[36]
Montney, Canada	Lower Triassic	0.03–8.2	2.64	[37]
Baling and Bendang Riang, Malaysia	Silurian–Devonian	0.73–24.6	6.71	[38]
Kubang Pasu, Malaysia	Lower Permian	1.01–19.65	5.74	[38]
Duwi Formation, El Sebaiya, Egypt	Late Campanian–early Maastrichtian	0.21–2.77	1.4	[39]
Kroh shale, Malaysia	Silurian–Devonian	0.13–8.56	1.71	Current study

Many studies have shown that TOC is one of the critical properties affecting the concentration of aromatic hydrocarbons in sediments [40]. They have shown a positive correlation between aromatic hydrocarbons and total organic carbon (TOC) content in sediments [41]. Therefore, the effect of TOC on aromatic and aliphatic hydrocarbons was investigated in this study. The absorbance of aromatic C=C stretching, aromatic C-H out-of-plane, aromatic C-H in-plane, and aliphatic =C–H bending was plotted vs. TOC to find any existing relationship (Figure 8). There was no correlation found between TOC and hydrocarbon functional groups, as the Kroh shale is overmature and its organic matter content has been exhausted due to the thermal maturity and the generation of hydrocarbons [42]. This might suggest that black shale, which is enriched in organic matter and mature enough to produce conventional oil and gas, has a more powerful absorption of aromatic hydrocarbon functional groups. Further investigation could be initiated with high TOC shale samples to understand this relationship better.

5.3. Hydrocarbon Compound Characterization Using E4/E6

The E4/E6 ratio is inversely related to the degree of condensation and the aromaticity of the humic substances and their degree of humification [43]. It is suggested that the values of the relationship E4/E6 for humic acid are smaller than 5.0 and between 6.0 and 8.0 for fulvic acids [43]. Most of the Kroh shale samples show an E4/E6 ratio lower than 5, indicating that the presence of humic acid dominates organic matter in the Kroh shale.

The E4 versus E6 plot for samples from the Kroh Formation black shales shows a weak relationship between the two parameters (Figure 9A). This correlation between E4 and E6 indicates that the supply of organic matter is not consistent in the environment throughout the deposition phase. A lower E4/E6 ratio evidenced a higher degree of aromaticity of humic acid in shale averaging 3.4 in the Kroh Formation. The E4/E6 ratio indicates the type and quality of humic matter [24]. The effect of TOC on the E4/E6 ratio was also investigated (Figure 9B). As humic acid is one of the components of organic matter, there might be a possibility that the abundance of humic acid could control the organic matter present in the Kroh shale. However, no correlation existed between the TOC and E4/E6 ratio (Figure 9B), indicating that humic acid alone is not the controlling factor of organic matter in the Kroh shale. Some other organic matter constituents also control the organic matter richness.

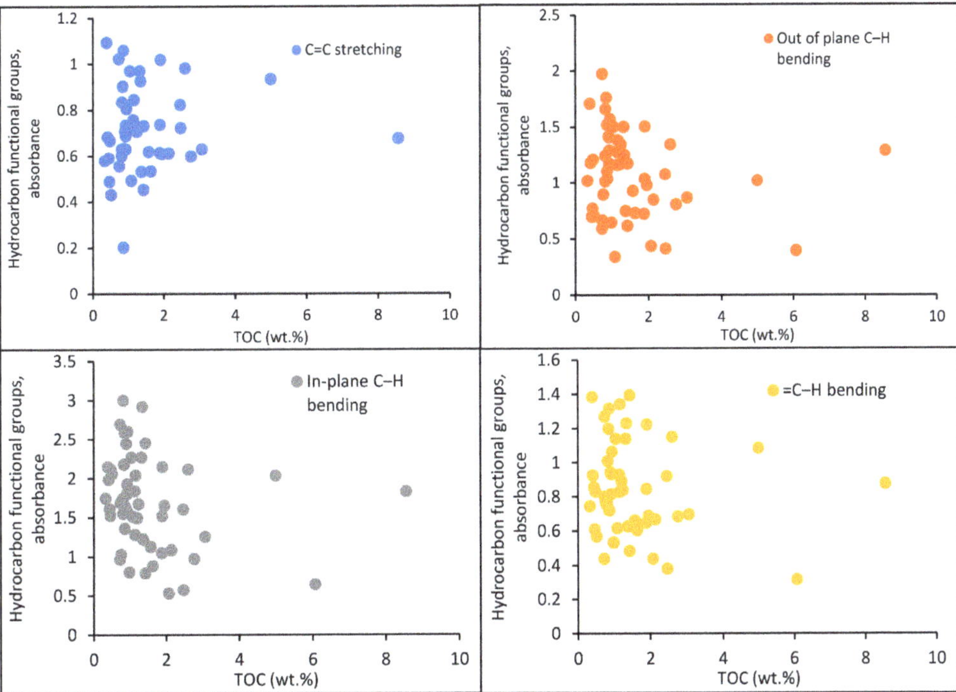

Figure 8. Relationship of TOC with hydrocarbon functional groups in the Kroh shales.

5.4. Depositional Environment of Organic Matter

Humic acid is one of the organic constituents of oil shale. It has been a major kerogen precursor, which can be a significant petroleum precursor. A large amount of humic acid was deposited in shale by the death of thick growth of vegetation; it then accumulated in large piles, which were then buried by rock and mudflows, as well as deposits of sand and silt. The weight of the overflow compacted or compressed out all of the moisture, and what remains today is a deposit of dried, prehistoric plant derivatives. As discussed in Section 4.2, only seven samples (21%) showed an E4/E6 ratio greater than 5 out of 33 studied samples, which showed the abundance of humic acid. Therefore, the presence of humic acid in the Kroh shale indicates that these shales contain plant-derived hydrophilic minerals which are exceedingly small compared to metallic minerals from the ground-up rocks and soil.

Some of the aromatic compounds found in crude oil and sediments are believed to have been derived from a modification of biologically produced compounds such as steroids and terpenoids. Steroids give rise to substituted phenanthrenes, and terpenoids produce alkyl naphthalenes. The processes by which higher plant triterpenoids in sediments are converted into aromatic hydrocarbons have been proposed to commence with the loss of the C-3 oxygen functionality, followed by sequential aromatization from the A ring through to the E ring.

Therefore, this process's ultimate products would be tetracyclic and pentacyclic aromatic hydrocarbons [44]. Alkyl naphthalenes are derived from various precursor compounds, and their composition changes with increasing thermal maturity [45–47]. However, alkyl naphthalenes are often abundant in oils and sedimentary organic matter that have undergone biodegradation or thermal cracking. Alkyl naphthalenes occur in terrestrial oils and rocks in higher concentrations than in marine oils and rocks, suggesting that their sourcing is mainly from terrestrial organic matter [48]. As discussed in Section 4.1, the Kroh shale samples comprise alkyl naphthalenes, which shows that the organic matter in these samples belongs to the terrestrial source. This interpretation is also supported by the presence of humic acid in the Kroh shales, as discussed above.

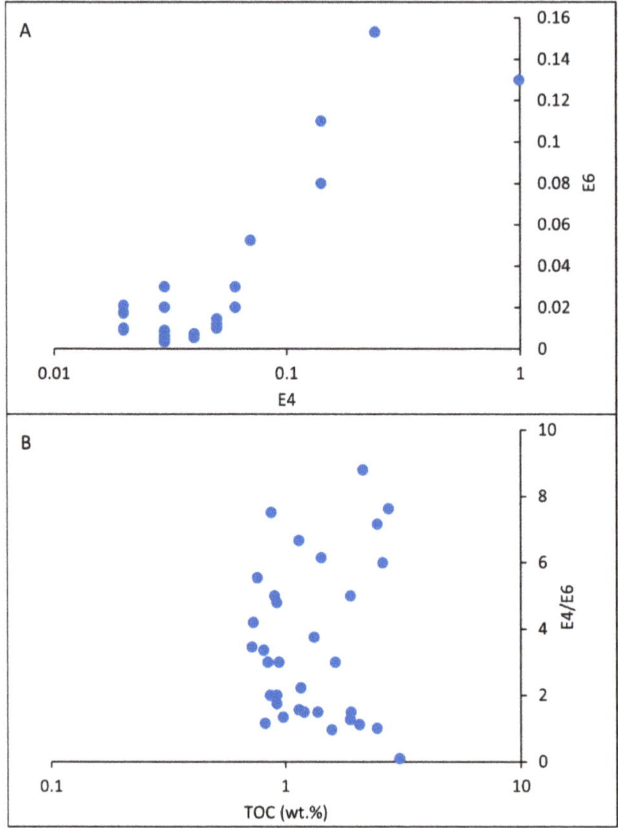

Figure 9. Relationship between (**A**) E4 and E6, (**B**) TOC and E4/E6.

6. Conclusions

The FTIR spectra results of the Kroh shale samples exhibited aromatic C=C stretching, aromatic C-H out-of-plane, aromatic C-H in-plane, and aliphatic =C–H bending hydrocarbon functional groups. The increment of the aliphatic hydrocarbon functional group with the aromatic hydrocarbon functional group indicates that the shale samples are thermally overmature. The absence of a correlation between TOC, aromatic, and aliphatic hydrocarbon functional groups indicates that the organic matter content of the Kroh shale has been exhausted due to thermal maturity and the generation of hydrocarbons in the past. The E4, E6, and E4/E6 ratios indicate that organic matter in the Kroh shale is rich in humic acid as compared to fulvic acid. The relationship of TOC with the E4/E6 ratio indicates that the enrichment of organic matter in the Kroh shale was not controlled by humic acid alone; however, it might be influenced by other constituents of organic matter for the organic matter richness. The depositional environment of organic matter in the Kroh shale was also investigated by hydrocarbon functional groups and the E4/E6 ratio. The abundance of aromatic hydrocarbon functional groups such as alkyl naphthalenes and humic acid indicates that the organic matter in these shales is terrestrial and contains plant-derived hydrophilic minerals.

Author Contributions: All authors have contributed to writing and editing this article. Writing original draft preparation, M.A.S. and H.T.G.; supervision, H.T.G.; writing—review and editing, O.R. and S.M.I.; Visualization, O.R. All authors have read and agreed to the published version of the manuscript.

Funding: The authors would like to acknowledge the financial support the PRF (Petroleum Research Fund) project "Advanced shale Gas Extraction Technology Using Electrochemical Methods" with Research Grant: (0153AB-A33) and Research Grant: YUTP-FRG (015LC0-325).

Data Availability Statement: Not applicable.

Conflicts of Interest: The authors declare no conflict of interest.

References

1. Myers, K.J.; Wignall, P.B. Understanding Jurassic organic-rich mudrocks—New concepts using gamma-ray spectrometry and palaeoecology: Examples from the Kimmeridge Clay of Dorset and the Jet Rock of Yorkshire. In *Marine Clastic Sedimentology*; Leggett, J.K., Zuffa, G.G., Eds.; Springer: Dordrecht, The Netherlands, 1987. [CrossRef]
2. Owusu, E.B.; Tetteh, G.M.; Asante-Okyere, S.; Tsegab, H. Error correction of vitrinite reflectance in matured black shales: A machine learning approach. *Unconv. Resour.* **2022**, *2*, 41–50. [CrossRef]
3. Tourtelot, H.A. Black shale—Its deposition and diagenesis. *Clays Clay Miner.* **1979**, *27*, 313–321. [CrossRef]
4. Naumann, D.; Meyers, R. *Encyclopedia of Analytical Chemistry*; John Wiley Sons: Chichester, UK, 2000; pp. 102–131.
5. Artz, R.R.E.; Chapman, S.J.; Robertson, A.H.J.; Potts, J.M.; Laggoun-Défarge, F.; Gogo, S.; Comont, L.; Disnar, J.-R.; Francez, A.-J. FTIR spectroscopy can be used as a screening tool for organic matter quality in regenerating cutover peatlands. *Soil Biol. Biochem.* **2008**, *40*, 515–527. [CrossRef]
6. Yokota, T.; Scriven, F.; Montgomery, D.S.; Strausz, O.P. Absorption and emission spectra of Athabasca asphaltene in the visible and near ultraviolet regions. *Fuel* **1986**, *65*, 1142–1149. [CrossRef]
7. Evdokimov, I.N.; Losev, A.P. Potential of UV-visible absorption spectroscopy for characterizing crude petroleum oils. *Oil Gas Bus.* **2007**, *1*, 1–21.
8. Eglinton, G. Applications of infrared spectroscopy to organic chemistry. In *An Introduction to Spectroscopic Methods for the Identification of Organic Compounds: Nuclear Magnetic Resonance and Infrared Spectroscopy*; Pergamon: Oxford, UK, 1970; pp. 123–143. [CrossRef]
9. Radke, M.; Rullkötter, J.; Vriend, S.P. Distribution of naphthalenes in crude oils from the Java Sea: Source and maturation effects. *Geochim. Cosmochim. Acta* **1994**, *58*, 3675–3689. [CrossRef]
10. Bastow, T.P.; Alexander, R.; Fisher, S.J.; Singh, R.K.; van Aarssen, B.G.K.; Kagi, R.I. Geosynthesis of organic compounds. Part V—Methylation of alkylnaphthalenes. *Org. Geochem.* **2000**, *31*, 523–534. [CrossRef]
11. Evdokimov, I.N.; Eliseev, N.Y.; Akhmetov, B.R. Assembly of asphaltene molecular aggregates as studied by near-UV/visible spectroscopy: II. Concentration dependencies of absorptivities. *J. Pet. Sci. Eng.* **2003**, *37*, 145–152. [CrossRef]
12. Maciel, G.E.; Bartuska, V.J.; Miknis, F.P. Correlation between oil yields of oil shales and 13C nuclear magnetic resonance spectra. *Fuel* **1978**, *57*, 505–506. [CrossRef]
13. Hutchison, C.S.; Tan, D.N.K. *Geology of Peninsular Malaysia*; The University of Malaya, The Geological Society of Malaysia: Kuala Lumpur, Malaysia, 2009; p. 479.
14. Gebretsadik, H.T.; Sum, C.W.; Talib, J.A. *Lithostratigraphy of Paleozoic Carbonates in the Kinta Valley, Peninsular Malaysia: Analogue for Paleozoic Successions*; ICIPEG 2016; Springer: Singapore, 2017; pp. 559–567.
15. Shoieb, M.A.; Gebretsadik, H.T.; Rahmani, O.; Ismail, M.S.; Ibad, S.M. Geochemical characteristics of the Silurian-Devonian Kroh black shales, Peninsular Malaysia: An implication for hydrocarbon exploration. *J. Geochem. Explor.* **2022**, *232*, 106891. [CrossRef]
16. Metcalfe, I. Tectonic evolution of the Malay Peninsula. *J. Asian Earth Sci.* **2013**, *76*, 195–213. [CrossRef]
17. Group, M.-T.W. Geology of the Pengkalan Hulu-Betong transect area along the Malaysia-Thailand border. *Geol. Pap.* **2009**, *7*, 1–84.
18. Griffiths, P.R. Fourier transform infrared spectrometry. *Science* **1983**, *222*, 297–302. [CrossRef]
19. Chen, Y.; Zou, C.; Mastalerz, M.; Hu, S.; Gasaway, C.; Tao, X. Applications of micro-fourier transform infrared spectroscopy (FTIR) in the geological sciences—A review. *Int. J. Mol. Sci.* **2015**, *16*, 30223–30250. [CrossRef]
20. McKelvy, M.L.; Britt, T.R.; Davis, B.L.; Gillie, J.K.; Lentz, L.A.; Leugers, A.; Nyquist, R.A.; Putzig, C.L. Infrared spectroscopy. *Anal. Chem.* **1996**, *68*, 93–160. [CrossRef]

21. Coates, J. Interpretation of infrared spectra, a practical approach. In *Encyclopedia of Analytical Chemistry*; John Wiley & Sons, Ltd.: Hoboken, NJ, USA, 2006. [CrossRef]
22. Stuart, B.H. *Infrared Spectroscopy: Fundamentals and Applications, Analytical Techniques in the Sciences*; Wiley: Hoboken, NJ, USA, 2004. [CrossRef]
23. Washburn, K.E.; Birdwell, J.E. Multivariate analysis of ATR-FTIR spectra for assessment of oil shale organic geochemical properties. *Org. Geochem.* **2013**, *63*, 1–7. [CrossRef]
24. Chen, Y.; Senesi, N.; Schnitzer, M. Information provided on humic substances by E4/E6 ratios. *Soil Sci. Soc. Am. J.* **1977**, *41*, 352–358. [CrossRef]
25. Schnitzer, M. Soil organic matter—The next 75 years. *Soil Sci.* **1991**, *151*, 41–58. [CrossRef]
26. Peters, K.E.; Cassa, M.R. Applied source rock geochemistry: Chapter 5: Part II. Essential elements. In *The Petroleum System—From Source to Trap*; American Association of Petroleum Geologists, AAPG: Tulsa, OK, USA, 1994; Volume 60, pp. 93–120.
27. Dow, W.G.; Pearson, D.B. Organic matter in Gulf Coast sediments. In Proceedings of the Offshore Technology Conference, Houston, TX, USA, 4 May 1975; pp. 85–94. [CrossRef]
28. Wiberley, S.E.; Gonzalez, R.D. Infrared spectra of polynuclear aromatic compounds in the CH stretching and out-of-plane bending regions. *Appl. Spectrosc.* **1961**, *15*, 174–177. [CrossRef]
29. Ibad, S.M.; Padmanabhan, E. Exploring the relationship between hydrocarbons with total carbon and organic carbon in black shale from Perak, Malaysia. *Pet. Coal* **2017**, *59*, 933–943.
30. Yang, G.-P. Polycyclic aromatic hydrocarbons in the sediments of the South China Sea. *Environ. Pollut.* **2000**, *108*, 163–171. [CrossRef] [PubMed]
31. Helgeson, H.C.; Richard, L.; McKenzie, W.F.; Norton, D.L.; Schmitt, A. A chemical and thermodynamic model of oil generation in hydrocarbon source rocks. *Geochim. Cosmochim. Acta* **2009**, *73*, 594–695. [CrossRef]
32. Tissot, B.P.; Welte, D.H. *Petroleum Formation and Occurrence*; Springer: Berlin/Heidelberg, Germany, 1984. [CrossRef]
33. Tan, J.; Horsfield, B.; Mahlstedt, N.; Zhang, J.; Boreham, C.J.; Hippler, D.; van Graas, G.; Tocher, B.A. Natural gas potential of Neoproterozoic and lower Palaeozoic marine shales in the Upper Yangtze Platform, South China: Geological and organic geochemical characterization. *Int. Geol. Rev.* **2015**, *57*, 305–326. [CrossRef]
34. Ross, D.J.K.; Bustin, R.M. Characterizing the shale gas resource potential of Devonian–Mississippian strata in the Western Canada sedimentary basin: Application of an integrated formation evaluation. *Am. Assoc. Pet. Geol. Bull.* **2008**, *92*, 87–125. [CrossRef]
35. Jarvie, D.M.; Hill, R.J.; Ruble, T.E.; Pollastro, R.M. Unconventional shale-gas systems: The Mississippian Barnett Shale of north-central Texas as one model for thermogenic shale-gas assessment. *Am. Assoc. Pet. Geol. Bull.* **2007**, *91*, 475–499. [CrossRef]
36. Du, X.; Song, X.; Zhang, M.; Lu, Y.; Lu, Y.; Chen, P.; Liu, Z.; Yang, S. Shale gas potential of the Lower Permian Gufeng Formation in the western area of the Lower Yangtze Platform, China. *Mar. Pet. Geol.* **2015**, *67*, 526–543. [CrossRef]
37. Egbobawaye, I.E. Petroleum Source-Rock Evaluation and Hydrocarbon Potential in Montney Formation Unconventional Reservoir, Northeastern British Columbia, Canada, Natural Resources. *Nat. Resour.* **2017**, *8*, 716. [CrossRef]
38. Ibad, S.M.; Padmanabhan, E. Methane sorption capacities and geochemical characterization of Paleozoic shale Formations from Western Peninsula Malaysia: Implication of shale gas potential. *Int. J. Coal Geol.* **2020**, *224*, 103480. [CrossRef]
39. Abou El-Anwar, E.; Salman, S.; Mousa, D.; Aita, S.; Makled, W.; Gentzis, T. Organic Petrographic and Geochemical Evaluation of the Black Shale of the Duwi Formation, El Sebaiya, Nile Valley, Egypt. *Minerals* **2021**, *11*, 1416. [CrossRef]
40. Li, B.; Feng, C.; Li, X.; Chen, Y.; Niu, J.; Shen, Z. Spatial distribution and source apportionment of PAHs in surficial sediments of the Yangtze Estuary, China. *Mar. Pollut. Bull.* **2012**, *64*, 636–643. [CrossRef]
41. Wu, F.; Xu, L.; Sun, Y.; Liao, H.; Zhao, X.; Guo, J. Exploring the relationship between polycyclic aromatic hydrocarbons and sedimentary organic carbon in three Chinese lakes. *J. Soils Sediments* **2012**, *12*, 774–783. [CrossRef]
42. González-Vila, F.J.; Amblès, A.; del Rıo, J.C.; Grasset, L. Characterisation and differentiation of kerogens by pyrolytic and chemical degradation techniques. *J. Anal. Appl. Pyrolysis* **2001**, *58*, 315–328. [CrossRef]
43. Cunha, T.J.F.; Novotny, E.H.; Madari, B.E.; Martin-Neto, L.; Rezende, M.O.d.O.; Canelas, L.P.; Benites, V.d.M. Spectroscopy characterization of humic acids isolated from Amazonian Dark Earth Soils (terra preta de Índio). In *Amazonian Dark Earths: Wim Sombroek's Vision*; Springer: Berlin/Heidelberg, Germany, 2009; pp. 363–372. [CrossRef]
44. Strachan, M.G.; Alexander, R.; Kagi, R.I. Trimethyl naphthalenes in crude oils and sediments: Effects of source and maturity. *Geochim. Cosmochim. Acta* **1988**, *52*, 1255–1264. [CrossRef]
45. Volkman, J.K.; Alexander, R.; Kagi, R.I.; Rowland, S.J.; Sheppard, P.N. Biodegradation of aromatic hydrocarbons in crude oils from the Barrow Sub-basin of Western Australia. *Org. Geochem.* **1984**, *6*, 619–632. [CrossRef]
46. Wang, X.; Cai, T.; Wen, W.; Ai, J.; Ai, J.; Zhang, Z.; Zhu, L.; George, S.C. Surfactin for enhanced removal of aromatic hydrocarbons during biodegradation of crude oil. *Fuel* **2020**, *267*, 117272. [CrossRef]
47. Cheng, X.; Hou, D.; Mao, R.; Xu, C. Severe biodegradation of polycyclic aromatic hydrocarbons in reservoired crude oils from the Miaoxi Depression, Bohai Bay Basin. *Fuel* **2018**, *211*, 859–867. [CrossRef]
48. Asahina, K.; Suzuki, N. Alkyl naphthalenes and tetralins as indicators of source and source rock lithology–Pyrolysis of a cadinane-type sesquiterpene in the presence and absence of montmorillonite. *J. Pet. Sci. Eng.* **2016**, *145*, 657–667. [CrossRef]

Article

Identification of Radioactive Mineralized Lithology and Mineral Prospectivity Mapping Based on Remote Sensing in High-Latitude Regions: A Case Study on the Narsaq Region of Greenland

Li He [1,2,*], Pengyi Lyu [1,3,*], Zhengwei He [1,3], Jiayun Zhou [4], Bo Hui [4], Yakang Ye [4], Huilin Hu [3], Yanxi Zeng [3] and Li Xu [3,4]

[1] State Key Laboratory of Geohazard Prevention and Geoenvironment Protection, Chengdu University of Technology, Chengdu 610059, China; hzw@cdut.edu.cn
[2] College of Tourism and Urban-Rural Planning, Chengdu University of Technology, Chengdu 610059, China
[3] College of Earth Sciences, Chengdu University of Technology, Chengdu 610059, China; abbey.wang@onestopwarehouse.com.au (H.H.); cissyzengxixi@sina.com (Y.Z.); xulicdzhs@163.com (L.X.)
[4] Institute of Multipurpose Utilization of Mineral Resources, China Geological Survey, Chengdu 610041, China; zhszjy@aliyun.com (J.Z.); huibo0728@foxmail.com (B.H.); yeyakang920617@sina.cn (Y.Y.)
* Correspondence: heli2020@cdut.edu.cn (L.H.); davislvpy1045@gmail.com (P.L.)

Citation: He, L.; Lyu, P.; He, Z.; Zhou, J.; Hui, B.; Ye, Y.; Hu, H.; Zeng, Y.; Xu, L. Identification of Radioactive Mineralized Lithology and Mineral Prospectivity Mapping Based on Remote Sensing in High-Latitude Regions: A Case Study on the Narsaq Region of Greenland. *Minerals* **2022**, *12*, 692. https://doi.org/10.3390/min12060692

Academic Editor: Amin Beiranvand Pour

Received: 17 May 2022
Accepted: 27 May 2022
Published: 30 May 2022

Publisher's Note: MDPI stays neutral with regard to jurisdictional claims in published maps and institutional affiliations.

Abstract: The harsh environment of high-latitude areas with large amounts of snow and ice cover makes it difficult to carry out full geological field surveys. Uranium resources are abundant within the Ilimaussaq Complex in the Narsaq region of Greenland, where the uranium ore body is strictly controlled by the Lujavrite formation, which is the main ore-bearing rock in the complex rock mass. Further, large aggregations of radioactive minerals appear as thermal anomalies on remote sensing thermal infrared imagery, which is indicative of deposits of highly radioactive elements. Using a weight-of-evidence analysis method that combines machine-learned lithological classification information with information on surface temperature thermal anomalies, the prediction of radioactive element-bearing deposits at high latitudes was carried out. Through the use of Worldview-2 (WV-2) remote sensing images, support vector machine algorithms based on texture features and topographic features were used to identify Lujavrite. In addition, the distribution of thermal anomalies associated with radioactive elements was inverted using Landsat 8 TIRS thermal infrared data. From the results, it was found that the overall accuracy of the SVM algorithm-based lithology mapping was 89.57%. The surface temperature thermal anomaly had a Spearman correlation coefficient of 0.63 with the total airborne measured uranium gamma radiation. The lithological classification information was integrated with surface temperature thermal anomalies and other multi-source remote sensing mineralization elements to calculate mineralization-favorable areas through a weight-of-evidence model, with high-value mineralization probability areas being spatially consistent with known mineralization areas. In conclusion, a multifaceted remote sensing information finding method, focusing on surface temperature thermal anomalies in high-latitude areas, provides guidance and has reference value for the exploration of potential mineralization areas for deposits containing radioactive elements.

Keywords: high latitudes; weak information; thermal anomalies; radioactive element deposits

1. Introduction

High-latitude regions lie between the 60° north and 60° south latitudes to the north and south poles of the Earth's surface, respectively, and receive the least solar radiation. Therefore, the climate is cold, and most areas are covered in snow and ice for long periods, which makes it difficult to carry out comprehensive geological field surveys. With remote sensing technology, it is possible to overcome the time constraints of field investigations and select multiple sources of remote sensing data for long time series analysis. Spectroscopic information from remote sensing can screen the diversity of the mineral spectrum, which depends on the physical interactions of electrons and molecular structures within the material [1–3]. Multispectral data and wave spectrum identification algorithms have made it possible for remote sensing technology to predict mineralized target areas [4–12]. The identification of the lithology based on remote sensing data automatically classified by computers can help quickly obtain geological background information of the target area in comparison to the long cycle time of a geological field survey. Lithological classification via machine learning not only fully utilizes the spectral and rock texture features among different rocks but also improves the lithological classification accuracy [13–20].

The Gardar igneous intrusions in southern Greenland are typically high in alkali elements, such as sodium, whereas the Ilimaussaq Complex, which was formed later in the magmatic intrusion system, has a high concentration of rare and radioactive elements [21,22]. Owing to the enrichment of radioactive elements, such as uranium and thorium, the earth heat flow generated will inevitably cause the enriched areas to exhibit extreme radioactivity; this radioactive heat can be detected by surface thermal anomalies [23–26]. Information on surface thermal anomalies can be obtained in various ways, and thermal infrared remote sensing technology is a widely used technique. Based on the information obtained from thermal anomalies, it is possible to interpret certain topographic changes (basement uplift and depression), volcanoes, hot springs, faults, etc. [27]. The use of thermal infrared remote sensing technology for geothermal resources has made it easier to develop resource-prospecting techniques. In the field of geology, this technology has been applied for decades, and it is widely used in large-scale geothermal resource surveys, mountain surveys, volcano early-warning systems, and earthquake prediction [28–32]. The introduction of remote sensing data, such as ASTER and Landsat TIRS, effectively increases the diversity of surface temperature inversion and more effectively traces radioactive minerals and geothermal resources, playing an important role in the field of geological and mineral exploration [33–37].

This paper aims to identify radioactive element enrichment areas and ore-bearing lithologies by remote sensing techniques and to study a method for predicting the favorability of mineralization of radioactive deposits at high latitudes using a weight-of-evidence model. The inversion of surface thermal anomalies from thermal infrared remote sensing data is conducted in the harsh Greenland Narsaq region, where areas of radioactive mineral enrichment are extracted. Machine learning techniques are also used to identify and classify regional lithologies enriched in radioactive elements. The integration of multi-source remote sensing information using the Weight of Evidence model can be effective in conducting mineral resource surveys in high latitude regions.

2. Study Area

2.1. Physical Geography

The study area is in the Narsaq region of Gardar Province in southern Greenland, ranging from 44°30′ to 46°30′ W in longitude and 60°45′ to 61°20′ N in latitude. The region has very few land-based road systems due to the extremely large number of bays (Figure 1). The region experiences a polar climate, with the average temperature in winter (January) being −6 °C, while the average temperature during the coastal summer (July) is 7 °C, with July and August having the highest temperatures of the year.

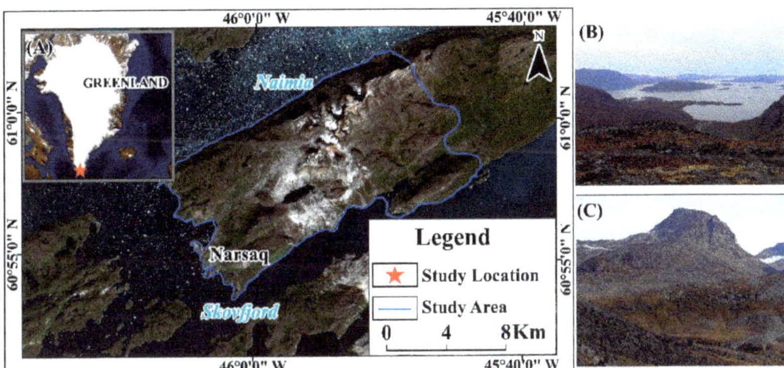

Figure 1. (**A**) Location map based on Sentinel-2's true color 432 band combinations; (**B,C**) A realistic view of the study area environment.

2.2. Geological Background

The geological setting of the study area is dominated by the southern parts of the Palaeo-Craton and the Palaeoproterozoic Ketilidian orogenic belts. The Mesoproterozoic Gardar igneous province crosses the Ketilidian orogenic belt. The province of Gardar is marked by the development of deposits of faulted, clastic, and volcanic rocks with high alkaline magmatic activity. The Gardar intrusive complex is dominated by differentiated silica-alumina rocks, including syenite, nepheline syenite, quartz syenite, and granite. Giant vein rocks are dominated by weakly alkaline gabbro and syenite gabbro, with faults developing parallel to rift valleys in formations affected by lithosphere stretching [38]. Within the Julianehab Granite, there are several east–northeast (NEE) oriented fault planes, where the lateral displacement along the fault planes is uncertain, but the vertical displacement is evident. The displacement faults incorporate NEE to north–east (NE) trending sinistral faults, as well as north–north-west (NNW) to north–north-east(NNE) trending r-dextral faults, forming conjugate faults (Figure 2) [22,39,40].

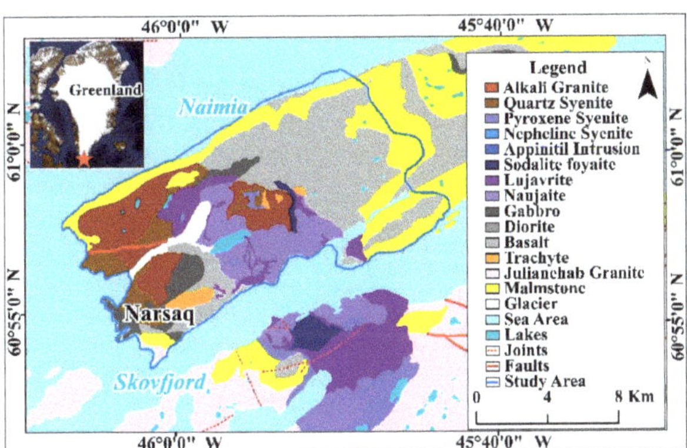

Figure 2. Geological map of the study area.

3. Material and Methods

3.1. Data and Pre-Processing

Several data sources were applied to meet the needs of the study (Table 1), including the following:

1. Visible light near the infrared (NIR) data of the Sentinel-2AB (S2AB) satellite;
2. Thermal infrared data of LANDSAT-8TIRS (LTRS) satellite;

3. Visible light near the infrared data of the Worldview-2 (WV-2) satellite;
4. ASTER GDEM 30 m spatial resolution ground elevation model data;
5. Measured data of the SVC HR-1024i full-spectrum ground object spectrometer.

Table 1. List of data used.

Data Type	Maximum Spatial Resolution (m)	Acquisition Time
Worldview-2	0.5	29 August 2017
Landsat 8 TIRS	15	29 July 2018, 7 August 2018, 26 August 2019
Sentinel-2	10	10 August 2019, 6 April 2020
ASTER GDEM V3	30	August 2019
SVC HR-1024i	-	July 2019

3.1.1. Visible NIR Remote Sensing Data

The visible NIR satellite remote sensing data were selected from S2AB and WV-2 satellite data. Among these, the main payload of the Sentinel satellite is the Multi-Spectral Imager (MSI), operating in the visible, near-infrared, and short-wave infrared spectral bands, with ground resolutions of 10 m, 20 m, and 60 m, respectively [41,42]. The WV-2 satellite is a high spatial resolution satellite data, capable of providing panchromatic images at 0.46 m and multispectral images at 1.8 m resolutions [43].

3.1.2. Thermal Infrared Remote Sensing Data

The LTRS data were chosen as a source of thermal infrared radiation information, whose thermal infrared sensor covers two thermal infrared bands, both of which have a resolution of 100 m in the wavelength range of 10.60–12.51 μm [44,45].

3.1.3. Topographic Surface Elevation Data

Topographic data were extracted using ASTER GDEM V3, a digital elevation model acquired and released by NASA's Earth observation satellite, named Terra, with a resolution of one arc-second (30 m), covering 99% of the global land surface from 83° N to 83° S [46].

3.1.4. Field Measurements of Feature Spectral Data

Field spectra were collected using the SVC HR-1024i (SVC, Poughkeepsie, NY USA) full spectrum spectroradiometer, which has a spectral measurement range of 350–2500 nm and a total of 1024 channels. The spectral resolution is 2.8 nm in the 350–1000 nm range, 3.6 nm in the 1000–1900 nm range, and 2.5 nm in the 1900–2500 nm range.

Remote sensing data pre-processing was carried out using the ENVI software (Version 5.6, ESRI, Redlands, CA, USA) [47], which provides multi-source remote sensing data with radiometric calibration, FLAASH atmospheric correction, geometric correction, and image enhancement as remote sensing image pre-processing steps. The measured spectral information was obtained from 50 rock samples collected at the mine site, and the rock spectra were collected in a dark room environment. As the spectral features of the rocks acquired from remote sensing images come from exposed rock surfaces, which are affected by weathering and other environmental factors, the rock samples were not ground to simulate the real conditions in the field, and a total of 128 valid spectra were collected (Figure 3).

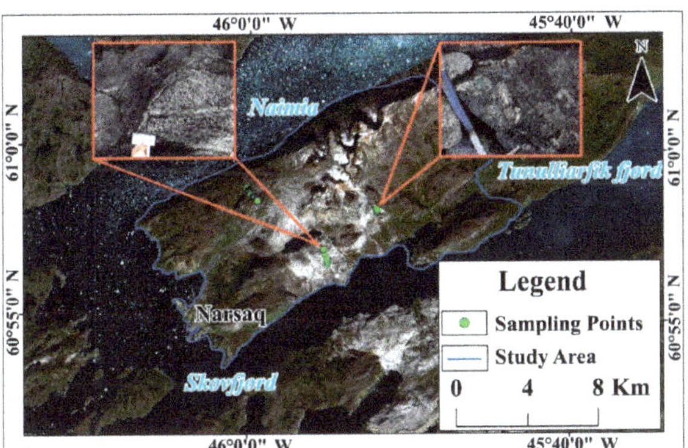

Figure 3. Distribution map of field spectral samples (the base image is a combination of the 432 bands of Sentinel-2).

3.2. Remote Sensing for Geological Background Information Extraction

The interpretation and investigation of geological background information are fundamental for the prediction of regional mineralization; during geological action, areas spatially located in geological-variation regions and marginal areas are often the sites of endogenous deposits. Significant deposits are often found at the junction of tectonic plates and are temporally associated with tectonic events, with the distribution of mineralization information roughly corresponding to the occurrence of tectonic anomalies. Furthermore, tectonics provides a good environment for the formation, storage, and transportation of deposits; the mapping of the geological base information will facilitate the understanding of the regional framework and the rapid tracing of mineralization prediction areas. Extraction was carried out from three perspectives: lithological, tectonic, and alteration information.

The interpretation of lithologies and formations in the study area was based on visual interpretation. Firstly, the ArcGIS and ENVI software were used to enhance the remote sensing image information, and the interpretability of the interpreted lithologies and structures was enhanced through optimal waveband analysis and image filtering. Directional features, which are important properties of linear constructions, were enhanced by directional filtering in the study area to identify linear constructions more intuitively. The image was enhanced using a 5 by 5 directional convolutional filtering method, and the image—after enhancement—exhibited extremely distinct linear features from north–north-east to north-east–east (45° range). Lithologies and formations smaller than the spatial resolution per image element are difficult to distinguish accurately and are often interpreted indirectly utilizing the topography, vegetation, water systems, etc.

For the extraction of alteration information, the study area used principal component analysis (PCA) based on the method proposed by Crosta scholars for the extraction of hydroxyl and iron-stained alterations. The PCA uses the multidimensional orthogonal linear variation of the interrelationships between variables, and the entire method is based on mathematical and statistical analyses. The method can reduce the dimensionality of remote sensing information, capture the spectral differences of features, and serve to enhance and compress the data while also removing correlations between information in the same region or the same remote sensing data band [48]. Secondary oxides are the most represented group of iron-stained alteration minerals, while only a small proportion of the other alteration minerals are primary. In the 2, 4, and 11 bands of Sentinel-2, the divalent and trivalent ions of iron have characteristic absorption valleys; therefore, these three bands were chosen as the main bands for iron-stained alteration extraction. Alteration minerals that contain hydroxyl or carbonate ions include chlorite and kaolinite. The spectra of the mineralized rocks have two unique features compared to the spectra of other rocks, where a slowly rising plateau forms at wavelengths of 1.0–1.4 μm, while the spectrum at

1.9–2.0 µm forms an extremely strong absorption valley, indicating absorption properties in the near-infrared band; therefore, bands 2, 8, and 12 of Sentinel-2 were chosen to extract the hydroxyl alteration.

3.3. The Support Vector Machines(SVM) Lithology Extraction Technique

SVMs are widely used in the field of geological rock identification and classification. As a method of machine learning, their core concept involves projecting data into a high-dimensional space, constructing an optimal hyperplane in the high-dimensional space, and using this optimal plane to classify different data. The object-oriented SVM classification method, which uses the object as the basic unit, is a classification method that combines multiple types of feature information, including spectrum, texture, shape, and topology information [49–52]. The method of classifying image units using SVMs (Figure 4) differs from those of other algorithms in that it minimizes a priori intervention and, therefore, presents the classification results objectively; in addition, it is efficient and stable [53,54].

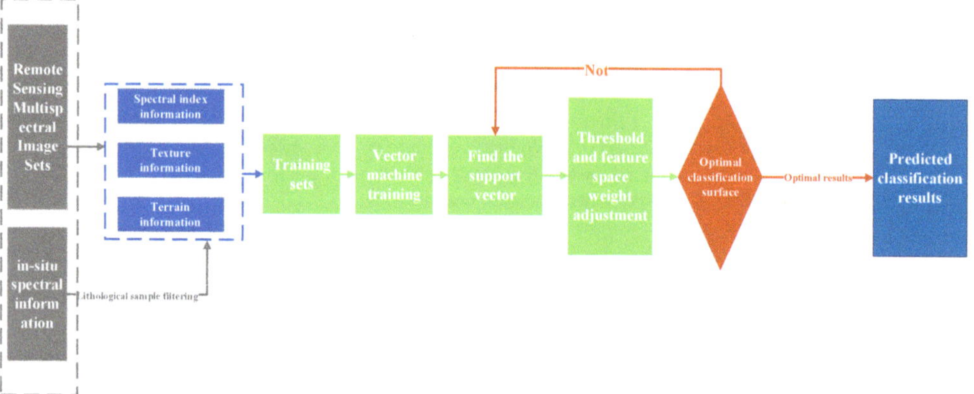

Figure 4. Support vector machine classification recognition process.

The SVM approach is considered to be a good method for classification extraction because it has high generalization performance and does not require prior knowledge, even if the dimensionality of the input space is high [55]. Intuitively, SVM algorithm extraction is based on finding a hyperplane, provided a set of points belong to either of the two classes, such that the proportion of points in the same class on the same side is maximized while also maximizing the distance between either class and the hyperplane [56] (Figure 5).

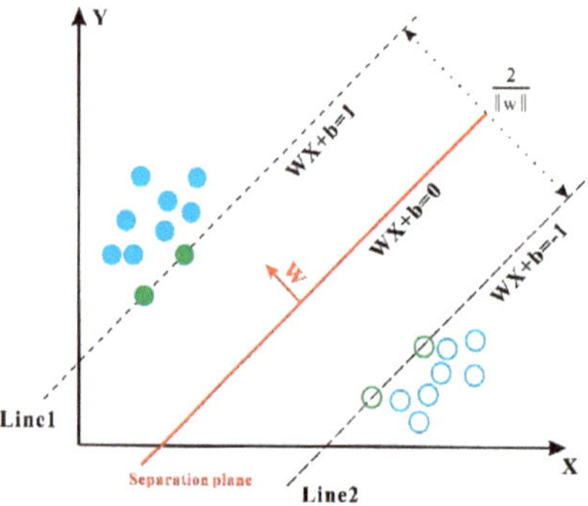

Figure 5. Support vector machine schematic.

A critical step in using non-linear SVMs is the selection of the kernel function, which performs a special spatial non-linear transformation, resulting in the projection of the training data into a high-dimensional feature space, which directly determines the dimensionality of the classification function. Ultimately, the optimal classification surface is found in the high-dimensional space, facilitating the classification calculations. SVM classification was carried out using ENVI 5.6; the radial basis kernel function (Gaussian radial function), which has high accuracy for classifying data, was chosen as the kernel function in this study, with the penalty parameter set to 100.

Owing to the narrowband and full spectral coverage of the SVC spectroradiometer, it is possible to effectively characterize the subtle spectral differences of rocks collected in the field. The measured feature spectral information is compared with the spectral information of remote sensing image elements to improve the accuracy and efficiency of training sample selection in remote sensing images. The measured spectra need to be resampled to the corresponding spectral resolution conditions of the remote sensing image when performing the comparison verification. The multivariate training element features are extracted using texture features, spectral index features, vegetation index features, and terrain features for different regions and levels of data limitation. The training samples obtained by the SVM method were all located within the field sampling work area and were analyzed by mineral rock identification. The rock samples collected covered four types of rocks: gabbro, Lujavrite, basalt, and Naujaite; Lujavrite, which is associated with the radioactive uranium ore, was analyzed by petrographic identification microscopy, and the collected samples all contained high mineral contents of eudialyte (Figure 6). Consequently, the SVM training samples were selected to create samples from these four lithologies, with a total of 653 samples (including 155 gabbro, 203 lujavrite, 105 basalt, and 190 naujaite).

(A)

(B)

Figure 6. (**A**) Lujavrite hand specimen; (**B**) microscopic photograph of Lujavrite.

3.4. Thermal Anomaly Information Extraction

Hydrothermal-type uranium deposits lead to surface thermal anomalies when they are formed and also have some influence on the geothermal flow in their vicinity after formation. Therefore, points with a high distribution of surface thermal anomaly values tend to be spatially coherent with uranium ores. Geothermal signatures are also used by some researchers as an indicator of hydrothermal uranium deposits, which are often closely related to their distribution in deep uranium exploration [25,57]. Surface temperature inversion is closely related to various resource and environmental processes on the Earth's surface. As an important physical parameter of the energy balance and circulation interchange processes between the Earth and the air, understanding the surface temperature has become an important facet of the field of quantitative remote sensing. In this study, the radiative transfer equation (RTE) method, which is well established and widely applicable, is used to invert the surface temperature in the study area, which has a solid physical basis due to its early development and high accuracy [58]. The RTE method is based on real-time atmospheric profile data, including humidity, temperature, and pressure, and uses radiant energy values obtained from individual thermal infrared bands observed by satellites—while removing atmosphere-related effects—to invert the surface temperature. RTE converts the thermal radiation values to surface temperatures after subtracting the influence of the atmospheric extinction coefficient as a parameter factor, whereby the atmospheric thermal radiation influence values can be obtained on the basis of atmospheric data (Table 2) [45,59]. The calculation is shown in Equation.

$$L_{\text{sensor}} = \left[\varepsilon B(T_s) + (1-\varepsilon)L_\downarrow\right]\tau + L_\uparrow \tag{1}$$

$$B(T_s) = \left[L_{\text{sensor}} - L_\uparrow - \tau(1-\varepsilon)L_\downarrow\right]/\tau\varepsilon \tag{2}$$

Table 2. Atmospheric profile parameters.

Data Type	Imaging Time	Atmospheric Transmissivity τ	Atmospheric Upward Radiation $L_\uparrow$ (w/m^2/sr/μm)	Atmospheric Downward Radiance $L_\downarrow$ (w/m^2/sr/μm)
	7 August 2018	0.95	0.30	0.53
Landsat TIRS10	29 July 2018	0.92	0.51	0.87
	26 August 2019	0.96	0.24	0.42

T_s is the surface temperature, in Kelvin; the blackbody radiance is denoted by B; L denotes the radiance, where the arrows pointing up and down represent the upward and downward radiance of the atmosphere; and the surface-specific radiance and the atmospheric transmittance in the thermal infrared band are denoted by ε and τ, respectively. Due to the continuous atmospheric profile, the atmospheric parameters vary at different altitudes, which also results in differences in atmospheric radiance. The two core parameters in the RTE algorithm are the atmospheric upward and downward radiation and the atmospheric transmittance parameters, for which the surface temperature is calculated using the following formula:

$$T_s = K_2 / \ln(1 + K_1/B(T_s)) \tag{3}$$

where K_2 and K_1 are constants that depend on the selected satellite metadata.

3.5. Remote Sensing Mineralization Prediction Based on the Weight of Evidence Methods

Agterberg proposed the Weight of Evidence Method (WofE), a geostatistical-based approach to mineralization prediction, using a Bayesian statistical analysis model [60]. The method aims to extract favorable areas (prospective areas) for mineralization, using geological information related to the formation of mineralization, overlaying and fusing such information, and analyzing it, which fully integrates AI technology, image analysis technology, and mathematical statistics technology. This approach is achieved by splitting all evidence layers into binary variables; in other words, evidence layers containing only '0' and '1' attributes, where '0' means that a single unit of evidence in the element layer does not exist (no ore), and '1' means that it does (contains ore). Assuming the number of units in the study area is expressed as S, the event element A is expressed as an element layer (hydrothermal alteration anomaly, mineral control structure, SVM classified lithology, radiothermal anomaly, etc.), and B is expressed as an ore-bearing unit. $P(B) = Area(B)/Area(S)$ denotes the prior probability of event B, where $Area()$ denotes the area. Bayesian statistical relations were introduced in the study area as the basis for the criterion, with A_i^+, A_i^- denoting the presence and absence of A_i favorable conditions, respectively, which divided the study area into four pooled parts, expressed as $B^+ \cap A_i^+$, $B^+ \cap A_i^-$, $B^- \cap A_i^+$, $B^- \cap A_i^-$. The posterior probability is calculated using the following formula:

$$(B \mid A_1 A_2 \cdots A_n) = e^{\sum_{i=0}^{n} W_j} / 1 + e^{\sum_{i=0}^{n} W_j} \tag{4}$$

For each evidence layer, it is necessary to introduce a contrast value C, $C = W^+ - W^-$, in order to express its correlation with the deposit or occurrence. The strength of the correlation is indicated by a significant C value, with a positive or negative C value representing a positive or negative relationship between the layer and the indicative mineralization. Studentized Index (SI) defined as:

$$SI = C / \sqrt{\delta^2(W^+) + \delta^2(W^-)} \tag{5}$$

The evidence elements in the layers were verified against each other in groups of two, the weights of the evidence elements were calculated, and the layers were combined statistically using superposition analysis to obtain the final posterior probability distribution of mineralization.

4. Results and Analysis

4.1. Remote Sensing for Geological Background Information Extraction

A total of 13 lithologies have been interpreted, including gabbro, syenite, ditroite, and Lujavrite; Lujavrite—containing steenstrupine and eudialyte—is the main ore mineral in the study area (Figure 7). Based on the tectonic features of the Narsaq area, interpretation markers were established to obtain the distribution pattern of lineaments and rings in the area. Four faults, nine rings, and 157 tectonic joints were interpreted (Figure 7). The mean linear orientation of the interpreted linear structures was analyzed using ArcGIS linear analysis, which calculated that the mean linear orientation of the linear structures across the study area is 62° (azimuthal), i.e., the tectonics in the study area—as a whole—are predominantly NEE oriented (Figure 8A). The strike rose diagram shows that the highest frequency of tectonics is between the north–north-east and north-east–east orientations (Figure 8B), which is spatially consistent with the distribution of the Southern Rift Zone.

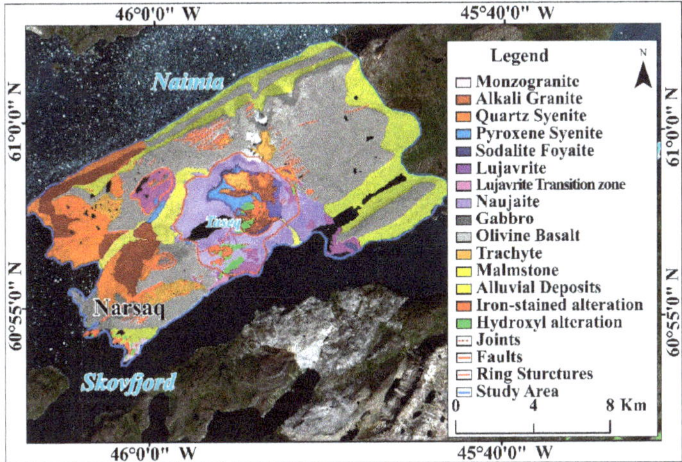

Figure 7. Interpreted map of remote sensing geological background information map (the base image is a combination of the 432 Sentinel-2 bands).

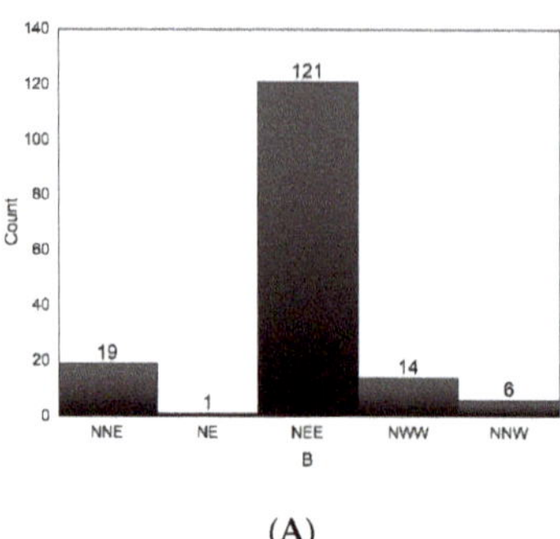
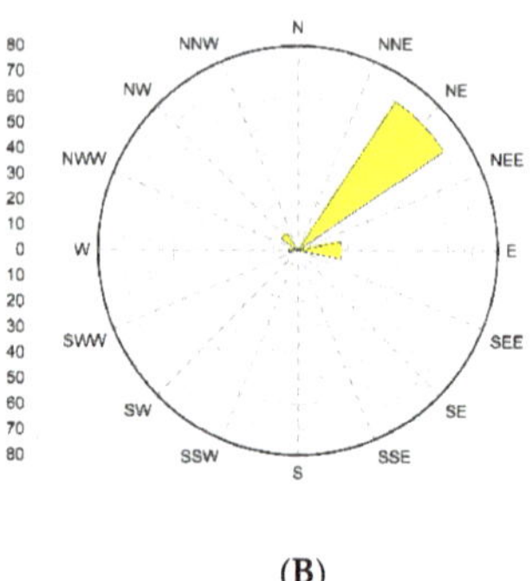

(A)   (B)

Figure 8. (**A**) Linear tectonic strike statistical histogram; (**B**) Linear tectonic strike rose diagram.

The hydroxy and iron-stained alteration is distributed in the center and lower parts of the study area, with the iron-stained alteration exhibiting a mass-like character and the hydroxy alteration showing a striped northeast spreading character. The two types of alteration information are mainly consistent with the location of surface outcrops of Naujaite and Lujavrite, which is indicative of the lithology.

4.2. Lithology Extraction Based on SVM

4.2.1. Lithological Information Enhancement and Analysis

Spectral absorption features were calculated using the IDL DISPEC software [1]. These features describe the shape of the spectrum, as reflected by the depth, width, area, and asymmetry (Table 3). The Lujavrite associated with the mineralization has the following characteristics (Figure 9): (i) a slowly rising plateau in the wavelength range of 1.0–1.3 μm; (ii) an extremely strong absorption valley in the spectrum of 1.9–2.0 μm, which indicates that the Lujavrite exhibits absorption properties in the near-infrared band. (iii) The absorption spectrum after continuum removal has a maximum absorption valley depth of 47.87 at 0.4 μm—the area enclosed by the envelope and the spectral curve is the largest here, and the diagnostic spectrum is located at 0.4 μm. It also shows strong absorption characteristics at 1.92 μm and 1.42 μm, as reflected in Table 3, which proves that it is influenced by the vibration of water molecules and the leap of hydroxyl ions.

4.2.2. Feature Information Extraction

In SVM lithology extraction, two major dimensions—texture information and topographic information—were used. The texture information was calculated using PCA and the greyscale formula matrix. WV-2 image data were used, whose spectrum covers the range of 0.4–1.04 μm; in this range, it is clear from the characteristic absorption in Table 3 that the Lujavrite has strong absorption properties in the coastal band and strong reflection properties in the near-infrared band. Further, the most informative bands of the image are concentrated in the true color band; therefore, bands one, two, five, and eight were selected for image enhancement using PCA (Figure 10B), and this image was used in the extraction of lithological PCA texture information. The texture information of WV-2 was also extracted using the greyscale formula matrix, and contrast (Figure 10C), angular-second-order moments (Figure 10D), homogeneity (Figure 10E), and phase dissimilarity (Figure 10F) were selected as feature statistics. The topographic relief was calculated using the slope information extracted from the DEM (Figure 10A). Before classification, the terrain and texture rasters were spatially resampled to ensure that the information had the same image size.

Table 3. Spectral absorption characteristics of lujavrite.

Wavelength (µ)	Depth	Wide	Area	Asymmetry
0.40	47.87	0.37	18.02	0.78
1.92	30.84	0.20	6.41	0.84
1.42	10.72	0.12	1.31	0.61
1.66	4.80	0.07	0.37	0.81
1.28	0.20	0.02	0.004	1

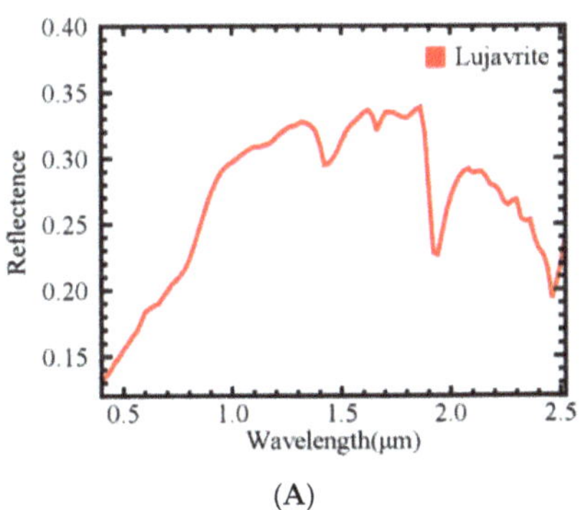

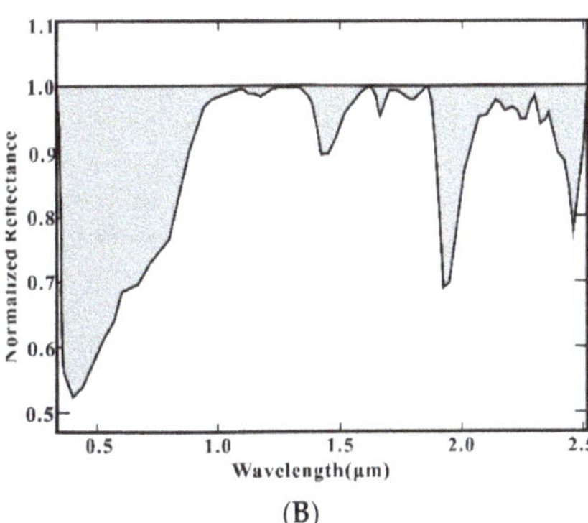

Figure 9. (**A**) Absorption characteristics of the continuum removal spectra of Lujavrite; (**B**) Characterization of the in situ spectra of Lujavrite.

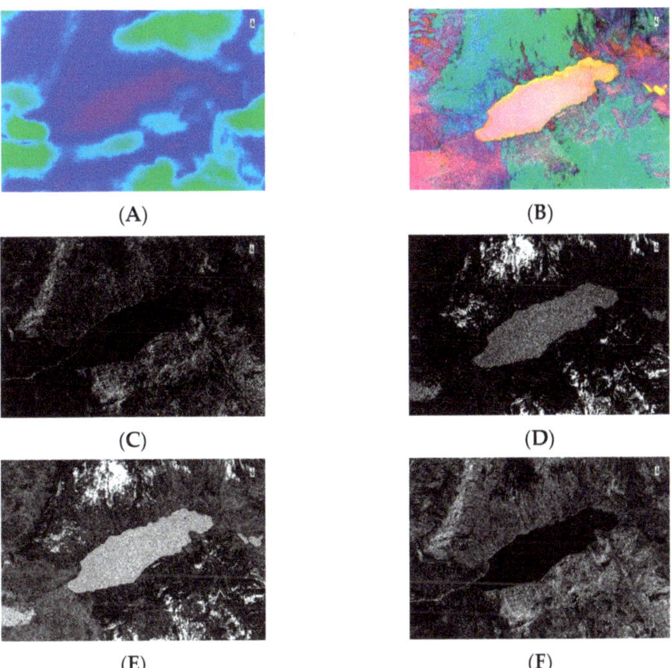

Figure 10. (**A**) Relief feature; (**B**) PCA feature; (**C**) Contrast feature; (**D**) Angular second-order moment feature; (**E**) Homogeneity feature; (**F**) Phase dissimilarity feature.

4.2.3. Results of SVM Lithology Classification

The classification images obtained by the SVM algorithm would show missing data in the classification patches, which were processed using majority/minority analysis to categorize the missing data into a category with a large percentage of surroundings; this helped eliminate the missing classification data. The geological map of the field survey and the remotely interpreted geological map of the area were used to compare and verify the classification results of the Lujavrite (Figure 7). From the classification results, patches with single texture and terrain information were significantly better classified than patches with complex information. In patches with complex feature classes and redundant terrain information, terrain features and texture features could not be accurately distinguished by the SVM algorithm. This is because high spatial resolution satellite data have a limited wavelength and low spectral resolution, making it difficult to distinguish between small diagnostic spectral information in the mixed image elements and reducing the accuracy of the algorithm's recognition. Lujavrite orthoclase is mainly clustered in the central and northern part of the study area and is distributed in bands (Figure 11).

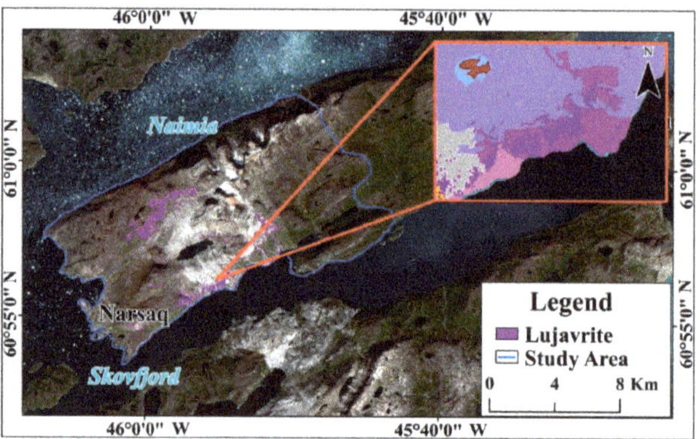

Figure 11. SVM Lithology classification map (the base image is a combination of Worldview-2's 321 bands).

The accuracy of the SVM lithology classification results was evaluated, as shown in Table 4. It is clear from the table that the classification user accuracy of the main mineralized lithology, namely the Lujavrite, is 89.57%; the overall accuracy of SVM classification is 87.75%, with a kappa coefficient of 0.84. After field route verification, dense grey Lujavrite was seen in the target route (Figure 12C), and the rocks were lined with oriented sodium–iron amphibole with a banded structure (Figure 12B), which is consistent with the region shown in the circle.

Table 4. Result accuracy evaluation table for classification using SVM.

Lithological Category	Lithological Category (Ground Truth Data)					
	Gabbro	Lujavrite	Basalt	Naujaite	Total	User Accuracy
Gabbro	134	4	9	8	155	86.45%
Lujavrite	4	189	1	17	211	89.57%
Basalt	13	0	94	9	116	81.03%
Naujaite	4	10	1	156	171	91.23%
Total	155	203	105	190	653	
Producer accuracy	86.45%	93.10%	89.52%	82.11%		

Overall Accuracy = 87.75%; Kappa coefficient = 0.84.

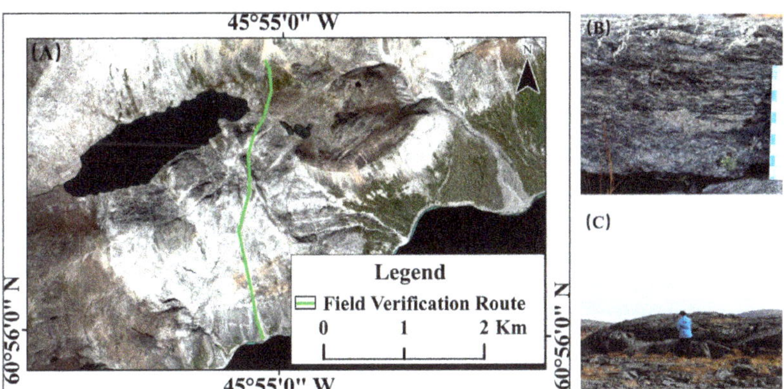

Figure 12. (**A**) Field validation route map (the base image is a combination of Sentinel-2's 432 bands); (**B,C**) Greyish-black dense Lujavrite.

4.3. Thermal Anomalies Extraction of Radioactive Minerals

Most of the uranium equivalents in the study area are greater than 4.3×10^{-6}, while the distribution pattern of greater than 5.2×10^{-6} is more consistent with the distribution pattern of nepheline syenite, proving that nepheline syenite is highly radioactive [61]. The arfvedsonite Lujavrite associated with rare earth-uranium mineralization is rich in elements such as U and Th, which are highly radioactive. The regional sandstone zone is composed mainly of feldspathic quartzite and contains radioactive minerals that have been subjected to low-pressure–high-temperature metamorphism and, thus, exhibit thermal anomalies in surface temperature. Over a given year, the average temperature in the study area is below 0 °C; the surface temperature is extremely low in winter due to the snow and ice cover, whereas water bodies are somewhat insulated, which leads to a lower surface temperature than the water body temperature. This masks the trace thermal anomalies of radioactive elements. The summer images were selected for use because there is less snow and ice cover during this season; further, the difference between the surface temperature of water bodies and land is significant, and the land surface temperature is greater than 0 °C. With Landsat's thermal infrared band, it is possible to effectively distinguish surface temperature differences and, thus, determine areas with radiothermal anomalies.

The Landsat TIRS 10 band was utilized for the surface temperature inversion using the RTE method. Thermal anomalies in the study area were mainly concentrated in the south-central part of the study area, where the overall surface temperature was low, and the average surface temperature in summer ranged from 5° to 16°, with the highest surface temperature values reaching 28° in some areas. Among them, there are three typical high-temperature areas (Figure 13): L1, L2, and L3 (Table 5). All three thermal anomalies are located in the vicinity of the southern rift zone, and the lithology of the high-temperature area is mainly alkaline rock body Naujaite and arfvedsonite Lujavrite; the arfvedsonite type ore is accompanied by uranium, thorium, and other elements, with obvious radioactive anomalies. The thermal anomalies exhibited a strong correlation with both faults and lithology in the area, which further suggests that surface temperature anomalies are indicative of mineralization. In conclusion, the thermal anomalies in the study area could mainly be found along the upper part of the Ilimaussaq Complex on the land margin, and the exposed lithology was arfvedsonite Lujavrite, exhibiting a blocky distribution of NE spreading. In some linear tectonically dense areas, the surface temperature values were significantly higher than those in the surrounding area, indicating that the surface temperature has some correlation with the tectonics. The distribution of thermal anomalies is somewhat indicative of the lithology, linear tectonics, and mineral distribution.

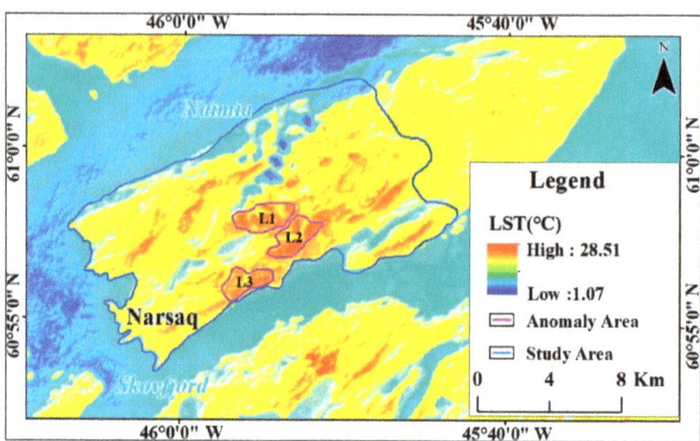

Figure 13. Surface temperature inversion map.

Table 5. Surface temperatures in areas of thermal anomalies.

Abnormal Area	Maximum Surface Temperature	Minimum Surface Temperature	Average Surface Temperature
L1	28.51	15.70	23.85
L2	27.08	14.59	22.76
L3	26.32	16.64	23.24
Study area	28.51	1.07	16.85

Through the official websites of the Geological Survey of Denmark and Greenland (De Nationale Geologiske Undersøgelser for Danmark og Grønland, GEUS), certain airborne radiometric data were selected for the Narsaq area: total uranium gamma radiation, uranium concentration (ppm), and thorium concentration (ppm) (Figure 14). These data were obtained from the GEUS South Greenland Regional Uranium Exploration Project (SYDURAN) [62], which used a helicopter-borne Scintrex GAD-6 for radiometric measurements. The correlation between this airborne radiation data and surface temperature inversion data was analyzed using the Spearman's correlation coefficient, and the thermal anomalies were found (Table 6) to be positively correlated with the total uranium gamma radiation, uranium, and thorium elements, with correlation coefficients of 0.63, 0.60, and 0.65, respectively. This further indicates that the surface temperature thermal anomalies are indicative of the presence of radioactive elements.

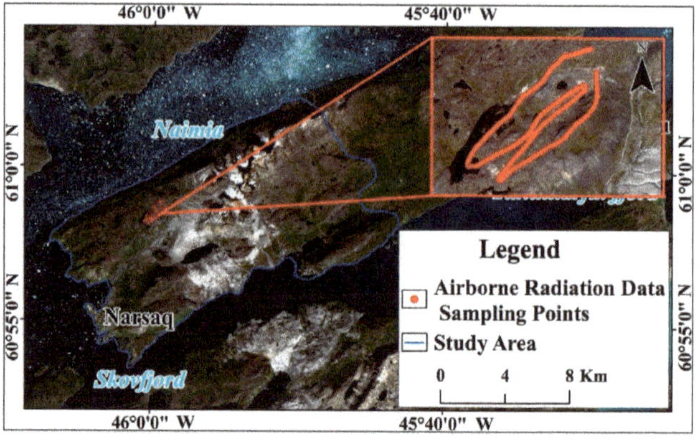

Figure 14. Map of sampling points for airborne radiation data.

Table 6. Correlations between thermal anomalies and radioactive elements.

Types	Spearman's Correlation Coefficient
Total uranium gamma radiation	0.63
U ppm	0.60
Th ppm	0.65

4.4. Mineral Prospectivity Mapping Based on Remote Sensing and Weight-of-Evidence Model

Mineralization is controlled by the formation lithology of a certain era, either directly or indirectly. Ore-bearing rock masses usually are more easily outcropped than the ore body, making the outcropping area larger. Ore-bearing rock masses closely related to the ore body are the marker bed for prospecting. The ore body can be delineated by tracking the ore-bearing rock masses [63]. Mineralization in the study area is mainly associated with Lujavrite, where black, dense, fine-grained arfvedsonite Lujavrite forms arfvedsonite ores, and mineralization elements such as rare earth elements and uranium are hosted in paragenetic minerals formed by the cooling and crystallization of magma. Research has shown that the mineralized minerals include steenstrupine, selenopatite, cerium phosphate sodalite, monazite, zirconium silicate minerals, etc. There are 13 types of mineralized minerals, among which the most important rare earth minerals are steenstrupine (5.58%), followed by monazite (0.09%), and sodium phosphorite, which are often found in agglomerates and contain associated uranium, thorium, and other elements; in regions where these are found, radioactive anomalies are very obvious [40,64–67]. By identifying such radioactive Lujavrite via remote sensing, the mineralization target area can be effectively traced. The main ore finding signatures in the study area are as follows: (1) Lithological and tectonic signatures—uranium-bearing minerals are concentrated in Lujavrite, among which the arfvedsonite type is the most important. The mineralization process is easily controlled by regional north-east tectonics, and tectonic activity often leads to strong deformations in the mineralized area, with the tectonic and hydrothermal alteration information output locations spreading north-eastwards, in parallel. (2) Thermal anomaly signatures—uranium ore is a radioactive mineral, and areas of radioactive thermal anomalies can be extracted in low-temperature areas using surface temperature thermal anomalies.

The study area was decomposed into 67,243 analysis units according to 10 m pixel units. The multi-layer raster data were imported for calculation using the ArcGIS geographic information analysis software developed by Esri.on. Through a comprehensive analysis of the aforementioned signatures, the four main elements of the weight of evidence were selected to include SVM machine learning lithological classification information, tectonic information, PCA hydrothermal alteration information, and surface thermal anomaly information. The comprehensive evaluation values corresponding to the four evidence elements were calculated through a priori probability analysis brought into the weight-of-evidence method (Table 7). A final probability map of favorable areas of mineralization in the study area was generated, with areas of high favorability values being spatially consistent with known mineralization in the study area (Figure 15).

Table 7. Statistical parameters for the binarization of evidence layers.

Evaluation Index Layer	W^+	$\sigma(W^+)$	W^-	$\sigma(W^-)$	C	SI	W
Thermal anomalies	4.44	1.05	−0.82	0.38	5.26	4.69	4.44
SVM	4.33	1.06	−0.68	0.36	5.01	4.47	4.33
Hydrothermal alteration	3.35	1.15	−0.20	0.28	3.55	2.98	3.35
Structural density	0.59	0.30	−1.07	0.58	1.67	2.52	0.59

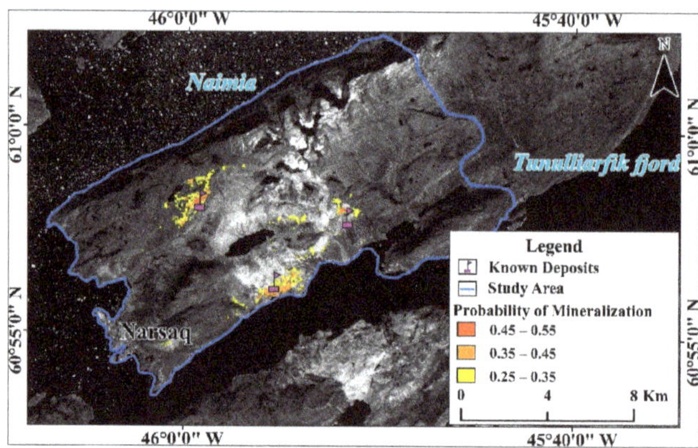

Figure 15. Map of projected potential mineralization areas (the base image was made using the Sentinel-2 panchromatic band).

The high probability area in the mineralization prediction map mainly covers the lithologies of Lujavrite and Naujaite. Combined with geological materials and field investigations, the uranium-bearing minerals are concentrated in the Lujavrite, with the arfvedsonite type being the most abundant. The mineralization is susceptible to regional north-east tectonic control, and tectonic activity often leads to the strong deformation of the mineralized area, with a parallel north-east spreading of tectonic and hydrothermal alteration information output locations. The alteration is dominated by alkaline alteration, enriching radioactive minerals in low-temperature areas; using surface temperature thermal anomalies can extract areas of radioactive thermal anomalies, and the average value of temperature anomalies in the predicted area is 23.24 °C (Figure 16).

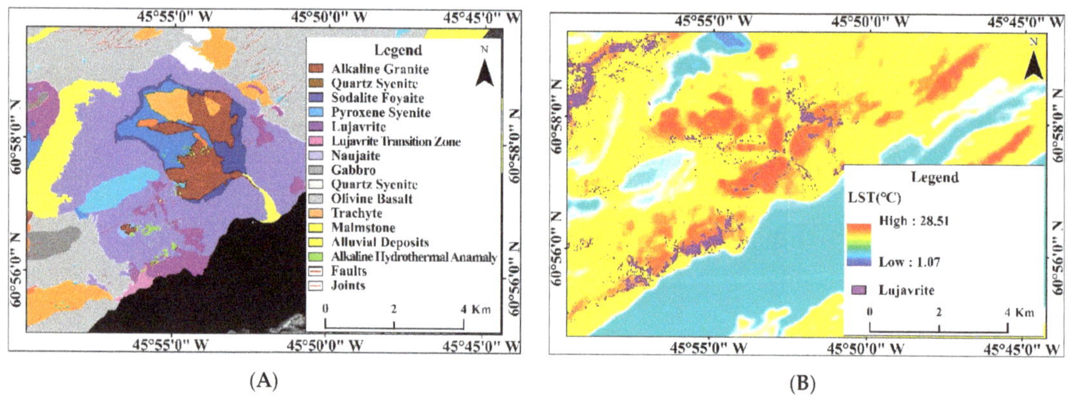

(A)　　　　　　　　　　　　　　　　　　　　(B)

Figure 16. (**A**,**B**) Mineral prospecting target area analysis map.

5. Discussions

High latitudes are heavily ice-covered, making it difficult for remote sensing to detect surface anomalies, with snow up to tens of meters thick completely covering any remote sensing information. However, in some areas, the snow and ice cover varies seasonally, as is the case in the southern Greenland region. The most significant advantage of remote sensing imagery is the multiplicity of data and the long time-series features, which facilitates the detection of geological phenomena irrespective of season or temperature. The use of remote sensing to detect geothermal heat is relatively diverse but is mainly carried out by detecting surface heat sources, such as volcanoes and hot springs [29,31,68,69]. The use of surface temperature inversion results to identify areas of high-temperature anomalies for the purpose of mapping the distribution of radioactive element enrichment zones is a novel method for undertaking geological mineralization surveys. In the harsh temperatures of the Narsaq region of Greenland, even small thermal anomalies can be captured by the thermal infrared sensor, which can be useful for identifying radioactive element enrichment zones in high latitudes. All types of data, including ASTER data, Sentinel Data-3, and Landsat TIRS data, there are limitations in terms of the resolution of the extracted surface temperature products, and there is a bias in the identification of specific geographical features [58,59,70,71]. With the SVM extraction method that used the in situ spectrum as a reference, the variation in the shape of the spectrum curve of the image elements within a rock unit, the variation in the position of the absorption valley, and the reflection peak (spectral difference), and the sample separability between the rock units affect the accuracy of the SVM classification. However, the SVM method was combined with the thermal anomaly inversion method to extract arfvedsonite Lujavrite-containing radioactive minerals in the region, and the two methods were used to corroborate the accuracy of the results. Further, Crosta's hydrothermal alteration information extraction technique [36,72,73] was utilized to extract relevant alteration information in the alkaline rock area, and the interpreted mineral control tectonic spreading characteristics were applied to the Lujavrite outcrop such that the mineralization characteristics could be optimized. This overlay analysis of multiple remote sensing data can increase the prediction accuracy while also solving the issue of predicting mineralization in areas of weak information. The geophysical and geochemical data in most areas of mineralization prediction are small-scale and do not have raster digitization. For small-range or large-scale mineralization studies, the accuracy of such data is severely lacking and, therefore, does not accurately reflect the geological and geochemical information of the area. The importance of such elements could not be measured while using this mineralization methodology. Such elements need to be refined in future studies by complementing them with large-scale studies. As the types of mineralization are not abundant in this article, it is not possible to build a sound statistical model, and more areas need to be studied. Although the use of remote sensing alone to support mineralization prediction is efficient and comprehensive, quantitative mathematical methods with multiple types of parameters should be used, and there is a need to add more geological anomaly information evaluation indicators to the research method to develop a more comprehensive method for mineralization prediction at larger scales.

6. Conclusions

A highly efficient and novel technical tool for regional mineralization investigations is proposed, which uses the remote sensing inversion of radiothermal anomalies in high-latitude areas. The study area is rich in radioactive minerals, and the average year-round temperature is below 0 °C, allowing weak thermal radiation to manifest through surface temperature anomalies. Landsat 8 thermal infrared data were used to invert the surface temperature using the RTE model to circle the high thermal anomaly area. The average surface temperature of the high thermal area was 23.28 °C, which was higher than the average temperature of the entire area, of 16.85 °C. By conducting Spearman correlation analyses with the airborne radiation data, a positive correlation with the uranium and thorium concentration and the correlation coefficients all exceeded 0.6, indicating that the thermal anomaly remote sensing inversion technique is a good indicator of low-temperature radioactive mineral enrichment areas. By establishing texture and topographic features, the SVM algorithm was used to identify the mineralized lithology of Lujavrite, with a classification accuracy of 89.57%; the classification results revealed that the Lujavrite was characterized by banded outcrops. Through the comprehensive analysis of remote sensing information, combined with metallogenic background information, the study area was deemed to be a favorable area for mineralization through the weight-of-evidence model, with high-value areas of mineralization potential overlapping well with known mineralization areas. The combination of remote sensing thermal anomaly information and rock interpretation methods in the Narsaq region of Greenland has, therefore, been validated for the analysis of mineralization in the region, and this integrated approach to remote sensing information can be extended to the prediction of mineralization in radiogenic high-altitude areas.

Author Contributions: Conceptualization, P.L., L.H. and Z.H.; methodology, P.L., L.H. and Z.H.; software, P.L.; validation, P.L., L.H. and Z.H.; formal analysis, P.L.; investigation, P.L., J.Z., B.H., Y.Y. and L.X.; resources, Z.H., J.Z. and B.H.; data curation, P.L., H.H. and Y.Z.; writing—original draft preparation, P.L., H.H. and Y.Z.; writing—review and editing, P.L., L.H. and Z.H.; visualization, P.L., L.H. and Z.H.; supervision, L.H. and Z.H.; project administration, Z.H. and J.Z. All authors have read and agreed to the published version of the manuscript.

Funding: This research received no external funding.

Data Availability Statement: Not applicable.

Acknowledgments: The authors would like to thank the following educational and research institutions for the infrastructure: Chengdu University of Technology (CDUT), Institute of Multipurpose Utilization of Mineral Resources, China Geological Survey. Without the support of these institutions this study could not be performed.

Conflicts of Interest: No potential conflict of interest was reported by the authors. The authors declare that they have no known competing financial interests or personal relationships that could have appeared to influence the work reported in this paper.

References

1. Hecker, C.; van Ruitenbeek, F.J.A.; van der Werff, H.M.A.; Bakker, W.H.; Hewson, R.D.; van der Meer, F.D. Spectral Absorption Feature Analysis for Finding Ore A tutorial on using the method in geological remote sensing. *IEEE Geosci. Remote Sens. Mag.* **2019**, *7*, 51–71. [CrossRef]
2. Hunt, G. Spectral Signatures of Particulate Minerals in the Visible and Near Infrared. *Geophysics* **1977**, *42*, 501–513. [CrossRef]
3. Clark, R.N.; King, T.V.V.; Klejwa, M.; Swayze, G.A.; Vergo, N. High spectral resolution reflectance spectroscopy of minerals. *Int. J. Rock Mech. Min. Sci. Geomech. Abstr.* **1991**, *28*, A217. [CrossRef]
4. Bierwirth, P.; Huston, D.; Blewett, R. Hyperspectral mapping of mineral assemblages associated with gold mineralization in the Central Pilbara, Western Australia. *Econ. Geol. Bull. Soc. Econ. Geol.* **2002**, *97*, 819–826. [CrossRef]
5. Kruse, F.A.; Perry, S.L.; Caballero, A. District-level mineral survey using airborne hyperspectral data, Los Menucos, Argentina. *Ann. Geophys.* **2006**, *49*, 83–92.
6. Rockwell, B.W.; Cunningham, C.G.; Breit, G.N.; Rye, R.O. Spectroscopic mapping of the White Horse alunite deposit, Marysvale Volcanic Field, Utah: Evidence of a magmatic component. *Econ. Geol.* **2006**, *101*, 1377–1395. [CrossRef]
7. Abulghasem, Y.A.; Akhir, J.B.M.; Hassan, W.F.W.; Samsudin, A.; Youshah, B.M. The use of remote sensing technology in geological investigation and mineral detection in Wadi shati, Libya. *Electron. J. Geotech. Eng.* **2012**, *17*, 1279–1291.
8. Amer, R.; Kusky, T.; Ghulam, A. Lithological mapping in the Central Eastern Desert of Egypt using ASTER data. *J. Afr. Earth Sci.* **2010**, *56*, 75–82. [CrossRef]
9. Xu, Y.; Chen, J.; Meng, P. Detection of alteration zones using hyperspectral remote sensing data from Dapingliang skarn copper deposit and its surrounding area, Shanshan County, Xinjiang Uygur autonomous region, China. *J. Vis. Commun. Image Represent.* **2019**, *58*, 67–78. [CrossRef]
10. Milovsky, G.A.; Makarov, V.P.; Troitsky, V.V.; Lyamin, S.M.; Orlyankin, V.N.; Shemyakina, E.M.; Gil, I.G. Use of Remote Sensing to Find a Localization Pattern of Gold Mineralization in the Central Part of the Ayan-Yuryakh Anticlinorium, Magadan Oblast. *Izv. Atmos. Ocean. Phys.* **2019**, *55*, 1389–1394. [CrossRef]
11. Yamaguchi, Y.; Kahle, A.B.; Tsu, H.; Kawakami, T.; Pniel, M. Overview of advanced spaceborne thermal emission and reflection radiometer (ASTER). *IEEE Trans. Geosci. Remote Sens.* **1998**, *36*, 1062–1071. [CrossRef]
12. Dehnavi, A.G.; Sarikhani, R.; Nagaraju, D. Image Processing and Analysis of Mapping Alteration Zones in environmental research, East of Kurdistan, Iran. *World Appl. Sci. J.* **2010**, *11*, 278–283.
13. Gupta, P.; Venkatesan, M. *Mineral Identification Using Unsupervised Classification from Hyperspectral Data*; Springer: Singapore, 2020.
14. Shayeganpour, S.; Tangestani, M.H.; Gorsevski, P.V. Machine learning and multi-sensor data fusion for mapping lithology: A case study of Kowli-kosh area, SW Iran. *Adv. Space Res.* **2021**, *68*, 3992–4015. [CrossRef]
15. Liu, Z.; Cao, J.; You, J.; Chen, S.; Lu, Y.; Zhou, P. A lithological sequence classification method with well log via SVM-assisted bi-directional GRU-CRF neural network. *J. Pet. Sci. Eng.* **2021**, *205*, 108913. [CrossRef]
16. Gu, Y.; Zhang, D.; Bao, Z.; Guo, H.; Zhou, L.; Ren, J. Lithology prediction of tight sandstone formation using GS-LightGBM hybrid machine learning model. *Bull. Geol. Sci. Technol.* **2021**, *40*, 224–234.

17. Mou, D.; Zhang, L.; Xu, C. Comparison of Three Classical Machine Learning Algorithms for Lithology Identification of Volcanic Rocks Using Well Logging Data. *J. Jilin Univ. Earth Sci. Ed.* **2021**, *51*, 951–956.
18. Kumar, C.; Chatterjee, S.; Oommen, T.; Guha, A. Automated lithological mapping by integrating spectral enhancement techniques and machine learning algorithms using AVIRIS-NG hyperspectral data in Gold-bearing granite-greenstone rocks in Hutti, India. *Int. J. Appl. Earth Obs. Geoinf.* **2020**, *86*, 15. [CrossRef]
19. Duan, Y.; Zhao, Y.; Ma, C.; Jiang, W. Lithology Identification Method Based on Multi -layer Ensemble Learning. *J. Data Acquis. Process.* **2020**, *35*, 572–581.
20. Han, Q.; Zhang, X.; Shen, W. Application of Support Vector Machine Based on Decision Tree Feature Extraction in Lithology Classification. *J. Jilin Univ. Earth Sci. Ed.* **2019**, *49*, 611–620.
21. Hutchison, W.; Finch, A.A.; Borst, A.M.; Marks, M.A.W.; Upton, B.G.J.; Zerkle, A.L.; Stueken, E.E.; Boyce, A.J. Mantle sources and magma evolution in Europe's largest rare earth element belt (Gardar Province, SW Greenland): New insights from sulfur isotopes. *Earth Planet. Sci. Lett.* **2021**, *568*, 117034. [CrossRef]
22. Steenfelt, A.; Kolb, J.; Thrane, K. Metallogeny of South Greenland: A review of geological evolution, mineral occurrences and geochemical exploration data. *Ore Geol. Rev.* **2016**, *77*, 194–245. [CrossRef]
23. Aydar, E.; Diker, C. Carcinogen soil radon enrichment in a geothermal area: Case of Guzelcamli-Davutlar district of Aydin city, western Turkey. *Ecotoxicol. Environ. Saf.* **2021**, *208*, 111466. [CrossRef] [PubMed]
24. Cui, Y.; Zhu, C.; Qiu, N.; Tang, B.; Guo, S. Radioactive Heat Production and Terrestrial Heat Flow in the Xiong'an Area, North China. *Energies* **2019**, *12*, 4608. [CrossRef]
25. Li, Q. *The Study of Deep Geothermal Features in Sichuan Basin*; Chengdu University of Technology: Chengdu, China, 1992.
26. Qing-yang, L.I.; Hui-rong, C.A.I.; Yan, C. The study and application of the relationship between the geothermal field and the deep uranium ore deposit. *Geol. China* **2010**, *37*, 198–203.
27. Wang, J.-h.; Zhang, J.-l.; Liu, D.-C. Discussion on the application potential of thermal infrared remote sensing technology in uranium deposits exploration. *World Nucl. Geosci.* **2011**, *28*, 32–41.
28. Salawu, N.B.; Fatoba, J.O.; Adebiyi, L.S.; Eluwole, A.B.; Olasunkanmi, N.K.; Orosun, M.M.; Dada, S.S. Structural geometry of Ikogosi warm spring, southwestern Nigeria: Evidence from aeromagnetic and remote sensing interpretation. *Geomech. Geophys. Geo-Energy Geo-Resour.* **2021**, *7*, 26. [CrossRef]
29. Gemitzi, A.; Dalampakis, P.; Falalakis, G. Detecting geothermal anomalies using Landsat 8 thermal infrared remotely sensed data. *Int. J. Appl. Earth Obs. Geoinf.* **2021**, *96*, 102283. [CrossRef]
30. Rodriguez-Gomez, C.; Kereszturi, G.; Reeves, R.; Rae, A.; Pullanagari, R.; Jeyakumar, P.; Procter, J. Lithological mapping of Waiotapu Geothermal Field (New Zealand) using hyperspectral and thermal remote sensing and ground exploration techniques. *Geothermics* **2021**, *96*, 102195. [CrossRef]
31. Xin, L.; Liu, X.; Zhang, B. Land surface temperature retrieval and geothermal resources prediction by remote sensing image: A case study in the Shijiazhuang area, Hebei province. *J. Geomech.* **2021**, *27*, 40–51.
32. Saibi, H.; Mia, M.B.; Bierre, M.; El Kamali, M. Application of remote sensing techniques to geothermal exploration at geothermal fields in the United Arab Emirates. *Arab. J. Geosci.* **2021**, *14*, 1251. [CrossRef]
33. Goldstein, B.A.; Hill, A.J.; Long, A.; Budd, A.R.; Holgate, F.; Malavazos, M. Hot dry rock geothermal exploration in Australia. *ASEG Ext. Abstr.* **2004**, *2004*, 1–4.
34. Wyborn, D.; Somerville, M. Prospective hot-dry-rock geothermal energy in Australia. GA Publication Research Newsletter. 1994. Available online: https://d28rz98at9flks.cloudfront.net/90424/ResearchNews_21_p001.pdf (accessed on 25 May 2022).
35. Coolbaugh, M.F.; Kratt, C.; Fallacaro, A.; Calvin, W.M.; Taranik, J.V. Detection of geothermal anomalies using Advanced Spaceborne Thermal Emission and Reflection Radiometer (ASTER) thermal infrared images at Bradys Hot Springs, Nevada, USA. *Remote Sens. Environ.* **2007**, *106*, 350–359. [CrossRef]
36. Aita, S.K.; Omar, A.E. Exploration of uranium and mineral deposits using remote sensing data and GIS applications, Serbal area, Southwestern Sinai, Egypt. *Arab. J. Geosci.* **2021**, *14*, 2214. [CrossRef]
37. Yousefi, M.; Tabatabaei, S.H.; Rikhtehgaran, R.; Pour, A.B.; Pradhan, B. Application of Dirichlet Process and Support Vector Machine Techniques for Mapping Alteration Zones Associated with Porphyry Copper Deposit Using ASTER Remote Sensing Imagery. *Minerals* **2021**, *11*, 1235. [CrossRef]
38. Marks, M.A.W.; Hettmann, K.; Schilling, J.; Frost, B.R.; Markl, G. The Mineralogical Diversity of Alkaline Igneous Rocks: Critical Factors for the Transition from Miaskitic to Agpaitic Phase Assemblages. *J. Petrol.* **2011**, *52*, 439–455. [CrossRef]
39. Steenfelt, A. High-technology metals in alkaline and carbonatitic rocks in greenland—Recognition and exploration. *J. Geochem. Explor.* **1991**, *40*, 263–279. [CrossRef]
40. Zhao, Y.; Lu, W.; Wang, A.; Lu, W. Research Progress on the Ilimaussaq Nb-Ta-U-REE Deposit, Greenland. *Geol. Sci. Techol. Inf.* **2013**, *32*, 9–17.
41. van der Werff, H.; van der Meer, F. Sentinel-2A MSI and Landsat 8 OLI Provide Data Continuity for Geological Remote Sensing. *Remote Sens.* **2016**, *8*, 883. [CrossRef]
42. Drusch, M.; Del Bello, U.; Carlier, S.; Colin, O.; Fernandez, V.; Gascon, F.; Hoersch, B.; Isola, C.; Laberinti, P.; Martimort, P.; et al. Sentinel-2: ESA's Optical High-Resolution Mission for GMES Operational Services. *Remote Sens. Environ.* **2012**, *120*, 25–36. [CrossRef]

43. Updike, T.; Comp, C. Radiometric use of WorldView-2 imagery. DigitalGlobe. 2010, p. 1. Available online: https://www.yumpu.com/en/document/read/43552535/radiometric-use-of-worldview-2-imagery-technical-note-pancroma (accessed on 25 May 2022).

44. Irons, J.R.; Dwyer, J.L.; Barsi, J.A. The next Landsat satellite: The Landsat Data Continuity Mission. *Remote Sens. Environ.* **2012**, *122*, 11–21. [CrossRef]

45. Jimenez-Munoz, J.C.; Sobrino, J.A.; Skokovic, D.; Mattar, C.; Cristobal, J. Land Surface Temperature Retrieval Methods From Landsat-8 Thermal Infrared Sensor Data. *IEEE Geosci. Remote Sens. Lett.* **2014**, *11*, 1840–1843. [CrossRef]

46. Abrams, M.; Crippen, R.; Fujisada, H. ASTER Global Digital Elevation Model (GDEM) and ASTER Global Water Body Dataset (ASTWBD). *Remote Sens.* **2020**, *12*, 1156. [CrossRef]

47. Jing, C.; Bokun, Y.; Runsheng, W.; Feng, T.; Yingjun, Z.; Dechang, L.; Suming, Y.; Wei, S. Regional-scale mineral mapping using ASTER VNIR/SWIR data and validation of reflectance and mineral map products using airborne hyperspectral CASI/SASI data. *Int. J. Appl. Earth Obs. Geoinf.* **2014**, *33*, 127–141. [CrossRef]

48. Pour, A.B.; Hashim, M. The application of ASTER remote sensing data to porphyry copper and epithermal gold deposits. *Ore Geol. Rev.* **2012**, *44*, 1–9. [CrossRef]

49. Brown, M.; Lewis, H.G.; Gunn, S.R. Linear spectral mixture models and support vector machines for remote sensing. *IEEE Trans. Geosci. Remote Sens.* **2000**, *38*, 2346–2360. [CrossRef]

50. Brown, M.; Gunn, S.R.; Lewis, H.G. Support vector machines for optimal classification and spectral unmixing. *Ecol. Model.* **1999**, *120*, 167–179. [CrossRef]

51. Brown, M.; Lewis, H. Support vector machines and linear spectral unmixing for remote sensing. In Proceedings of the International Conference on Advances in Pattern Recognition (ICAPR 98), Plymouth, UK, 23–25 November 1998; pp. 395–404.

52. Lackner, M.; Conway, T.M. Determining land-use information from land cover through an object-oriented classification of IKONOS imagery. *Can. J. Remote Sens.* **2008**, *34*, 77–92. [CrossRef]

53. Tsang, I.W.; Kwok, J.T.; Cheung, P.M. Core vector machines: Fast SVM training on very large data sets. *J. Mach. Learn. Res.* **2005**, *6*, 363–392.

54. Xiao, A.; Zhao, W.; Hu, D.; Liu, L.; Li, J. Remote sensing information extraction based on object-oriented and support vector machines. *Sci. Surv. Mapp.* **2010**, *35*, 154–157.

55. Pasolli, E.; Melgani, F.; Tuia, D.; Pacifici, F.; Emery, W.J. SVM Active Learning Approach for Image Classification Using Spatial Information. *IEEE Trans. Geosci. Remote Sens.* **2014**, *52*, 2217–2233. [CrossRef]

56. Cortes, C.; Vapnik, V. Support-vector networks. *Mach. Learn.* **1995**, *20*, 273–297. [CrossRef]

57. Room, C.A.o.S.G. *Introduction to Mine Geothermal*; China Coal Industry Publishing House: Beijing, China, 1981.

58. Duan, S.; Ru, C.; Li, Z.; Wang, M.; Xu, H.; Li, H.; Wu, P.; Zhan, W.; Zhou, J.; Zhao, W.; et al. Reviews of methods for land surface temperature retrieval from Landsat thermal infrared data. *J. Remote Sens.* **2021**, *25*, 1591–1617.

59. Sekertekin, A.; Bonafoni, S. Land Surface Temperature Retrieval from Landsat 5, 7, and 8 over Rural Areas: Assessment of Different Retrieval Algorithms and Emissivity Models and Toolbox Implementation. *Remote Sens.* **2020**, *12*, 294. [CrossRef]

60. Bonham-Carter, G.F.; Agterberg, F.P.; Wright, D.F. Weight of evidence modeling: A new approach to mapping mineral potential. In *Statistical Applications in the Earth Sciences*; Wiley: Hoboken, NJ, USA, 1989; pp. 171–183.

61. Riisager, P.; Rasmussen, T.M. Aeromagnetic survey in south-eastern Greenland: Project Aeromag 2013. *Geol. Surv. Den. Greenl. Bull.* **2014**, *31*, 63–66. [CrossRef]

62. Armour-Brown, A.; Tukiainen, T.; Wallin, B. *South Greenland Uranium Exploration Programme*; Final Report; Groenlands Geologiske Undersoegelse: Copenhagen, Denmark, 1982; Volume 14.

63. Xiaoliang, F.; Mingjie, W.; Chengmin, W.E.N.; Huihua, Z. The preliminary study of the exploration potential of the Liwu copper deposit and its surrounding areas, western Sichuan. *Sediment. Geol. Tethyan Geol.* **2007**, *27*, 9–13.

64. Petersen, O.V.; Gault, R.A.; Balic-Zunic, T. Odintsovite from the Ilimaussaq alkaline complex, South Greenland. *Neues Jahrb. Mineral.-Mon.* **2001**, 235–240.

65. Petersen, O.V. List of all minerals identified in the Ilímaussaq alkaline complex, South Greenland. *Geol. Greenl. Surv. Bull.* **2001**, *190*, 25–33. [CrossRef]

66. Petersen, O.V.; Johnsen, O.; Micheelsen, H.I. Turkestanite from the Ilimaussaq alkaline complex, South Greenland. *Neues Jahrb. Mineral.-Mon.* **1999**, 424–432.

67. Quan, X.; Liu, C.; Zhao, Y. The evaluation of uranium mineral resources potential in Greenland. *Geol. Bull. China* **2019**, *38*, 1071–1079.

68. Peleli, S.; Kouli, M.; Marchese, F.; Lacava, T.; Vallianatos, F.; Tramutoli, V. Monitoring temporal variations in the geothermal activity of Miocene Lesvos volcanic field using remote sensing techniques and MODIS—LST imagery. *Int. J. Appl. Earth Obs. Geoinf.* **2021**, *95*, 102251. [CrossRef]

69. Hewson, R.; Mshiu, E.; Hecker, C.; van der Werff, H.; van Ruitenbeek, F.; Alkema, D.; van der Meer, F. The application of day and night time ASTER satellite imagery for geothermal and mineral mapping in East Africa. *Int. J. Appl. Earth Obs. Geoinf.* **2020**, *85*, 101991. [CrossRef]

70. Sanchez-Aparicio, M.; Andres-Anaya, P.; Del Pozo, S.; Laguela, S. Retrieving Land Surface Temperature from Satellite Imagery with a Novel Combined Strategy. *Remote Sens.* **2020**, *12*, 277. [CrossRef]

71. Romaguera, M.; Vaughan, R.G.; Ettema, J.; Izquierdo-Verdiguier, E.; Hecker, C.A.; van der Meer, F.D. Detecting geothermal anomalies and evaluating LST geothermal component by combining thermal remote sensing time series and land surface model data. *Remote Sens. Environ.* **2018**, *204*, 534–552. [CrossRef]
72. Crosta, A.P.; De Souza, C.R.; Azevedo, F.; Brodie, C. Targeting key alteration minerals in epithermal deposits in Patagonia, Argentina, using ASTER imagery and principal component analysis. *Int. J. Remote Sens.* **2003**, *24*, 4233–4240. [CrossRef]
73. Chen, C.; Leng, C.; Si, G. Comprehensive Metallogenic Prediction Based on GIS and Analytic Hierarchy Process: A Case Study of Kumishi Region in Xinjiang. *Gold Sci. Technol.* **2020**, *28*, 213–227.

minerals

MDPI

Fusion of Lineament Factor (LF) Map Analysis and Multifractal Technique for Massive Sulfide Copper Exploration: The Sahlabad Area, East Iran

Aref Shirazi [1], Ardeshir Hezarkhani [1,*] and Amin Beiranvand Pour [2,*]

1   Department of Mining Engineering, Amirkabir University of Technology (Tehran Polytechnic), Tehran 1591634311, Iran; aref.shirazi@aut.ac.ir
2   Institute of Oceanography and Environment (INOS), University Malaysia Terengganu (UMT), Kuala Nerus 21030, Terengganu, Malaysia
*   Correspondence: ardehez@aut.ac.ir (A.H.); beiranvand.pour@umt.edu.my (A.B.P.); Tel.: +98-21-6454-2968 (A.H.); +60-9-668-3824 (A.B.P.); Fax: +98-21-6640-5846 (A.H.); +60-9-669-2166 (A.B.P.)

Abstract: Fault systems are characteristically one of the main factors controlling massive sulfide mineralization. The main objective of this study was to investigate the relationship between fault systems and host lithology with massive sulfide copper mineralization in the Sahlabad area, South Khorasan province, east of Iran. Subsequently, the rose diagram analysis, Fry analysis, lineament factor (LF) map analysis and multifractal technique were implemented for geological and geophysical data. Airborne geophysical analysis (aeromagnetometric data) was executed to determine the presence of intrusive and extrusive masses associated with structural systems. Accordingly, the relationship between the formation boundaries and the fault system was understood. Results indicate that the NW-SE fault systems are controlling the lithology of the host rock for copper mineralization in the Sahlabad area. Hence, the NW-SE fault systems are consistent with the main trend of lithological units related to massive sulfide copper mineralization in the area. Additionally, the distance of copper deposits, mines and indices in the Sahlabad area with fault systems was calculated and interpreted. Fieldwork results confirm that the NW-SE fault systems are entirely matched with several massive sulfide copper mineralizations in the area. This study demonstrates that the fusion of lineament factor (LF) map analysis and multifractal technique is a valuable and inexpensive approach for exploring massive sulfide mineralization in metallogenic provinces.

Keywords: fault system analysis; massive sulfide copper exploration; airborne geophysical analysis; Fry analysis; multifractal technique

Citation: Shirazi, A.; Hezarkhani, A.; Pour, A.B. Fusion of Lineament Factor (LF) Map Analysis and Multifractal Technique for Massive Sulfide Copper Exploration: The Sahlabad Area, East Iran. *Minerals* **2022**, *12*, 549. https://doi.org/10.3390/min12050549

Academic Editor: Behnam Sadeghi

Received: 18 March 2022
Accepted: 26 April 2022
Published: 28 April 2022

Publisher's Note: MDPI stays neutral with regard to jurisdictional claims in published maps and institutional affiliations.

1. Introduction

The spatial distribution of mineral reserves is controlled by various parameters [1,2]. The most important controllers of ore mineralization distribution at regional scale are host-rock lithology, intrusive or extrusive masses and structural systems [3,4]. Therefore, to identify areas of mineral potential, determining the relationship between mineralization and structural features is of great importance [5–7]. Numerous studies analyzed the relationship between structural features controlling mineralization and the spatial distribution of mineral resources [7–11]. The purpose of these studies was to extract exploratory keys to identify new high-potential areas [8]. The relationship between structural features such as the fault system and ore mineralization has been identified through various methods such as fractal analysis, fault density mapping and combining this information with geochemical layers and remote sensing data [9–13]. In the regions where field information has not been collected, remote sensing surveys are used to identify lineaments such as faults and fractures [13–15]. Simultaneous use of aeromagnetic data analysis with remote sensing data helps to generate an accurate structural map [16–20]. In the case of structural field data collected and mapped by geologists, the analysis of behavior and impact of structural systems, especially major faults in the spatial distribution of mineralization, could be performed using analytical methods such as fault density mapping and the analytic hierarchy process (AHP) decision method in combination with other information layers [21]. Multifractal analysis, rose diagram analysis and Fry analysis were used in fault interpretation and determining the relationship between mineralization and the fault system [22–26].

Massive sulfide mineralization is typically associated with regional fault systems, which are documented in many regions such as the Main Urals Fault (MUF), South Urals, the Selwyn Basin, Canada and the North Australian Craton [27–29]. The Sahlabad area located in South Khorasan province, east of Iran, has a large number of copper mines, deposits and indices (Figure 1A). Mesgaran ore deposit is one of the biggest copper mineralizations (Cyprus-type massive sulfide) in the study area, in which the regional fault system acts as a controlling structural factor for copper mineralization [30–32]. According to the volume, extent and trend of their distribution in the region, to identify the mineralization potentials of copper in this area, structural controlling factors and their relationship with mineralization zones need to be determined.

In the present study, in order to analyze the relationship between the fault system and massive sulfide copper mineralization in the Sahlabad area, rose diagram analysis, Fry analysis, multifractal technique and lineament factor (LF) map were implemented. To investigate the host-rock lithological trend, aeromagnetic data analysis was also used. Thus, the main faults controlling the host lithology trend and playing a key role to determine the spatial distribution of mineralization were identified. Finally, fractal analysis was used to extract more detailed characteristics of the relationship between the fault system and the mineralization distribution in the area. Consequently, high-potential areas were categorized in terms of control by the fault system and the relationship of each mineralization point with the map of the nearest community of high-intensity LFs. This approach provides innovative and valuable information about the fault systems controlling massive sulfide copper mineralization in the study area. The main objectives of this investigation were: (1) to provide a rose diagram analysis for fault systems in the region including major faults, minor faults, inferred faults, thrust faults; (2) to apply Fry analysis to the spatial distribution of mineralization points in the region; (3) to perform airborne magnetometric analysis to identify deep faults controlling the host lithology trend; and (4) to generate a lineament factor (LF) map and concentration–area (C-A) fractal analysis to classify different LF communities.

2. Geology of the Study Area

The Sahlabad area is located in the east of Iran and South Khorasan province. It is positioned between longitudes 59°30′ E to 60° E and latitudes 32° to 32°30′ N (Figure 1A,B). The study area is completely located in the flysch belt and ophiolite melange in the Sistan structural zone of eastern Iran [33]. This structural zone is situated between the Nehbandan fault (in the west) and Harirod fault (in the east) and is 800 km long and 200 km wide [34,35]. Based on the geological map of Sahlabad, the regional faults of the area are divided into four categories: major faults, minor faults, inferred faults and thrust faults. This zone has undergone evolutionary stages from oceanic to continental crust and is one of the derivations of the "young Tethys" type [35–37]. In this area, igneous, metamorphic and sedimentary lithological units related to the Late Cretaceous to Neogene are exposed [38]. The Sahlabad area is entirely located in the flysch and colored melange belt of eastern Iran. The geological formations observed in the area include rocks with the characteristics of this belt, which are attributed to the Upper Cretaceous and Lower Tertiary, and the volcanic cover and younger Tertiary sediments [36].

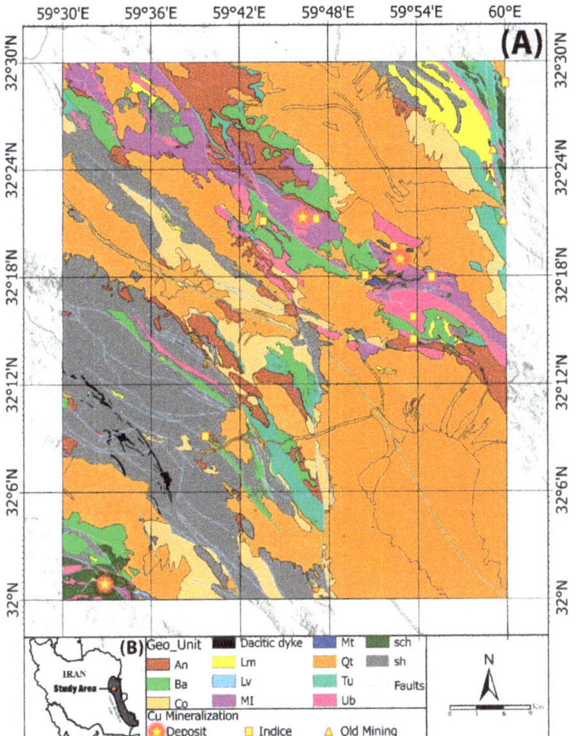

Figure 1. (**A**) Geological map of Sahlabad area (scale of 1:100,000) (modified [39]). Abbreviations: An = Andesite, Ba = Basalt, Co = Conglomerate, Dd = Dacitic dyke, Lm = Limestone, Lv = Listwanite, Ml = Ophiolite melange, Mt = Metadiabase, Qt = Quaternary sediments, Tu = Tuff, Ub = Ultrabasic rocks, Sch = Schist, Sh = Shale and sandstone. (**B**) Geographical location of Sistan structural zone in Iran.

2.1. Regional Tectonics

The Sahlabad area belongs to the ophiolitic melanges and flysch belts of eastern Iran and is located in the Lut structural block. The main trend of the belt is north-south, which gradually changes to the southeast-northwest. The intense folding of the flysch deposits and the irregular structure of the melange complexes indicate high compaction in the area [40]. The most severe crustal deformation has occurred at the southwestern tip of the region, where a narrow zone of thrust and metamorphosis (metamorphic ophiolitic melanges) indicates the close connection of the flysch and ophiolitic melanges belt to the Lut structural block. Folding and crushing, along with tilting, which results in a random mixing of different types of rock, characterize the ophiolitic melange complexes of the region [41]. Narrow and intense folding and longitudinal faulting with post-Middle Eocene (Oligocene) age have affected the volcanic and sedimentary formations of the Paleogene. The Cretaceous-era Zar-Kooh mountain flysches are trusted on the Eocene sediments of Bezo Mountain in a southwesterly direction [37,38]. The uniform tectonical motions create a system of parallel mountain ranges and the depressions between them that characterize the current topography. Neogene deposition in the depressions has led to moderate folding and minor faults [33,36]. Andesites and basalts, as representative of the youngest volcanic rocks (Neogene, probably Early Quaternary), show soft tilt (with a slope of about 20°) in the lower units and with a semi-real position in the higher units [34,35].

2.2. Copper Mineralization in the Study Area

Due to the diversity of lithology consisting of ultrabasic, alkaline-based volcanic, intermediate and acidic rock units, metamorphic rocks, listwanites and other rock units in the Sahlabad area, there are mineralization potentials for copper, gold, nickel, chromium and magnesite. Old mining activity and excavations have been reported in the study area. Copper mineralization in the study region (mines, deposits and indices) was investigated and classified from various reports obtained from exploratory studies in the Sahlabad area, such as geological map reports, economic geology reports, preliminary and detailed exploration reports of mineral areas, etc., [39,42–46]. The location of copper mines, deposits and indices in the Sahlabad area are marked on the geology map (Figure 1A). Copper mineralizations such as malachite, chalcopyrite and chalcocite were observed and documented in the study area. Figure 2A–F show polished sections of copper mineralizations selected from the Mesgaran deposit, Chah-Rasteh deposit and Zahri deposit. Classified information about 14 copper mineralization zones in the Sahlabad area is presented in Table 1.

Table 1. Classified information of 14 copper mineralization points in Sahlabad area.

Row	Copper Mineralization Name	Anomaly Center Coordinates		Anomaly Area (Km²)	Alteration	Lithology (Host Rock)	Cu Dominant Mineral
		Longitude (E)	Latitude (N)				
1	Mesgaran Deposit	59°52′49″	32°18′58″	8	Phy + Arg + Pp + Chl + Qtz	Ba + Anb	Cpy + Mch
2	Chah-Rasteh Deposit	59°46′15″	32°21′19″	4	Phy + Arg + Pp + Chl + Cab	An + Anb	Ch + Mch
3	Zahri Deposit	59°32′52″	32°00′50″	2	Phy + Arg + Pp + Hem	Ub + Sch	Cpy + Ch + Mch
4	Kasrab Abandoned Mine	59° 59′45″	32°21′05″	3.8	Phy + Arg + Pp + Sep	Ub	Mch
5	Cheshme-Zangi Abandoned Mine	59°59′08″	32°25′02″	2.5	Phy + Arg + Pp + Silicification	Limestone shale + Listwanite	Cpy + Mch
6	Shir-Shotor Indice	59°53′50″	32°14′28″	1	Arg + Pp + Sep	An + Serpentinite (Ub)	Mch + Az
7	Dastgerd Indice	59°43′39″	32°21′03″	2	Arg + Pp + Sep + Hem	Harzburgite	Mch
8	Torshaab Indice	59°59′56″	32°28′48″	5	Phy + Arg + Pp + Hem + Lm	Sch	Mch + Az
9	Chah-Anjir Indice	59°53′37″	32°15′44″	2	Pp + Sep	Serpentinite (Ub)	Mch + Az
10	Zargaran Indice	59°47′09″	32°21′14″	1	Phy + Arg + Pp + Lm + Goe + Hem	An + Db	Mch + Az
11	West Mesgaran Indice	59°52′26″	32°19′36″	1.5	Arg + Pp + Hem + Lm	Mtd	Cpy + Mch + Az
12	Mirsimin Indice	59° 54′58″	32°17′53″	9	Arg + Pp + Hem	Db	Cpy + Mch + Az

Table 1. *Cont.*

Row	Copper Mineralization Name	Anomaly Center Coordinates		Anomaly Area (Km²)	Alteration	Lithology (Host Rock)	Cu Dominant Mineral
		Longitude (E)	Latitude (N)				
13	Kuharod Indice	59°50′31″	32°18′01″	1	Phy + Arg + Pp + Hem	Db	Mch
14	Barghan Indice	59° 39′38″	32°09′05″	2	Arg + Pp + Lm + Geo + Hem	Db + Limestone	Mch

Abbreviations: Cpy = Chalcopyrite, Py = Pyrite, Mch = Malachite, Ch = Chalcocite, Az = Azorite, Ba = Basalt, An = Andesite, Anb = Andesite-Basalt, Ub = Ultrabasic, Sch = Schist, Db = Diabase, Mtd = Metadiabase, Chl = Chlorite Alteration, Qtz = Quartz Alteration, Cab = Carbonate Alteration, Pp = Propylitic Alteration, Arg = Argillic Alteration, Phy = Phyllic Alteration, Sep = Serpentine Alteration, Hem = Hematite Alteration, Lm = Limonite Alteration, Goe = Goethite Alteration.

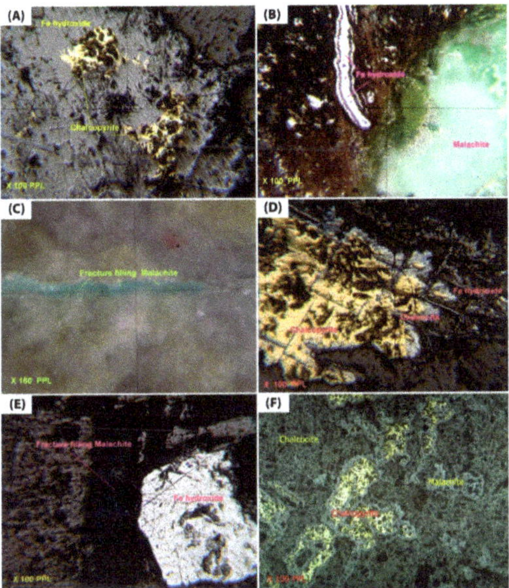

Figure 2. Selected polished sections prepared from collected samples of copper mineralizations in the Sahlabad area. (**A**) Mesgaran Deposit: Chalcopyrite and Fe-hydroxide; (**B**) Mesgaran Deposit:

Malachite and Fe-hydroxide; (**C**) Chah-Rasteh Deposit: Fracture Filling Malachite; (**D**) Chah-Rasteh Deposit: Chalcocite, Fe-hydroxide and Chalcopyrite; (**E**) Zahri Deposit: Fracture Filling Malachite and Fe-hydroxide; (**F**) Zahri Deposit: Chalcopyrite, Malachite and Chalcocite.

3. Materials and Methods

3.1. Geology and Geophysical Data

Geological data, including lithological map of area, structural features (fault system and lineaments) and location of copper ore deposits, old mines and indices were collected from the reports of the Geological Survey of Iran (GSI) as well as the Ministry of Industry, Mines and Trade of Iran [39,42,47]. Network data of 7.5 km of Iranian airborne magnetometry between 1974 and 1976 were commissioned by the Geological Survey of Iran (GSI) by the American company Aero Service, one of the largest companies active in the field of airborne geophysics at that time. The distance between the flight lines was 7.5 km, the fixed flight altitude was 300 m above the ground, and the distance between the vertical flight control lines was 40 km. The aircraft used to record this data was a twin-engine aircraft with a cesium vapor magnetometer mounted on it (with sensitivity of 0.002 nT). This data collection was performed in 62 separate flight blocks and was presented at an acceptable level in terms of quality [48]. An overview of methodological flowchart is presented in Figure 3.

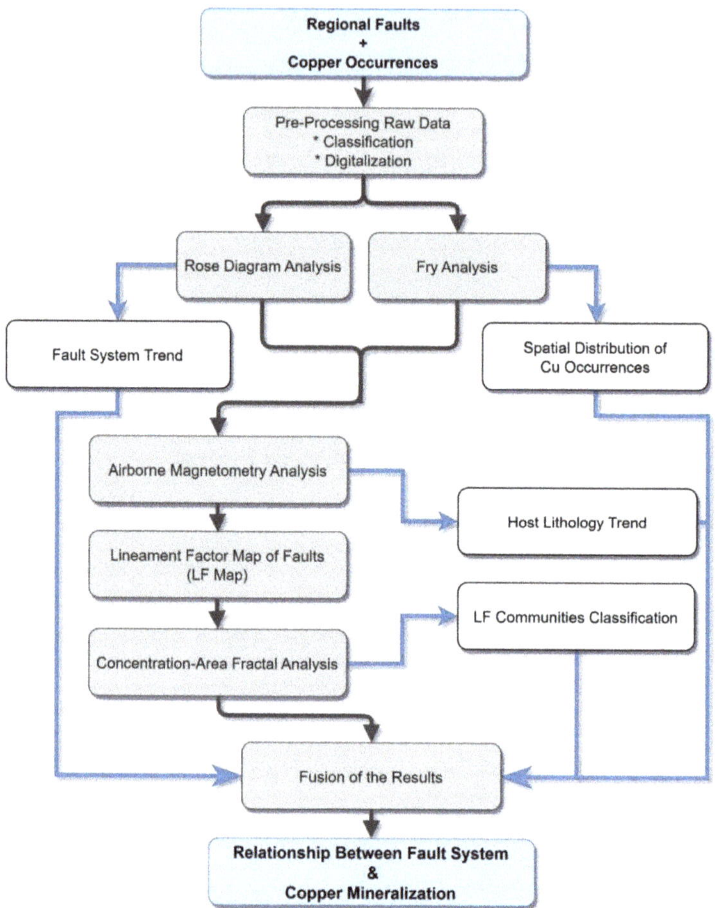

Figure 3. An overview of methodological flowchart used in this study.

3.2. Rose Diagram Analysis

Rose diagram is a type of circular histogram used to display directional data and the repetition rate of each category. This diagram is used in structural geology to show the trend of faults, fractures, lineaments and dykes [49,50]. In this study, rose diagram analysis was used to investigate the trend of faults in the Sahlabad region, which was subsequently compared to copper mineralization information as well as the host lithology trend. By analyzing the trend of faults and lithology, as well as the trend of copper mineralization in the area, it is possible to find out the effect of controlling faults [51,52].

3.3. Fry Analysis

Fry analysis is a complementary method in structural geological studies, which can be used to study the distribution of mineralization in a region and its relationship with linear structures. In other words, the application of Fry analysis method is useful in linear and directional analysis. This analysis is used to investigate the patterns of mineralization dispersion at the regional scale and also to describe mineralization zones, such as the direction of mineralization, for high-grade zones and the distribution of grade at a deposit scale [53–55]. Spatial distribution of mines, deposits and mineral indices is affected by factors such as formation environment, host rocks and other mineralization factors as well as structural controllers such as faults. Considering the importance of information about the spatial distribution of mineralizations, which is an important factor in regional exploration and mineral potential detection, in this study, the role of structural controllers in the spatial distribution of copper mineralization in Sahlabad area was investigated [56–58].

3.4. Airborne Magnetometry Analysis

Airborne magnetic data of Sahlabad region were isolated from these data and were used after corrections. In this study, gradient tensor method was used to analyze airborne magnetic data. The purpose of analyzing airborne magnetic data is to identify the position and trend of intrusive masses and to investigate their relationship with regional faults. There are various methods for analyzing magnetometric geophysical data, which use gradient analysis to detect geological lineaments. Some methods use only dx and dy horizontal gradients or only dz vertical gradients. However, in the gradient tensor method, horizontal gradients and vertical gradients are used simultaneously (dx, dy and dz). It provides more accurate and acceptable results in detecting lines on the border of magnetic anomalies [6,16]. For this purpose, using the gradient tensor method, a map of the residual magnetic intensity was prepared, and the faults associated with these masses were investigated.

3.5. Concentration–Area (C-A) Fractal Analysis

Fractal is a geometric structure that is obtained by enlarging each part of this structure in a certain proportion to the original structure. In other words, a fractal is a structure whose every part is the same as its whole. Fractals are seen from the same distance and closeness. This feature is called self-similarity [59,60]. Fractals are one of the most important tools in computer graphics and can be used in many ways [61,62]. The purpose of concentration–area fractal analysis is to examine the parameters related to the concentration and the area occupied by it. An exponential equation is given below for the aggregation of materials or fractal properties.

$$A(\geq v) \propto v^{-\alpha} \tag{1}$$

$A(\geq v)$ is the cumulative area enclosed by contours whose corresponding degree is greater than or equal to v. The value of α represents dimension of fractal corresponding to different amplitudes [63,64]. In this study, in order to classify the results of the lineament factor (LF) map, concentration–area (C–A) fractal analysis was used. The result of this analysis is the presentation of different groups that have different degrees of importance in the control of mineralization by faults.

4. Analysis and Results

4.1. Rose Diagram Analysis

In this study, in order to study the trend of faults in the area, rose diagrams of faults were generated. Rose diagrams for each type of fault are shown in Figure 4. A rose diagram of all the faults in the area is shown in Figure 5. The distance between the classes is 5 degrees; the average direction angle of faults is 129.8° (230.2°) with a confidence interval of 2.9° (95%). Figure 5 shows the frequency percentage of faults in the extended intervals. Faults are divided into three categories based on frequency percentage: low frequency, medium and frequent, which are distinguished by blue, yellow and red colors, respectively. As shown in the diagram, the main direction of the faults in the area is in the range of 115 to 135 degrees, which can be understood that the main extension of the faults is northwest-southeast.

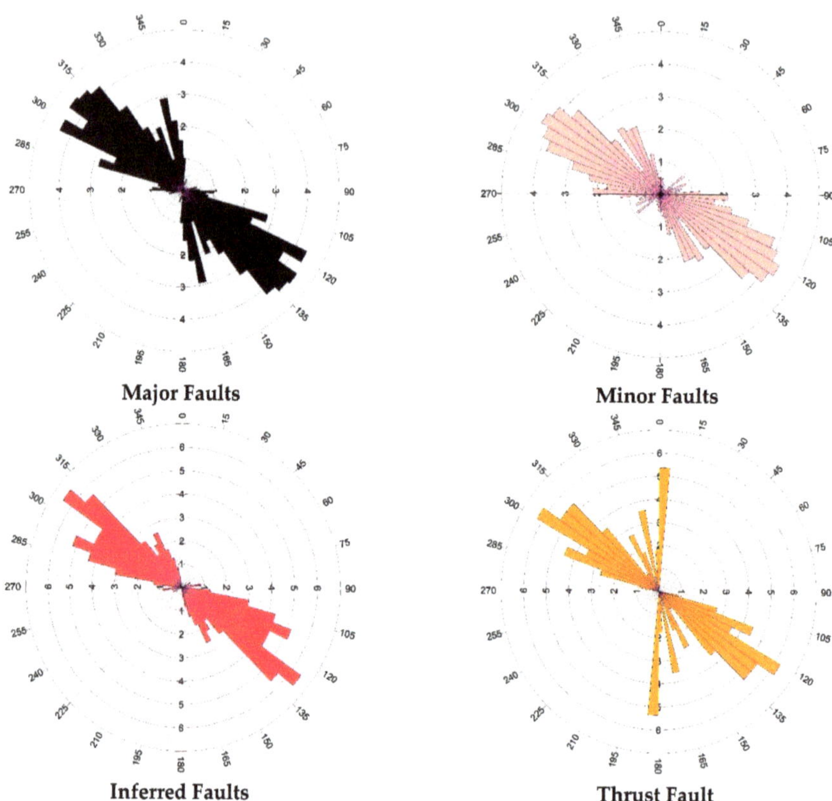

Figure 4. Rose diagram of faults by each of the types in the study area. The exact locations of the faults according to the colors assigned in the rose diagrams are shown in Figure 8 as a map.

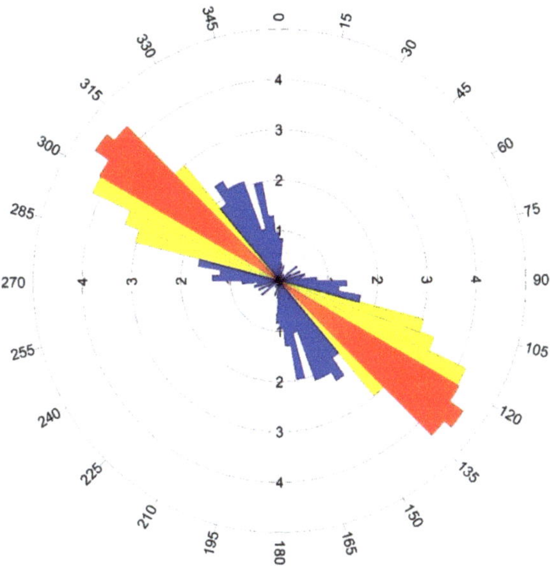

Figure 5. Rose diagram of all faults in the study area. Faults are divided into three categories based on frequency percentage: low frequency, medium and frequent, which are distinguished by blue, yellow and red colors, respectively.

4.2. Fry Analysis

After analyzing the rose diagram of regional faults and detecting the trend of the fault system in the region, which was identified as northwest-southeast. Fry analysis was performed to determine the mineralization trend of copper in the Sahlabad region. One of the main purposes of this study was to compare the trend of the fault system and the mineralization trend in the area. The locations of 14 mines, ores and mineral indices of copper in the Sahlabad area were drawn as dots on a separate layer, and then Fry analysis was performed on it. The result of this analysis is presented in Figure 6.

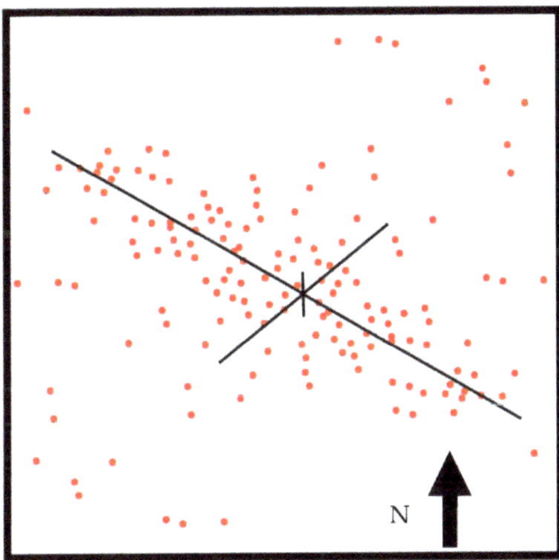

Figure 6. Fry analysis related to copper mineralization dispersion in the study area.

Based on Figure 6 as the result of Fry analysis, the mineralization trend of copper in the Sahlabad area is mostly northwest-southeast. By comparing the result of Fry analysis and Rose diagram analysis of faults in the Sahlabad area, it is evident that the mineralization trend in the area conforms to the dominant trend of faults. Therefore, in order to identify areas with high potential for copper mineralization, the study of faults is of great importance and is considered as a valuable exploratory key. Considering that the trend of copper mineralization is coincident with the trend of the Sahlabad fault system, in order to confirm the control of mineralization by faults, other trends such as host mineralization lithology and hydrothermal alterations should be examined.

4.3. Lithology Trend Analysis

Using airborne magnetic data of the Sahlabad area, a residual map of the magnetic intensity was produced. Figure 7 shows the residual magnetic map of the Sahlabad area. Results show that there is a magnetic dipole with a northwest-southeast trend, which according to the geological map of the area, is related to basaltic, andesitic, granite and ophiolite melange units in the area. In this regard, there are effects of serpentinization and a high probability of alteration effects due to the proximity of carbonate masses with basic and ultrabasic masses. According to airborne geophysical evidence and the geological map, there is a possibility of copper mineralization in this area, especially in the central parts of the area.

In this analysis, the purpose of analyzing airborne magnetic data was to identify faults that are associated with intrusive masses in the area. In other words, these faults, in addition to having an older formation time than other faults, also play a major role in controlling the lithological trend in the area. Using the gradient tensor method and intensity magnetic field map of the area, geological lineaments related to intrusive masses were identified. Figure 7 shows the geological lineament obtained from the magnetic field intensity map of the Sahlabad area. The lineaments identified by this method are in accordance with the faults in the geological map (see Figure 1). Since the main purpose of the magnetometric study was to study intrusive masses, the faults shown in Figure 7 are considered deep faults that have defined the boundary and trend of igneous masses.

The main fault trend, which extends from the northwest to the east of the map, generally defines the boundary of the ophiolitic melange unit with basaltic, ultrabasic and andesitic massifs. Expansion and formation of ophiolitic melanges in the Sahlabad region (northwest-southeast) have occurred in the direction of this fault. Therefore, it can be considered as the main fault that controls the lithology trend in the area. The faults north of the map are also located at the boundary of basaltic and andesitic units, and thus the elongation of these massifs is evident along the faults. Other faults that are shorter than the others also show control over the extension of intrusive masses in the area. It is noteworthy that because the faults were identified based on airborne magnetic analysis, the boundary of the intrusive masses in the area plays an essential role in the final result. Therefore, they are clearly shown linearly and based on the boundaries of geological units.

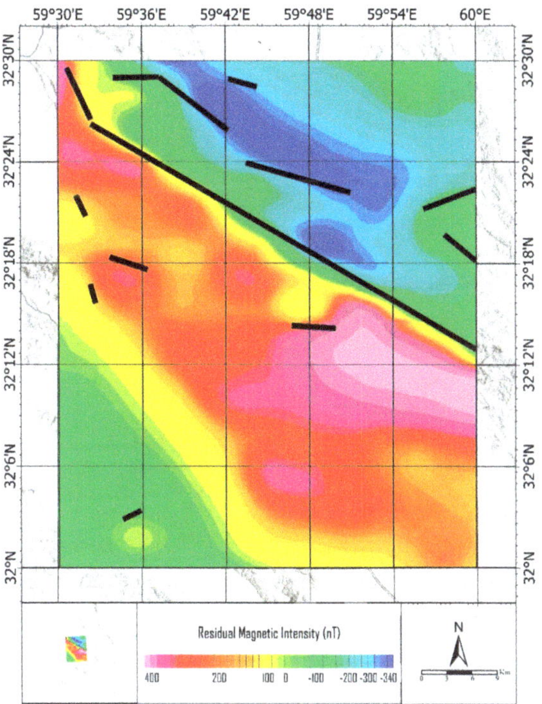

Figure 7. Map of changes in residual magnetic field intensity and faults identified by airborne magnetometry in the Sahlabad area.

4.4. Lineament Factor (LF) Map Analysis

The lineament factor (LF) map scores various parameters related to faults based on their degree of importance and finally shows the areas that are important in terms of fault activity. The parameters used in this study were: (i) frequency of faults, (ii) length of the faults and (iii) number of fault intersections. Initially, the network of the Sahlabad area was divided into 100-square-meter cells in order to study the faults and draw an LF map using the RockWorks software package. The scores of these factors were considered from top to bottom 1, 2 and 3, respectively [51,65]. The frequency of fault intersections plays an important role in the formation of magmatic and hydrothermal deposits because these intersections create a suitable space for mineralization in the bedrock [66]. However, the length of faults is also an important factor in the formation of hydrothermal deposits and leads to fluid conduction. The frequency of faults including structures before and after mineralization is the least important among the mentioned factors [67,68]. The lineament factor map is presented in Figure 8. According to the lineament factor map, based on the above-mentioned scores, the effect of fault control on copper mineralization in the Sahlabad area is shown. The importance of fault control in copper mineralization, from gray (lowest) to red (highest), is shown in the LF map. In order to determine the threshold of the impact of faults on copper mineralization, grouping was performed using fractal methods.

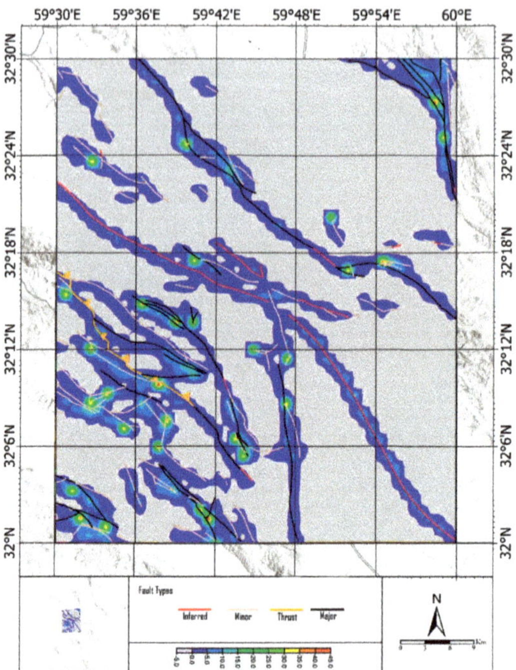

Figure 8. Lineament factor (LF) map of Sahlabad area. The faults in the map are divided into four categories, inferred, minor, thrust and major, which are marked with red, cream, yellow and black colors, respectively. Faults were identified using information extracted from the geological map (Geological Survey of Iran (GSI)).

LF Map Classification by Fractal Modeling

Based on the map presented in Figure 8, the fractal diagram of the concentration–area (C–A) of the faults was produced. According to the LF values shown in Figure 8, the area associated with lower LF values to higher LF values was calculated using Surfer software. Then, based on the concentration–area (C–A) fractal methodology, logarithmic values were examined and are shown in Figure 9. The C–A fractal diagram is shown in Figure 9. The diagram shows the multifractal nature of faults in the Sahlabad area. The results of the fractal classification of faults are presented in Table 2.

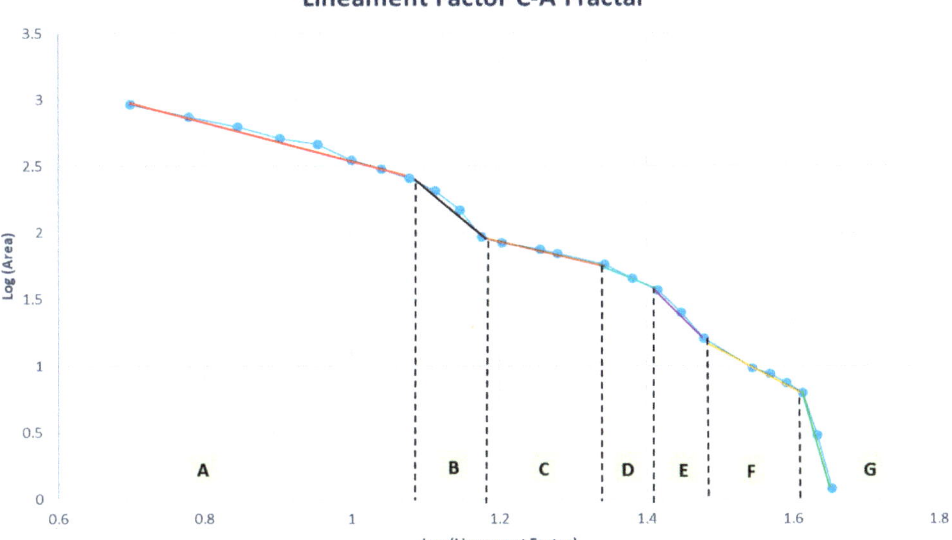

Figure 9. Concentration–area (C–A) fractal diagram of faults in the Sahlabad area. The trend change is indicated by colored lines, and each sub-community is marked with the letters A to G, respectively.

Table 2. Range of lineament factor values obtained from the concentration–area (C–A) fractal model of faults in the Sahlabad area.

Communities	Background		Medium Intensity			High Intensity	
Sub-Communities	A	B	G	D	E	F	G
LF Threshold	Less than 12	12–15	15–22	22–26	26–30	30–41	More than 41

According to Table 2, three communities and seven sub-communities were identified in terms of the LF concentration of faults in the area. The first community is the background in which the cell counts in this class are calculated from 12 to 25. The second community shows the average intensity of the LF concentration of faults in which the range of cell values is 15 to 30. The third community, which is introduced as the community of a high-intensity concentration of LFs, includes values above 30. Now, based on the map presented in Figure 8, the important areas in terms of fault activity can be easily distinguished and studied. The distance of each of the existing copper mineralizations (mines, ores and indices) from the regional faults and the LF high-intensity community of the faults is presented in Table 3.

Table 3. Copper mineralization distance from regional faults and LF high-intensity community.

Distance from LF High-Intensity Community (Km)	Distance from Regional Faults (Km)	Copper Mineralization	Row
1.9	1.1	Mesgaran Deposit	1
4.1	0.95	Chah-Rasteh Deposit	2
0.45	Coincident	Zahri Deposit	3
4.3	Coincident	Kasrab Abandoned Mine	4
Coincident	Coincident	Cheshme-Zangi Abandoned Mine	5
2.79	Coincident	Shir-Shotor Indice	6
3.67	1.64	Dastgerd Indice	7
1.8	Coincident	Torshaab Indice	8
1.52	0.8	Chah-Anjir Indice	9
3.5	1.49	Zargaran Indice	10
1.34	1.25	West Mesgaran Indice	11
Coincident	Coincident	Mirsimin Indice	12
1	Coincident	Kuharod Indice	13
1.59	Coincident	Barghan Indice	14
1.99	0.51	Average Distance (Km)	

4.5. Field Evidence

In order to conduct a field check, some points were selected as the target. These points are typically andesite-basalt and ultrabasic rocks, which are the host rocks of copper mineralization in the Sahlabad area. Generally, surface exposures of copper mineralization in the form of malachite and azurite were observed in the faults and fractures associated with andesitic-basaltic outcrops. An overview of copper mineralization in the faults and fracture zone is shown in Figure 10A–D. The reason for choosing these points as control points was the presence of andesite-basalt and ultrabasic rock units and the conformity of this lithology on one of the parts of the community in the high-intensity LF map (see Figure 8). Moreover, these zones are in line with the copper mineralization trend (analyzed by Fry analysis) on a regional scale. As shown in Figure 10, copper oxide mineralization is widespread in the outcrops of these areas.

Figure 10. *Cont.*

Figure 10. View of copper mineralization (malachite-azurite) points in fracture zone and andesitic-basaltic bedrock. (**A–C**) Malachite-azurite mineralization in field survey (high-intensity LF); (**D**) an overview of fracture zone in field check points.

5. Discussion

The exploration of massive sulfide mineralization involves specific, robust and tailored exploration techniques, which can be further developed using geology and geophysical data. Massive sulfide deposits are diligently related to low-angle detachment faults [27,29]. They are typically hosted in various altered ultramafic rocks (tectonic melange) and are enriched in Au, Ag, Co, Cu, Zn, Ni [28]. Because of their complex tectonic settings, these deposits are difficult to explore. In this study, using analytical methods such as the rose diagram, Fry analysis, lineament factor (LF) map, multifractal technique and aeromagnetic data analysis, the regional trend of faults systems and the trend of massive sulfide copper mineralization in host rock were investigated in the Sahlabad area, South Khorasan province, east of Iran. Due to the boundary of lithologies of mineralization host rock in the Sahlabad area, which can be seen from the residual magnetic field map and geological map, the fault system has played an important role in orienting the host lithology.

The concentration–area (C-A) multifractal technique, which was applied on the lineament factor (LF) map, divided the fault lineament factor community into seven sub-communities. In order to simplify the results, these sub-communities were divided into three general communities, which are background, medium intensity and high intensity. Then, in order to investigate the relationship between copper occurrences (deposits, mines and indices) with the fault system, the distances of these anomalies to the LF high-intensity community of faults were measured. According to Table 3, about 60% of these anomalies are coincident with the fault system, and on average, all copper occurrences in the Sahlabad area are within 500 m of regional faults and 2 km from the LF high-intensity community. The main development of the present study, compared to previous studies, is the fusion of airborne (aeromagnetic) geophysical data with the regional fault system information derived from geological data. It is worth mentioning that before this research, no study had been conducted on the relationship between the fault system and copper mineralization in the Sahlabad area. The NW-SE fault systems are, along with the main trend of lithological units, related to massive sulfide copper mineralization in the area. Field evidence established that the NW-SE fault systems are matched with a number of massive sulfide copper mineralizations.

6. Conclusions

In this study, the relationship between the fault system and copper mineralization in the Sahlabad area, South Khorasan province, east of Iran, was identified. The lineament factor (LF) map was generated, and multifractal analysis was implemented. The main achievements of this research are:

- In general, the trend of faults at the regional scale is northwest-southeast, which is consistent with the trend of lithology units related to mineralization.
- Based on the classified information related to faults in mines, deposits and copper indices of the Sahlabad area, it is observed that in most cases, mineralization has taken place at the fault systems that have a trend perpendicular to the faults in the area.
- Studies on airborne magnetometric data indicate that the faults identified by this method are faults associated with intrusive masses, and thus the faults control the lithology trend in the area.
- Overall, it can be said that the faults in the area control the bedrock lithology and the source of massive sulfide copper mineralization in the region, while the regional faults (on a mining scale) in mines, deposits and indices control the mineralization in the region.

The distance of copper mineralization in the Sahlabad area from regional faults and also from the community of high-intensity lineament factors (LFs) is on average 500 m and 2 km, respectively. It is noteworthy that a number of mineralizations correspond exactly to the regional faults as well as the high-intensity linear factor community.

In conclusion, the approach developed in this study is a valuable and inexpensive tool for exploring massive sulfide mineralization in metallogenic provinces.

Author Contributions: All authors have contributed to writing and editing this article. Writing original draft preparation, A.S. and A.H.; supervision, A.H. and A.B.P.; writing—review and editing, A.B.P. and A.S. All authors have read and agreed to the published version of the manuscript.

Funding: This research received no external funding.

Data Availability Statement: Not applicable.

Acknowledgments: We would appreciate Department of Mining and Metallurgy Engineering Amirkabir University of Technology (Tehran Polytechnic). This study is part of the research activity carried out during the first author's PhD research at Department of Mining and Metallurgy Engineering Amirkabir University of Technology (Tehran Polytechnic). The Institute of Oceanography and Environment (INOS), Universiti Malaysia Terengganu (UMT) is also acknowledged for providing facilities during editing, rewritten and re-organizing the manuscript.

Conflicts of Interest: The authors declare no conflict of interest.

References

1. Hoggard, M.J.; Czarnota, K.; Richards, F.D.; Huston, D.L.; Jaques, A.L.; Ghelichkhan, S. Global distribution of sediment-hosted metals controlled by craton edge stability. *Nat. Geosci.* **2020**, *13*, 504–510. [CrossRef]
2. Prior, Á.; Benndorf, J.; Mueller, U. Resource and grade control model updating for underground mining production settings. *Math. Geosci.* **2021**, *53*, 757–779. [CrossRef]
3. Baseri, F.; Nezafati, N. Structural and Chemical Controllers of the North and Northwest of Torud Based on Involved Fluid Studies, Structural and Geochemical Analyses. *Teh. Glas.* **2021**, *15*, 339–349. [CrossRef]
4. Tuduri, J.; Chauvet, A.; Barbanson, L.; Labriki, M.; Dubois, M.; Trapy, P.-H.; Lahfid, A.; Poujol, M.; Melleton, J.; Badra, L. Structural control, magmatic-hydrothermal evolution and formation of hornfels-hosted, intrusion-related gold deposits: Insight from the Thaghassa deposit in Eastern Anti-Atlas, Morocco. *Ore Geol. Rev.* **2018**, *97*, 171–198. [CrossRef]
5. Malaekeh, A.; Ghassemi, M.R.; Afzal, P.; Solgi, A. Fractal Modeling and Relationship between Thrust Faults and Carbonate-Hosted Pb-Zn Mineralization in Alborz Mountains, Northern Iran. *Geochemistry* **2021**, *81*, 125803. [CrossRef]
6. Ghasempoor, F.; Heyhat, M.R.; Khatib, M.M. Investigation of the Relationship between Faults and Joints with Copper Mineralization, Using Remote Sensing and Magnetism in Vorezg Region (East of Ghaen-Birjand). *J. Tecton.* **2020**, *4*, 35–57.
7. Shirazy, A.; Shirazi, A.; Hezarkhani, A. *Behavioral Analysis of Geochemical Elements in Mineral Exploration: Methodology and Case Study*; LAP LAMBERT Academic Publishing: Saarbrücken, Germany, 2020.
8. Shirazy, A.; Shirazi, A.; Hezarkhani, A. *Advanced Integrated Methods in Mineral Exploration*; LAP LAMBERT Academic Publishing: Saarbrücken, Germany, 2022; p. 160.

9. Haddad-Martim, P.M.; Carranza, E.J.M.; de Souza Filho, C.R. The fractal nature of structural controls on ore formation: The case of the iron oxide copper-gold deposits in the Carajás Mineral Province, Brazilian Amazon. *Econ. Geol.* **2018**, *113*, 1499–1524. [CrossRef]

10. Sun, T.; Xu, Y.; Yu, X.; Liu, W.; Li, R.; Hu, Z.; Wang, Y. Structural controls on copper mineralization in the Tongling Ore District, Eastern China: Evidence from spatial analysis. *Minerals* **2018**, *8*, 254. [CrossRef]

11. Zhang, L.C.; Bai, Y.; Zhu, M.T.; Huang, K.; Peng, Z.D. Regional heterogeneous temporal–spatial distribution of gold deposits in the North China Craton: A review. *Geol. J.* **2020**, *55*, 5646–5663. [CrossRef]

12. Nadoll, P.; Sośnicka, M.; Kraemer, D.; Duschl, F. Post-Variscan structurally-controlled hydrothermal Zn-Fe-Pb sulfide and F-Ba mineralization in deep-seated Paleozoic units of the North German Basin: A review. *Ore Geol. Rev.* **2019**, *106*, 273–299. [CrossRef]

13. Aita, S.K.; Omar, A.E. Exploration of uranium and mineral deposits using remote sensing data and GIS applications, Serbal area, Southwestern Sinai, Egypt. *Arab. J. Geosci.* **2021**, *14*, 2214. [CrossRef]

14. Dasgupta, S.; Mukherjee, S. Remote sensing in lineament identification: Examples from western India. In *Developments in Structural Geology and Tectonics*; Elsevier: Amsterdam, The Netherlands, 2019; Volume 5, pp. 205–221.

15. Moradpour, H.; Rostami Paydar, G.; Pour, A.B.; Valizadeh Kamran, K.; Feizizadeh, B.; Muslim, A.M.; Hossain, M.S. Landsat-7 and ASTER remote sensing satellite imagery for identification of iron skarn mineralization in metamorphic regions. *Geocarto Int.* **2020**, *35*, 1–28. [CrossRef]

16. Eldosouky, A.M.; El-Qassas, R.A.; Pour, A.B.; Mohamed, H.; Sekandari, M. Integration of ASTER satellite imagery and 3D inversion of aeromagnetic data for deep mineral exploration. *Adv. Space Res.* **2021**, *68*, 3641–3662. [CrossRef]

17. Riahi, S.; Bahroudi, A.; Abedi, M.; Aslani, S.; Elyasi, G.R. Integration of airborne geophysics and satellite imagery data for exploration targeting in porphyry Cu systems: Chahargonbad district, Iran. *Geophys. Prospect.* **2021**, *69*, 1116–1137. [CrossRef]

18. Uwiduhaye, J.d.A.; Ngaruye, J.C.; Saibi, H. Defining potential mineral exploration targets from the interpretation of aeromagnetic data in western Rwanda. *Ore Geol. Rev.* **2021**, *128*, 103927. [CrossRef]

19. Shebl, A.; Csámer, Á. Reappraisal of DEMs, Radar and optical datasets in lineaments extraction with emphasis on the spatial context. *Remote Sens. Appl. Soc. Environ.* **2021**, *24*, 100617. [CrossRef]

20. Innocent, A.J.; Chidubem, E.O.; Chibuzor, N.A. Analysis of aeromagnetic anomalies and structural lineaments for mineral and hydrocarbon exploration in Ikom and its environs southeastern Nigeria. *J. Afr. Earth Sci.* **2019**, *151*, 274–285. [CrossRef]

21. Hedayat, B.; Ahmadi, M.E.; Nazerian, H.; Shirazi, A.; Shirazy, A. Feasibility of Simultaneous Application of Fuzzy Neural Network and TOPSIS Integrated Method in Potential Mapping of Lead and Zinc Mineralization in Isfahan-Khomein Metallogeny Zone. *Open J. Geol.* **2022**, *12*, 215–233. [CrossRef]

22. Nicolis, O. Review chapter on mineral exploration section. In *Methods and Applications in Petroleum and Mineral Exploration and Engineering Geology*; Elsevier: Amsterdam, The Netherlands, 2021; pp. 231–235.

23. Nicolis, O. Statistical analysis of geological faults for characterizing mineral deposits. In *Methods and Applications in Petroleum and Mineral Exploration and Engineering Geology*; Elsevier: Amsterdam, The Netherlands, 2021; pp. 285–293.

24. Khosravi, V.; Shirazi, A.; Shirazy, A.; Hezarkhani, A.; Pour, A.B. Hybrid Fuzzy-Analytic Hierarchy Process (AHP) Model for Porphyry Copper Prospecting in Simorgh Area, Eastern Lut Block of Iran. *Mining* **2022**, *2*, 1–12. [CrossRef]

25. Talebi, H.; Mueller, U.; Peeters, L.J.; Otto, A.; de Caritat, P.; Tolosana-Delgado, R.; van den Boogaart, K.G. Stochastic Modelling of Mineral Exploration Targets. *Math. Geosci.* **2022**, *54*, 593–621. [CrossRef]

26. Shahbazi, S.; Ghaderi, M.; Afzal, P. Prognosis of of gold mineralization phases by multifractal modeling in the Zehabad epithermal deposit, NW Iran. *Iran. J. Earth Sci.* **2021**, *13*, 31–40.

27. Betts, P.G.; Giles, D.; Lister, G. Aeromagnetic patterns of half-graben and basin inversion: Implications for sediment-hosted massive sulfide Pb–Zn–Ag exploration. *J. Struct. Geol.* **2004**, *26*, 1137–1156. [CrossRef]

28. Patten, C.G.; Coltat, R.; Junge, M.; Peillod, A.; Ulrich, M.; Manatschal, G.; Kolb, J. Ultramafic-hosted volcanogenic massive sulfide deposits: An overlooked sub-class of VMS deposit forming in complex tectonic environments. *Earth-Sci. Rev.* **2021**, *224*, 103891. [CrossRef]

29. Melekestseva, I.Y.; Zaykov, V.; Nimis, P.; Tret'Yakov, G.; Tessalina, S. Cu–(Ni–Co–Au)-bearing massive sulfide deposits associated with mafic–ultramafic rocks of the Main Urals Fault, South Urals: Geological structures, ore textural and mineralogical features, comparison with modern analogs. *Ore Geol. Rev.* **2013**, *52*, 18–36. [CrossRef]

30. Agharezaei, M.; Hezarkhani, A. Delineation of geochemical anomalies based on Cu by the boxplot as an exploratory data analysis (EDA) method and concentration-volume (CV) fractal modeling in Mesgaran mining area, Eastern Iran. *Open J. Geol.* **2016**, *6*, 1269–1278. [CrossRef]

31. Shirazi, A.; Shirazy, A.; Karami, J. Remote sensing to identify copper alterations and promising regions, Sarbishe, South Khorasan, Iran. *Int. J. Geol. Earth Sci.* **2018**, *4*, 36–52.

32. Shirazy, A.; Hezarkhani, A.; Shirazi, A.; Timkin, T.V.; Voroshilov, V.G. Geophysical explorations by resistivity and induced polarization methods for the copper deposit, South Khorasan, Iran. *Bull. Tomsk. Polytech. Univ. Geo Assets Eng.* **2022**, *333*, 99–110. [CrossRef]

33. Tirrul, R.; Bell, I.; Griffis, R.; Camp, V. The Sistan suture zone of eastern Iran. *Geol. Soc. Am. Bull.* **1983**, *94*, 134–150. [CrossRef]

34. Reyre, D.; Mohafez, S. *A First Contribution of the NIOC-ERAP Agreements to the Knowledge of Iranian Geology*; Editions Technip: Paris, France, 1972.

35. Nabavi, M. *An Introduction to the Iranian Geology*; Geological Survey of Iran: Tehran, Iran, 1976; Volume 110.

36. Samani, B.; Ashtari, M. The development of geology in Sistan and Baluchestan region. *Q. J. Earth Sci.* **1991**, *4*, 26–40.
37. Carey, S.W. *The Expanding Earth*; Elsevier: Amsterdam, The Netherlands; Oxford, UK; New York, NY, USA, 1976; p. 488.
38. Alavi, M. Sedimentary and structural characteristics of the Paleo-Tethys remnants in northeastern Iran. *Geol. Soc. Am. Bull.* **1991**, *103*, 983–992. [CrossRef]
39. Navai, I. *Geological Map of Sahlabad Area (on Sacale 1:100,000)*; Geological Survey of Iran (GSI): Tehran, Iran, 1974.
40. Ghorbani, M. Geological Setting and Crustal Structure of Iran. In *The Geology of Iran: Tectonic, Magmatism and Metamorphism*; Springer: Berlin/Heidelberg, Germany, 2021; pp. 1–22.
41. Ghorbani, M. A summary of geology of Iran. In *The Economic Geology of Iran*; Springer: Berlin/Heidelberg, Germany, 2013; pp. 45–64.
42. Monazami Bagherzade, R. *Economic Geology Report (Hammer Exploration) Sahlabad Sheet (Scale 1:100,000)*; Geological Survey and Mineral Exploration of Northeast Iran: South Khorasan Province, Iran, 2001; p. 20.
43. Samimi Namin, M. *Report on Potential Detection and Preliminary Exploration of Iron, Gold and Copper in Sarbisheh City*; Madankav Consulting Engineers Co.: Tehran, Iran, 2006; p. 238.
44. Zaman, Z.K. *Detailed Exploration Operation Report of Kooh Kheiri Copper Deposit*; Industries and Mines Organization of South Khorasan Province, Iran: South Khorasan Province, Iran, 2012.
45. Kav, P.K. *Detailed Exploration Report of Mesgaran Copper Deposit*; Parsi Kan Kav Eng. Co.: Soth Khorasan Province, Iran, 2013; p. 477.
46. Jalili, F. *Detailed Exploration Operation Report of Zahri Copper Deposite*; Industries and Mines Organization of South Khorasan Province: South Khorasan Province, Iran, 2014.
47. GSI. *Report of Systematic Geochemical Explorations in the Sahlabad Area (Sheet on Scale 1:100,000—Geochemistry of Stream Sediments)*; Geological Survey of IRAN (GSI): Tehran, Iran, 2001; p. 644.
48. Friedberg, J.L.; Yousefi, E. *Aeromagnetic Map of Iran: Quaderangle, No. D7, Based on Airborne Survey by Aero Service Corporation under Contract to the Geological Survey of Iran*; Geological Survey of Iran: Tehran, Iran, 1978.
49. Hill, J.S. Post-Orogenic Uplift, Young Faults, and Mantle Reorganization in the Appalachians. Ph.D. Thesis, The University of North Carolina at Chapel Hill, Chapel Hill, NC, USA, 2018.
50. Akgün, E.; İnceöz, M.; Manap, H.S. Geological lineament analyses application to a fault segment on the east Anatolian fault zone. In Proceedings of the International Multidisciplinary Scientific GeoConference: SGEM, Albena, Bulgaria, 28 June–6 July 2019; Volume 19, pp. 547–552.
51. Ahmadfaraj, M.; Mirmohammadi, M.; Afzal, P.; Yasrebi, A.B.; Carranza, E.J. Fractal modeling and fry analysis of the relationship between structures and Cu mineralization in Saveh region, Central Iran. *Ore Geol. Rev.* **2019**, *107*, 172–185. [CrossRef]
52. Wu, R.; Chen, J.; Zhao, J.; Chen, J.; Chen, S. Identifying Geochemical Anomalies Associated with Gold Mineralization Using Factor Analysis and Spectrum–Area Multifractal Model in Laowan District, Qinling-Dabie Metallogenic Belt, Central China. *Minerals* **2020**, *10*, 229. [CrossRef]
53. Maanijou, M.; Daneshvar, N.; Alipoor, R.; Azizi, H. Spatial Analysis on Gold Mineralization in Southwest Saqqez Using Point Pattern, Fry and Fractal Analyses. *Geotectonics* **2020**, *54*, 589–604. [CrossRef]
54. Behyari, M.; Rahimsouri, Y.; Hoseinzadeh, E.; Kurd, N. Evaluating of lithological and structural controls on the barite mineralization by using the remote sensing, Fry and fractal methods, Northwest Iran. *Arab. J. Geosci.* **2019**, *12*, 167. [CrossRef]
55. Nguemhe Fils, S.C.; Mimba, M.E.; Nyeck, B.; Nforba, M.T.; Kankeu, B.; Njandjock Nouck, P.; Hell, J.V. GIS-Based Spatial Analysis of Regional-Scale Structural Controls on Gold Mineralization Along the Betare-Oya Shear Zone, Eastern Cameroon. *Nat. Resour. Res.* **2020**, *29*, 3457–3477. [CrossRef]
56. Amirihanza, H.; Shafieibafti, S.; Derakhshani, R.; Khojastehfar, S. Controls on Cu mineralization in central part of the Kerman porphyry copper belt, SE Iran: Constraints from structural and spatial pattern analysis. *J. Struct. Geol.* **2018**, *116*, 159–177. [CrossRef]
57. Habibkhah, N.; Hassani, H.; Maghsoudi, A.; Honarmand, M. Application of numerical techniques to the recognition of structural controls on porphyry Cu mineralization: A case study of Dehaj area, Central Iran. *Geosyst. Eng.* **2020**, *23*, 159–167. [CrossRef]
58. Zheng, D.; Ni, C.; Zhang, S.; Chen, Z.; Zhong, J.; Zhu, J.; Li, Y.; Yan, Y. Significance of the spatial point pattern and Fry analysis in mineral exploration. *Arab. J. Geosci.* **2020**, *13*, 883. [CrossRef]
59. Anguera, J.; Andújar, A.; Jayasinghe, J.; Chakravarthy, V.; Chowdary, P.; Pijoan, J.L.; Ali, T.; Cattani, C. Fractal antennas: An historical perspective. *Fractal Fract.* **2020**, *4*, 3. [CrossRef]
60. Karaca, Y.; Baleanu, D. A novel R/S fractal analysis and wavelet entropy characterization approach for robust forecasting based on self-similar time series modeling. *Fractals* **2020**, *28*, 2040032. [CrossRef]
61. Babič, M.; Mihelič, J.; Cali, M. Complex network characterization using graph theory and fractal geometry: The case study of lung cancer DNA sequences. *Appl. Sci.* **2020**, *10*, 3037. [CrossRef]
62. Sadeghi, B.; Cohen, D.R. Category-based fractal modelling: A novel model to integrate the geology into the data for more effective processing and interpretation. *J. Geochem. Explor.* **2021**, *226*, 106783. [CrossRef]
63. Nazarpour, A. Application of CA fractal model and exploratory data analysis (EDA) to delineate geochemical anomalies in the: Takab 1: 25,000 geochemical sheet, NW Iran. *Iran. J. Earth Sci.* **2018**, *10*, 173–180.
64. Saadati, H.; Afzal, P.; Torshizian, H.; Solgi, A. Geochemical exploration for lithium in NE Iran using the geochemical mapping prospectivity index, staged factor analysis, and a fractal model. *Geochem. Explor. Environ. Anal.* **2020**, *20*, 461–472. [CrossRef]

65. Asadi, H.H.; Sansoleimani, A.; Fatehi, M.; Carranza, E.J.M. An AHP–TOPSIS predictive model for district-scale mapping of porphyry Cu–Au potential: A case study from Salafchegan area (central Iran). *Nat. Resour. Res.* **2016**, *25*, 417–429. [CrossRef]
66. Berger, B.R.; Ayuso, R.A.; Wynn, J.C.; Seal, R.R. Preliminary model of porphyry copper deposits. *US Geol. Surv. Open-File Rep.* **2008**, *1321*, 55.
67. Pirajno, F. *Hydrothermal Processes and Mineral Systems*; Springer Science & Business Media: Berlin/Heidelberg, Germany, 2008.
68. Carranza, E.J.M. Controls on mineral deposit occurrence inferred from analysis of their spatial pattern and spatial association with geological features. *Ore Geol. Rev.* **2009**, *35*, 383–400. [CrossRef]

 *minerals*

MDPI

Article

Depth Estimation of Sedimentary Sections and Basement Rocks in the Bornu Basin, Northeast Nigeria Using High-Resolution Airborne Magnetic Data

Stephen E. Ekwok [1], Ogiji-Idaga M. Achadu [2], Anthony E. Akpan [1], Ahmed M. Eldosouky [3,*], Chika Henrietta Ufuafuonye [1], Kamal Abdelrahman [4] and David Gómez-Ortiz [5]

[1] Applied Geophysics Programme, Department of Physics, University of Calabar, Calabar 540271, Cross River State, Nigeria; styvnekwok@unical.edu.ng (S.E.E.); anthonyakpan@unical.edu.ng (A.E.A.); hufuafuonye@gmail.com (C.H.U.)
[2] Department of Geology, University of Calabar, Calabar 540271, Cross River State, Nigeria; martinsachadu@unical.edu.ng
[3] Department of Geology, Suez University, Suez 43518, Egypt
[4] Department of Geology & Geophysics, College of Science, King Saud University, P.O. Box 2455, Riyadh 11451, Saudi Arabia; khassanein@ksu.edu.sa
[5] Department of Biology and Geology, Physics and Inorganic Chemistry, ESCET, Universidad Rey Juan Carlos, Móstoles, 28933 Madrid, Spain; david.gomez@urjc.es
* Correspondence: ahmed.eldosouky@sci.suezuni.edu.eg

Citation: Ekwok, S.E.; Achadu, O.-I.M.; Akpan, A.E.; Eldosouky, A.M.; Ufuafuonye, C.H.; Abdelrahman, K.; Gómez-Ortiz, D. Depth Estimation of Sedimentary Sections and Basement Rocks in the Bornu Basin, Northeast Nigeria Using High-Resolution Airborne Magnetic Data. *Minerals* 2022, 12, 285. https://doi.org/10.3390/min12030285

Academic Editor: Frank Wendler

Received: 12 January 2022
Accepted: 22 February 2022
Published: 24 February 2022

Abstract: This study involves the use of high-resolution airborne magnetic data to evaluate the thicknesses of sedimentary series in the Bornu Basin, Northeast Nigeria, using three depth approximation techniques (source parameter imaging, standard Euler deconvolution, and 2D GM-SYS forward modelling methods). Three evenly spaced profiles were drawn in the N-S direction on the total magnetic intensity map perpendicular to the regional magnetic structures. These profiles were used to generate three 2-D models. The magnetic signatures were visually assessed to determine the thickness of depo-centres and the position of intrusions. The thicknesses of sedimentary series based on source parameter imaging results are approximately ranged 286 to 615 m, 695 to 1038 m, and 1145 to 5885 m for thin, intermediate, and thick sedimentation, respectively. Similarly, the standard Euler deconvolution result shows thin (130 to 917 m), intermediate (1044 to 1572 m), and thick (1725 to 5974 m) sedimentation. The magnetic model of Profile 1, characterized by two major breaks, shows that the igneous intrusions and basement rocks are covered by sediments with thickness varying from 300 to <3500 m, while Profile 2 has a maximum estimated depth value of about 5000 m at the southern part. Furthermore, the Profile 3 model shows sediment thicknesses of 2500 and 4500 m in the northern and southern flanks of the profile, respectively. The maximum sediment thickness value from the various depth estimation methods used in this study correlate relatively well with each other. Furthermore, the anomalous depth zone revealed by the 2D forward models coincides with the locality of the thick sedimentation revealed by the source parameter imaging and standard Euler-deconvolution (St-ED) methods. The maximum depth values obtained from the various depth estimation methods used in this study correlated strongly with each other. The widespread occurrence of short-wavelength anomalies in the southern part of the study area as indicated by the jagged nature of the magnetic signature was validated by the analytic signal and upward-continuation results. Generally, it was observed that the southern part of the research area is characterized by thick sedimentation and igneous intrusions.

Keywords: aeromagnetic; Bornu Basin; Precambrian; basement depth

1. Introduction

Potential field techniques have various successful applications in exploration geophysics [1–10]. One of the most essential applications of the magnetic dataset is to define the location and depth of magnetic bodies. Conventionally, it is often used to determine sedimentary thicknesses for oil and gas exploration purposes. The false solution problems associated with different depth approximation techniques can be controlled by combining two or more depth determination procedures [11,12], or enhancing the signal/noise ratio through proper evaluation of the derivatives of the field [13].

In this research, vital depth estimation methods like source-parameter imaging (SPI) [14,15], standard-Euler deconvolution (St-ED) [16], and 2D-GM-SYS modelling code [17] in the Oasis–Montaj software (Geosoft Inc., Toronto, ON, Canada) are used for the interpretation of airborne magnetic data collected in the Bornu Basin, Northeast Nigeria. The SPI and St-ED techniques are non-dependent on assumptions about the geologic model [18,19]. Hence, the applications of these techniques have made the procedure of magnetic data interpretation significantly simpler [20]. The use of these depth estimation methods in this study to the same magnetic sources will considerably enhance the reliability of depth solutions. Furthermore, analytic signal [20] and upward-continuation (UP-C) [21–23] filters will be applied to image the location and source of the main tectonics that caused the buried deformations and geologic structures [24,25] within the study area.

Modern geoscience investigations in the Nigerian inland basins are centred on gravity, seismic, magnetic, paleoclimatography, geochemistry, aerial photography, source rocks and rock facies evaluations that have been studied by several researchers [11,26]. At the reconnaissance stage, the basement framework, depositional centres [27], and depth solutions [11,28] can be discerned from high resolution airborne magnetic data. Several researchers [11,23–27,29–33] have investigated the thickness of sediments in the Cretaceous inland basins of Nigeria. In their separate studies, several depth solutions (1.5–12 km) were estimated involving different depth to basement procedures [11]. Remarkably, these investigations have led to the detection of commercial hydrocarbon in the Cretaceous inland Anambra basin. This discovery has further triggered geoscientists' interest to properly evaluate the hydrocarbon prospects of the Nigerian Benue Trough.

Modern airborne (magnetic) data were collected (between 2005 and 2010) by Fugro Airborne-Services, Canada was used in this research. The object of this study which involves the assessment of sediment thicknesses, mapping of basin topography, and delineation of the spatial spread of magnetic sources were resolved using these magnetic datasets. These data can further be employed in the demarcation of regional surficial geologic boundaries [26], mineral assessment programs [11], mapping of hydrothermally modified rocks [27,34–37], edges of sources [38], geothermal potentiality [39], and geologic structures such as faults, fractures, dykes, sills, etc. [27,40]. Furthermore, it can be applied in hydrocarbon exploration [27], archaeological investigations, and unexploded ordnance (UXO) detection [41]. Recently, the readily available high-speed and robust computer programs have made the processes of magnetic data correction, enhancement, modelling, and interpretation easier. Furthermore, the advancements in technology have made it increasingly possible to produce more details, and delineate elusive magnetic anomalies. However, the associated inverse problem of magnetic data is often ill-posed, therefore making the solution unstable and uncertain [42–44]. Nonetheless, a reliable solution to an ill-posed problem can be obtained by combining accurate geologic information with recent innovative magnetic data enhancement and modelling procedures [11].

The Bornu Basin, often described as part of the Upper Benue Trough, is a fraction of the Chad Basin [45]. The basin has witnessed extensive geological and geophysical studies for hydrocarbon, coal and minerals, and groundwater resources [11]. The detection of commercial hydrocarbon in the neighboring Republics of Niger and Chad in the 1970s has further oil and gas exploration activities in the Bornu Basin [46] and other Nigerian inland basins in the last forty years. The geology [45,47,48], stratigraphy [49,50], tectonics and tectono-sedimentary framework of the Nigerian sector of the Chad Basin [24,25,28,51] have been extensively investigated and properly documented. These investigations have indicated a relatively thick succession of sedimentary series overlying coexistent igneous intrusions within the horst/graben structures of the Bornu Basin [23]. The preliminary results triggered the drilling of twenty-three exploratory wells by the Nigerian National Petroleum Cooperation (NNPC) that revealed evidence of gas accumulation [52].

2. Geologic Setting of the Study Area

The investigated region is located at the Northeast Nigeria frontiers with the Republics of Chad, Niger, and Cameroon. It is situated between longitude 11°30′ E and 14°00′ E and latitude 12°00′ N and 14°00′ N.

The Bornu Basin (Figure 1) is characterized by an elevation ranging from 200–500 m above sea level [15]. It is often described as an interior sag basin [53] and is a portion of the Chad basin [25]. The Chad basin is composed of two coeval rift systems Central African Rift System (CARS) and West African Rift System (WARS) [54] that are physically disconnected but genetically related. The origin of the Central and West African Rift Systems is essentially attributed to the breakup of Gondwana and the opening of the South-Atlantic Ocean and the Indian Ocean at about 120–130 Ma [55].

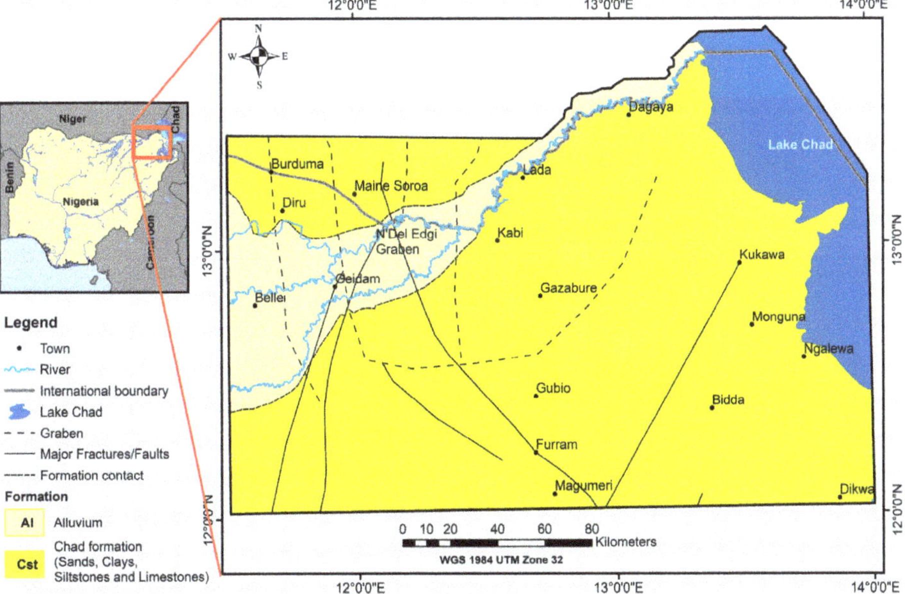

Figure 1. Geologic map of the study area.

The origin and tectonic structure of the Bornu Basin occurred in the evolution period of the WARS [25]. The regional structure and tectonic evolution of the Cretaceous to Recent rift basins of Niger, Chad, and the Central African Republic have been properly studied and well documented [25,55]. Geophysical and geological analyses of data indicate a complex sequence of Cretaceous grabens spanning from the Benue Trough to the southwest [23]. These datasets reveal near-surface intrusive bodies in the horst/graben structures as well as a relatively thick sedimentary section [25].

Refs. [50,56], and others reported the stratigraphic settings of the Southern part of the Chad basin (Figure 2). The Precambrian basement of the basin is directly overlain by continental, sparsely fossiliferous, poorly sorted, and medium to coarse-grained, feldspathic sandstones described as the Bima Formation (Sandstone). Overlying the Bima Formation is the transitional calcareous deposit called Gongila Formation [47,57]. It is composed mainly of sandstone and calcareous shale deposits [23]. This formation shows the beginning of marine incursion into the Chad Basin [58]. Marine transgression in the Albian got to its peak in the Turonian resulting in the deposition of ammonite-rich, bluish-black, open-marine Fika Shale Formation [58] which continued into Senonian. In the Maastrichtian, the regressive depositional Gombe Sandstone comprised of intercalations of siltstones, ironstones, and shales was deposited in a deltaic/estuarine environment.

AGE	FORMATION	LITHOLOGY	THICKNESS (m)	SEDIMENTTHICKNESS FROM SEISMIC DATA (m)	MEAN THICKNESS (m)	DEPOSITIONAL ENVIRONMENT
Pliocene, Pleistocene	Chad Formation	Clay, Sand	Not investigated	800	400	Continental
		(Unconformity)				
Palaeocene	Kerri-Kerri Formation	Coarse Sandstone, Clay stone, Sandstone	Not investigated		130	Continental
		(Unconformity)				
Maastrichtian	Gombe Sandstone	Shale, Sandstone, Siltstone,	Not investigated	0-1000	315	Deltaic Estuarine
Turonian-Santonian	Fika Shale	Blue-Black Shale	840-1453	0-900	430	Marine
Turonian	Gongilla Formation	Sandstone, Shale	162-420	0-800	420	Marine, Estuarine
Cenomanian	Bima Formation	Sandstone	716-850	2000	3050	Continental
		(Unconformity)				
Crystalline Basement						

Figure 2. The stratigraphic succession, average thicknesses of formations and thicknesses recorded in the studied wells in the Nigerian sector of the Chad Basin [47,57].

A phase of extensional deformation in the Chad basin occurred between the Late Maastrichtian to the end of the Cretaceous period. This extension deformation created an elongated graben system that trends in the Northeast-Southwest direction. The associated remnant basin that followed this tectonic deformation created a depositional site for the Tertiary Kerri-Kerri Formation that overlies the Cretaceous sedimentary series unconformably [57]. The continental (lacustrine) deposits of the Chad Formation were deposited unconformably over the Kerri-Keri Formation in the Pleistocene and perhaps, in the Pliocene period. The central and southern parts of the Bornu Basin have witnessed extensive volcanic activity towards the end of the Tertiary, and even recently [59]. Currently, dunes heap up in the Chad basin. The youngest deposits that blanket some parts of the south and southwestern flanks of Lake Chad are the river alluvium and deltaic and lagoonal clay flats [59].

3. Materials and Methods

3.1. Data Acquisition

In the middle of 2005 and 2010, Fugro-Airborne Surveys (FAS), Canada, under contract to the Central of Nigeria and supervised by the Nigerian-Geological-Survey Agency (NGSA) collected airborne magnetic data in Nigeria. The data were acquired with Flux-Adjusting Surface Data Assimilation System (FASDAS) with a terrain clearance of between 0.08–0.1 km, tie line spacing of 0.5 km, and flight-line spacing of 0.1 km along 826,000 lines. The flight lines and tie-lines were oriented in the northwest-southeast and northeast-southwest directions, respectively. The tie-lines orientation was carefully and deliberately designed to run across the major geological strike. The 10th generation International Geomagnetic Reference Field (IGRF) version 4.0 was applied by FAS on the compiled dataset to remove IGRF. In magnetic data reduction practice, IGRF, which is easily available and universally accepted, provides consistency in data reduction procedure [60]. Processed and corrected high-resolution aero-magnetic data are very appropriate for mineral evaluation programs, basin framework, hydrocarbon, and hydrogeological explorations, and delineation of the regional geologic boundary [26,61–67]. The airborne magnetic data employed in this investigation were corrected and processed to a total magnetic intensity map (Figure 3) which were gridded, saved, and displayed in Oasis Montaj Geosoft in colour raster format.

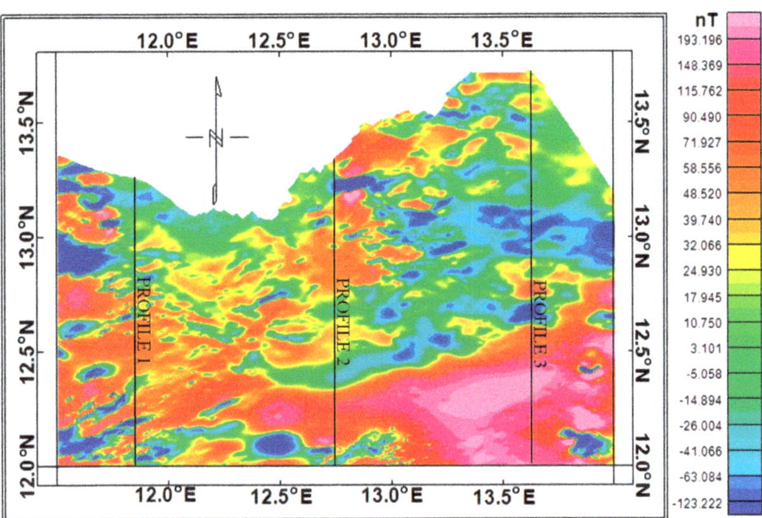

Figure 3. Total magnetic intensity map with the profile lines used for modelling.

3.2. Data Processing and Modelling

FAS, Canada carried out all the required airborne magnetic data filtering and correction processes. The local datum-transformation and projection method were applied to input magnetic data into World Geodetic System 84 (WGS 84) and Universal Transverse Mercator coordinate system at zone 32 of Northern hemisphere (UTM-32N), respectively. The data file was loaded into source parameter imaging, Euler deconvolution, MAGMAP, and 2-D GM-SYS extensions which generated control files for the various procedures. These processes were comprehensively applied to determine sedimentary thicknesses, define the basement framework, and map the location and the spatial distributions of the main intrusions in the investigated area. The source parameter imaging (SPI) technique [15] output is typically a map from which sediment thicknesses can be estimated [19]. This method evaluates the properties of the second vertical derivative and analytic signal responses [14,19,68]. Just like Euler deconvolution, depth approximation does not depend on any assumptions about the geologic model [19]. Furthermore, the geologic model can be determined correctly from the analysis, and the depth solutions are independent of the magnetic inclination and declination. Consequently, it is not necessary to apply a reduction-to-pole input grid [15]. The magnetic data interpretation procedure is made considerably easier with the availability of correct information on the local geology [15].

The approximations of depth using the SPI method are ordinarily from the wavelength of the analytic signal. Refs. [14,21,69] defined the analytic signal $A_1(x, z)$ as

$$A_1(x,z) = \frac{\partial M(x,z)}{\partial x} - j\frac{\partial M}{\partial z} \tag{1}$$

where, $i = \sqrt{-1}$, $M(x, z)$ is the magnitude of the anomalous total magnetic field, z and x are the Cartesian coordinates for the vertical and the horizontal directions perpendicular to the strike, respectively.

The standard-Euler deconvolution (St-ED) method relies on Euler's homogeneity equation:

$$(x - x_0)\frac{\partial T}{\partial x} + (y - y_0)\frac{\partial T}{\partial y} + (z - z_0)\frac{\partial T}{\partial Z} = N(B - T) \tag{2}$$

where (x_0, y_0, z_0) is the position of the magnetic field of which total field is observed at (x, y, z), while B is the value of the regional value of the total field. The degree of homogeneity N is interpreted as a structural index [16]. Unlike several other computer-aided procedures before it, St-ED does not adopt any particular geologic model. In addition, the method which can be applied directly to gridded data is interpreted even when the geology cannot be suitably described by dikes or prisms [69].

Two-dimensional forward modelling [70] involving the GM-SYS tool of Oasis Montaj [42] was used to evaluate depth to the basement and basin framework of the area. The GM-SYS profile is a program for computing magnetic and gravity responses from a cross-section of geologic models [71]. The application of algorithms described by [72] enabled the forward modeling procedure to create a hypothetical geologic model and compute the magnetic/gravity responses based on [73,74]. Generally, the subsurface is partitioned into two layers that are, top sedimentary series and basal basement rock. All points on the profile used for modelling in the GM-SYS platform have values obtained from total magnetic intensity, magnetic susceptibility, geographic coordinates, elevation, inclination, declination, and depth to the basement (generated from source parameter imaging database). The gridded data used for modelling were obtained along with three profiles, drawn mainly in the N-S direction on the total magnetic intensity gridded data (Figure 3). The profile placements were determined after carrying outsource parameter imaging and standard Euler operations. These processes were intended to make sure the profile lines were drawn across the depositional centers within the study area. All operations in this investigation were carried out applying codes obtainable in Oasis-Montaj version 7.0.1 (OL) (2008).

The analytic signal (AS) filter ([14,73,74] generates peak responses over magnetic anomalies. This technique is usually applied at low magnetic latitude because of the in-built problem associated with the reduction-to-pole filter. References [14,20,68] indicated that the amplitude of the AS can be obtained from the three orthogonal derivatives of the magnetic field as:

$$\left| ASIG_{(X,Y)} \right| = \sqrt{\left(\frac{\partial A}{\partial x}\right)^2 + \left(\frac{\partial A}{\partial y}\right)^2 + \left(\frac{\partial A}{\partial z}\right)^2} \tag{3}$$

where, A is the observed magnetic field.

Upward-continuation (UP-C) filter is used in evaluating the regional magnetic structures emanating from deeply buried magnetic field sources. The equation of the wavenumber domain filter to generate upward continuation [67] is essential:

$$F(\omega) = e^{-hw} \tag{4}$$

where, h is the continuation height. This function drops progressively with increasing wave-number, decreasing the higher wavenumbers more severely, and so creating a map in which more regional anomalies dominate [60].

4. Results

The SPI and St-ED methods are data enhancement processes for evaluating the positions and depth of magnetic bodies [11]. These methods are suitable for delineating isolated and multiple magnetic source geometries [67], and susceptibility disparity [26]. Wide-ranging colours (pink-blue) showing various depths and locations of different magnetic bodies within the subsurface are displayed by the SPI and St-ED gridded maps (Figure 4). The colour legend bar is described by a negative sign indicating depth measurement from the Earth's surface downward [15]. The SPI (Figure 4a) indicates shallow (red-pink), intermediate (yellow-red) and deep (lemon green-blue) depth ranges of 286 to 615 m, 695 to 1038 m, and 1145 to 5885 m, respectively. Similarly, the standard Euler deconvolution (Figure 4a) reveals the depth to shallow (130 to 917 m), intermediate (1044 to 1572 m), and deep (1725 to 5974 m) magnetic sources characterized by red-pink, yellow-red and lemon green-blue, respectively. The wide range of depth to deep magnetic sources obtained from Figure 4 explains the undulant nature of the underlying basement surface. From the results (Figure 4), zones described by yellow-pink colour (intermediate-high magnetization) are recognized to be localized residual magnetizations associated with ferruginous sediments, horst/graben structures, near-surface igneous intrusions, and related baked sediments of the area [23]. In the late Tertiary, and even in recent times [59], reported extensive volcanic activities in the southern and central parts of the basin. Reference [58] described the area to be blanketed by sand dunes, river alluvium, and deltaic and clay sediments.

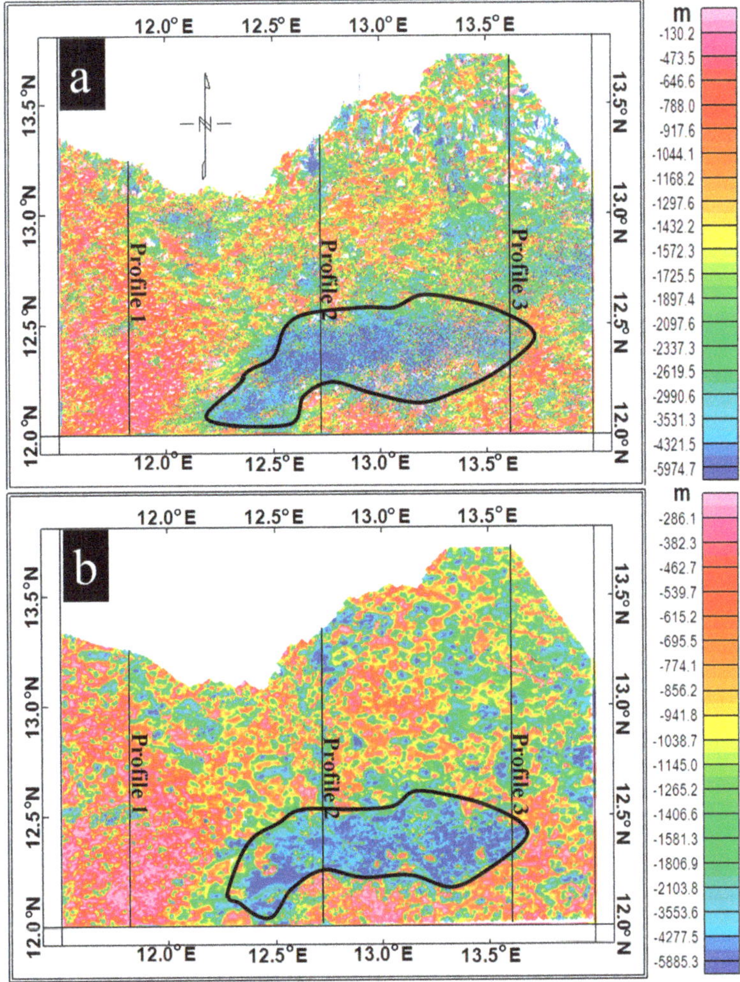

Figure 4. (**a**) Standard Euler deconvolution (structural index = 1.0; max. % depth tolerance = 15.0, window size = 10) and (**b**) source parameter imaging maps.

The magnetic signature of Profile 1 (Figure 5), which runs across part of the western flank of the study area in the N-S direction (Figure 3), is characterized by two major breaks around the middle and southern parts of the model. These breaks signify weak zones within the Precambrian basement that created openings for igneous intrusions. These zones of magnetic signature disorders show the occurrence of magnetic anomalies previously revealed by the depth determination techniques (Figure 4). The prevalence of magnetic structures observed along profile 1 indicate the widespread invasion of the Cretaceous-Recent sediments and underlying basement rocks by basaltic lavas, mafic and felsic intrusives [23,25,55,59], whose sizes range from dyke-like structures to massive granitic structures that may perhaps link to generate massive structures [11,28]. In general, Profile 1 model shows that the igneous intrusions and basement rocks are blanketed by sedimentary series with thickness varying from ~300 to <3500 m.

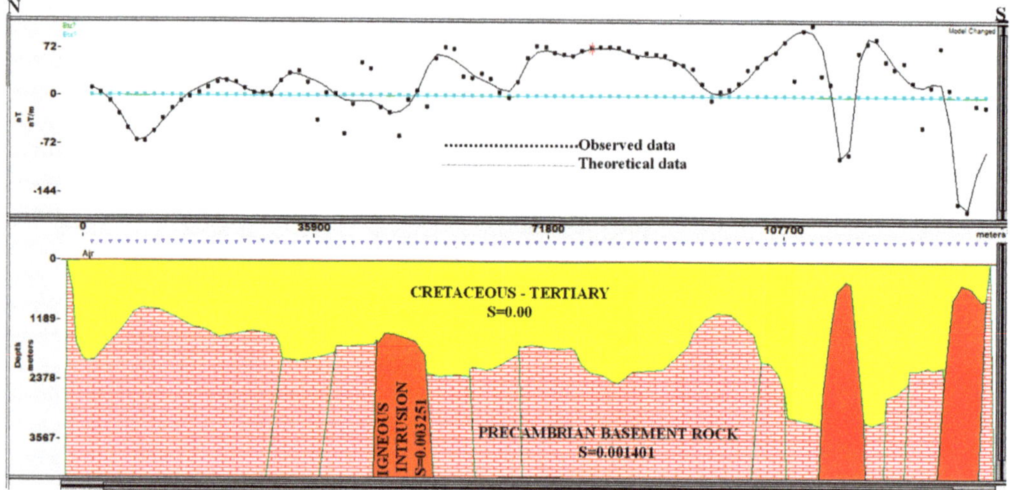

Figure 5. 2-D total magnetic intensity model obtained at Profile 1 showing thickness of sediments, basement and igneous rocks, and their respective susceptibility values.

Furthermore, the model obtained at the centre that runs in the N-S direction (Profile 2, Figure 6) is defined by having one serration towards the northern part of the magnetic signature. This is an indication of magnetic intrusion that penetrated the weak zone within the basement and overlying sediments. The southern end of the model is characterized by an anomalous depth [25] value of about 5000 m. This region coincides with the thick sediments zone previously circumscribed by a black polygon in Figure 4.

Profile 3 is situated at the eastern flank runs in the N-S direction (Figure 3). It is characterized by jagged and smooth magnetic signature at the southern and northern ends, respectively (Figure 7). The serrated pattern of the curve at the southern part shows multiple block faults within the basement and overlying sediments caused by tectonic events [25,49,55]. In general, the model shows sediments cover of about 2500 and 4500 m in the northern and southern flanks of the profile, respectively. The region with thick sediments falls under the southern flank of the model. This anomalous depth region that runs east-west coincides with the locality of the thick sedimentation revealed by Figures 4 and 6.

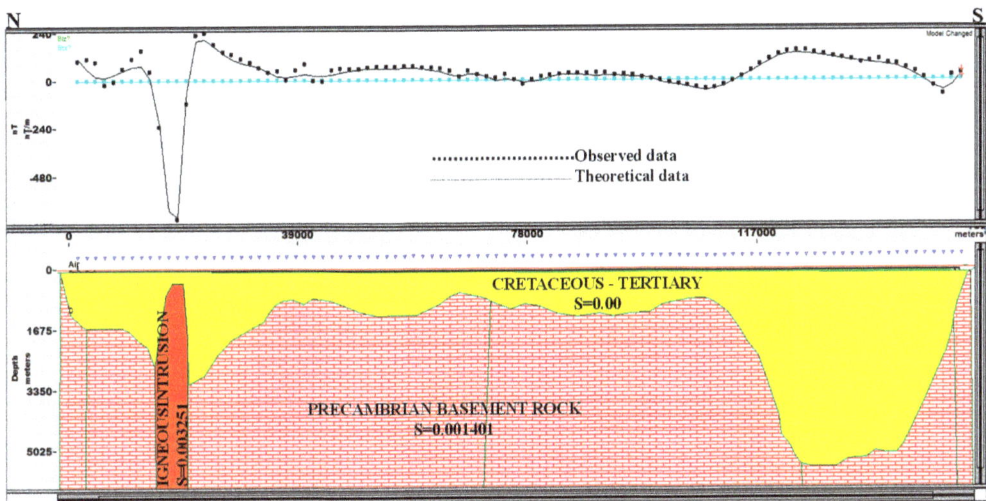

Figure 6. 2-D total magnetic intensity model obtained at Profile 2 showing thickness of sediments, basement and igneous rocks, and their respective susceptibility values.

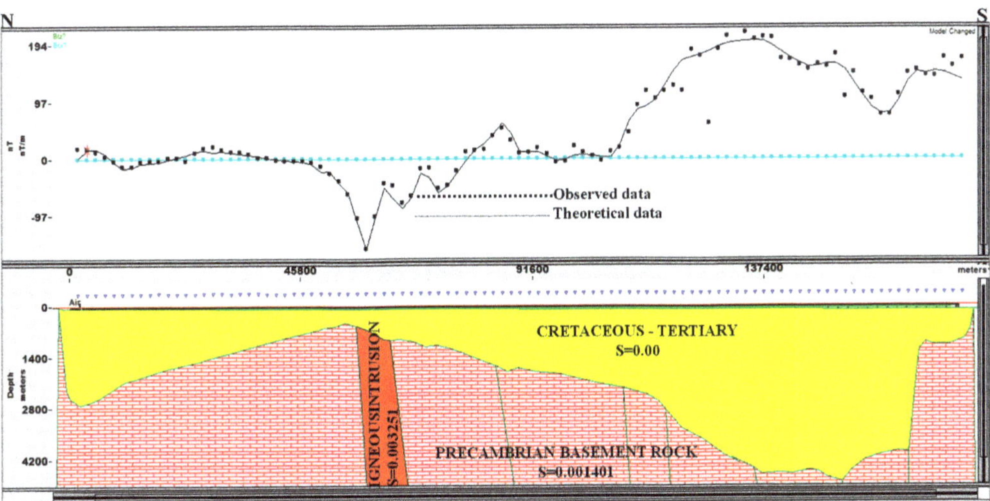

Figure 7. 2-D total magnetic intensity model obtained at Profile 3 showing thickness of sediments, basement and igneous rocks, and their respective susceptibility values.

Analytic signal (AS) filter [21] creates peak responses over distinctive magnetic bodies. The amplitude of magnetization mostly generated by the AS is independent of the direction of the magnetic body [21]. Figure 8a outlined low (0.003421–0.008461 nT/m), intermediate (0.009898–0.017579 nT/m) and high (0.019647–0.093958 nT/m) magnetizations described by blue, lemon green-yellow and red-pink colours correspondingly. The delineated regions of high magnetizations (red-pink colours) labeled A, B and C correlate with the sites of near-surface igneous intrusions (Figure 8a) occurring alongside the horst/graben structures in the Bornu Basin described by [23,25]. Similarly, Figure 8a delineated the U-like structure dominated by low-intermediate magnetizations (blue-yellow colours).

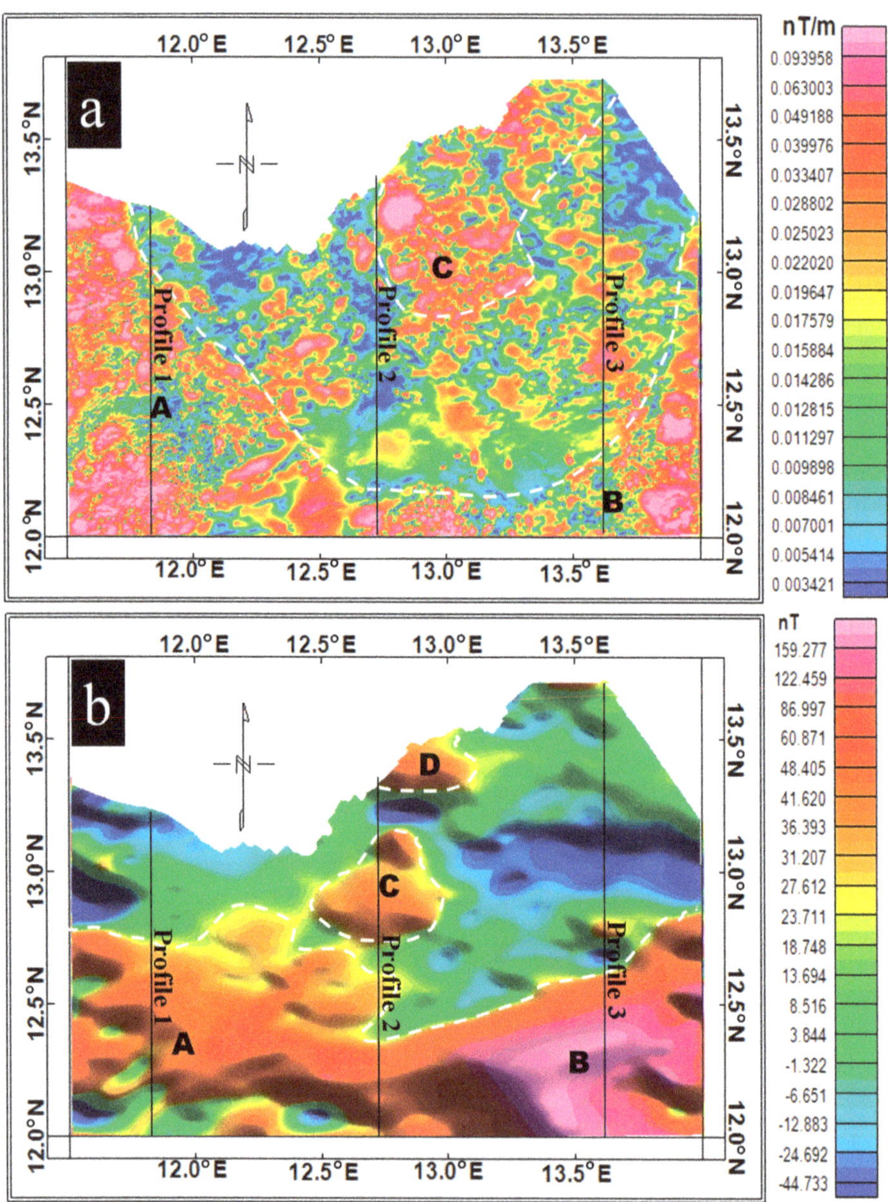

Figure 8. (**a**) Analytic signal and (**b**) total magnetic intensity data upward continued 5000 m maps.

To reveal the main igneous intrusions that caused near-surface magnetizations, geologic structures, baking of basement and overlying sedimentary materials, the upward continuation filter [20,75] was applied to the magnetic data. The data were upward continued to 5000 m (Figure 8b) to eliminate magnetic effects emanating from high-frequency magnetic bodies. Figure 8b displays a deeply seated ridge-like intrusive structure (labeled as A-B dominated by red-pink colours) that straddles the southern part of the study area in the E-W direction. Furthermore, Figure 8b showed igneous intrusions in the central and the northern parts of the investigated area labeled as C and D, respectively. The location of these igneous validates the findings of [59]. Furthermore, A, B, C, and D igneous intrusions match with the anticlinal structures of the study area that are sandwiched by synclines (well-defined by lemon green-blue colours). Additionally, Figure 8b showed a distinct deeply buried E-W weak zone well defined by blue colour at the eastern flank of the area. Similar features observed at the northwestern flank correlate with the location of N'Dgel Edgi graben that extends into Nigeria from the Niger Republic [24]. These regions characterized by intermediate magnetizations (Figure 8a) are dominated by intermediate-thick sedimentations (Figure 4).

5. Discussion

The Bornu Basin, which is part of the Upper Benue Trough [45], is described by [53] as an interior sag basin that is genetically interrelated to the Central African Rift System and West African Rift System [54]. The area is characterized by sedimentary thicknesses in the range of about 130 to 5974 m from SPI, St-ED, and 2-D modeling results (Figures 4–7). The respective peak depth values of 5885 m, 5974 m, and 5500 m revealed from the various depth determination methods used, strongly agree with each other. The depth results match relatively well with the depth estimation results obtained by previous researchers in the Bornu Basin [25,30,32,33,49]. Similarly, studies carried out in the genetically and structurally connected Lower Benue Trough, involving potential field data and various depth estimation methods [11,26,27,31,34,59,76] show strong correlation with depth solutions of the Bornu Basin. However, [29,47] reported contrary sedimentary thicknesses of about 10,000 m and 9000 m, in the Bornu Basin and Lower Benue Trough, respectively.

Additionally, an anomalous sedimentary thickness (Figure 4) represented by a polygon sits on igneous intrusions revealed by the UP-C magnetic map (Figure 8b). Documented geophysical and geological works revealed that the investigated area is dominated by horst/graben structures, block faults, and associated igneous intrusions [24,25,55]. However, the findings of [23] and others indicated that these structures are blanketed by somewhat thick sediments. Their reports motivated the Nigerian National Petroleum Corporation (NNPC) to drill some exploratory wells within the area characterized by thick sedimentation. The exploration exercise revealed evidence of gas accumulation [52] which may perhaps be caused by igneous intrusions [56,58] and related enhanced geothermal gradient. The 2-D models with sediment thicknesses in the range of ~300 to ~5500 m (Figures 5–7) obtained from the three profiles (Figure 3) are characterized by jagged magnetic signatures. The serrated pattern of the curves indicates multiple block faults within the basement, and widespread invasion of igneous intrusions into overlying Cretaceous sediments [23,25,55,59]. These igneous invasions were detected by SPI and St-ED (Figure 4), as well as analytic signal and upward continuation enhancements (Figure 8). Geologic structures caused by intrusions in such regions are the potential pathway for hydrothermal fluid migration and mineralization in rift environments [18,34,64,77–79] like the Bornu Basin. However, Figure 8b revealed E-W anticlinal structures (red-pink colour) adjoined by a trough-like structure (lemon green-blue colour) located at the southern and northern parts of the study area, respectively. The northern part is characterized by two isolated domal high magnetization structures and E-W weak zone (blue colour) separating them. The western end of this weak zone runs into N'Dgel Edgi graben, Niger Republic [25], where commercial hydrocarbon is exploited. Generally, the northern flank of the study area is characterized by intermediate depths (Figure 5) and magnetizations (Figure 8). Hence, further oil and gas surveys involving drilling of exploratory wells and seismic reflection techniques should be shifted to this area defined by insignificant relics of post-Cretaceous tectonic events [26].

6. Conclusions

Modern high-resolution aero-magnetic data measured between 2005 and 2010 were used to infer thicknesses of the sedimentary sections of Northeastern Nigeria. Various depth estimation methods like source parameter imaging (SPI), Standard-Euler deconvolution (St-ED), and 2D GM-SYS forward modelling were used in this investigation. It was generally observed from the various results obtained that the southern part of the study area is dominated by thick sedimentary series coexistent with igneous intrusions. The SPI result revealed 286 to 615 m, 695 to 1038 m, and 1145 to 5885 m for thin, intermediate, and thick sedimentation, respectively. Likewise, the St-ED result indicated values of 130 to 917 m, 1044 to 1572 m, and 1725 to 5974 m for thin, intermediate, and thick sedimentation correspondingly. A 2D forward model of Profile 1 is described by two major breaks, indicating that the igneous intrusions and basement rocks are blanketed by sediments with thicknesses ranging from about 300 to <3500 m while the 2D model of Profile 2 has an approximated maximum depth value of about 5000 m at the southern flank. Furthermore, the Profile 3 model shows sedimentary thickness in the range of 2500 and 4500 m in the northern and southern ends of the profile, respectively. The anomalous depth region detected by the 2D forward models matches with the location of the thick sedimentary cover identified by the SPI and St-ED techniques. The maximum depth values obtained from the different depth approximation procedures employed in this investigation matched closely with each other. In addition, the forward models revealed that the underlying basement framework is undulant, and the weak zones intruded by igneous rocks. Further hydrocarbon investigations involving the seismic reflection method and exploratory wells should be shifted to the northern flank characerized by shallow-intermediate depths and magnetizations. On the whole, the extensive occurrence of igneous intrusions in the southern part of the study area have caused severe fracturing and faulting of the basement, and overlying sediments. Related geologic structures caused by tectonic events serve as a potential pathway for hydrothermal fluid migration and mineralization accumulation. High resolution ground gravity and electromagnetic studies should be employed in the exploration of igneous rock related minerals.

Author Contributions: Conceptualization, S.E.E. and O.-I.M.A.; methodology, O.-I.M.A.; software, S.E.E.; validation, A.E.A., S.E.E. and C.H.U.; formal analysis, C.H.U.; investigation, A.E.A.; resources, S.E.E.; data curation, C.H.U.; writing—original draft preparation, S.E.E., O.-I.M.A. and A.M.E.; writing—review and editing, S.E.E., O.-I.M.A. and A.M.E.; visualization, S.E.E., O.-I.M.A. and A.M.E.; supervision, S.E.E., A.E.A., D.G.-O. and A.M.E.; project administration, S.E.E.; funding acquisition, K.A. All authors have read and agreed to the published version of the manuscript.

Funding: This research was funded by [King Saud University, Riyadh, Saudi Arabia] grant number [RSP-2022/351] And The APC was funded by [Researchers Supporting Project number (RSP-2022/351), King Saud University, Riyadh, Saudi Arabia].

Data Availability Statement: Data available on request from the authors.

Acknowledgments: Deep thanks and gratitude to the Researchers Supporting Project number (RSP-2022/351), King Saud University, Riyadh, Saudi Arabia for funding this research article.

Conflicts of Interest: There are no potential conflict exist in the paper. There is no declaration of interest.

References

1. Koné, A.Y.; Nasr, I.H.; Traoré, B.; Amiri, A.; Inoubli, M.H.; Sangaré, S.; Qaysi, S. Geophysical Contributions to Gold Exploration in Western Mali According to Airborne Electromagnetic Data Interpretations. *Minerals* **2021**, *11*, 126. [CrossRef]
2. Fu, J.; Jia, S.; Wang, E. Combined Magnetic, Transient Electromagnetic, and Magnetotelluric Methods to Detect a BIF-Type Concealed Iron Ore Body: A Case Study in Gongchangling Iron Ore Concentration Area, Southern Liaoning Province, China. *Minerals* **2020**, *10*, 1044. [CrossRef]
3. Mehanee, S.; Zhdanov, M. A quasi-analytical boundary condition for three-dimensional finite difference electromagnetic modeling. *Radio Sci.* **2004**, *39*, RS6014. [CrossRef]
4. Mehanee, S.; Golubev, N.; Zhdanov, M.S. Weighted regularized inversion of magnetotelluric data. In *SEG Technical Program Expanded Abstracts 1998*; Society of Exploration Geophysicists: Tulsa, OK, USA, 1998; pp. 481–484. [CrossRef]

5. Biswas, A.; Mandal, A.; Sharma, S.P.; Mohanty, W.K. Delineation of subsurface structures using self-potential, gravity, and resistivity surveys from South Purulia Shear Zone, India: Implication to uranium mineralization. *Interpretation* **2014**, *2*, T103–T110. [CrossRef]
6. Biswas, A.; Rao, K. Interpretation of Magnetic Anomalies over 2D Fault and Sheet-Type Mineralized Structures Using Very Fast Simulated Annealing Global Optimization: An Understanding of Uncertainty and Geological Implications. *Lithosphere* **2021**, *2021*, 2964057. [CrossRef]
7. Essa, K.S.; Mehanee, S.; Elhussein, M. Magnetic Data Profiles Interpretation for Mineralized Buried Structures Identification Applying the Variance Analysis Method. *Pure Appl. Geophys.* **2020**, *178*, 973–993. [CrossRef]
8. Gan, J.; Li, H.; He, Z.; Gan, Y.; Mu, J.; Liu, H.; Wang, L. Application and Significance of Geological, Geochemical, and Geophysical Methods in the Nanpo Gold Field in Laos. *Minerals* **2022**, *12*, 96. [CrossRef]
9. Liu, Y.; Na, X.; Yin, C.; Su, Y.; Sun, S.; Zhang, B.; Ren, X.; Baranwal, V.C. 3D joint inversion of airborne electromagnetic and magnetic data based on local Pearson correlation constraints. *IEEE Trans. Geosci. Remote Sens.* **2022**. [CrossRef]
10. Mehanee, S.A. A new scheme for gravity data interpretation by a faulted 2-D horizontal thin block: Theory, numerical examples and real data investigation. *IEEE Trans. Geosci. Remote Sens.* **2022**. [CrossRef]
11. Ekwok, S.E.; Akpan, A.E.; Ebong, E.D.; Eze, O.E. Assessment of depth to magnetic sources using high resolution aero-magnetic data of some parts of the Lower Benue Trough and adjoining areas, Southeast Nigeria. *Adv. Space Res.* **2021**, *67*, 2104–2119. [CrossRef]
12. Salem, A.; Ravat, D. A combined analytic signal and Euler method (AN-EUL) for automatic interpretation of mag-netic data. *Geophysics* **2003**, *68*, 1952–1961. [CrossRef]
13. Davis, K.; Li, Y. Enhancement of depth estimation techniques with amplitude analysis. In *SEG Technical Program Expanded Abstracts 2009*; Society of Exploration Geophysicists: Tulsa, OK, USA, 2009; pp. 908–912. [CrossRef]
14. Mehanee, S.; Essa, K.S.; Diab, Z.E. Magnetic Data Interpretation Using a New R-Parameter Imaging Method with Application to Mineral Exploration. *Nonrenew. Resour.* **2020**, *30*, 77–95. [CrossRef]
15. Thurston, J.B.; Smith, R.S. Automatic conversion of magnetic data to depth, dip, and susceptibility contrast using SPITM method. *Geophysics* **1997**, *62*, 807–813. [CrossRef]
16. Reid, A.B.; Allsop, J.M.; Granser, H.; Millett, A.J.; Somerton, I.W. Magnetic interpretation in three dimensions using Euler deconvolution. *Geophysics* **1990**, *55*, 80–91. [CrossRef]
17. Northwest Geophysical Associates (NGA), Inc. GM–SYS Gravity/Magnetic Modeling Software, User's Guide, Version 4.9. 2004. pp. 1–101. Available online: http://pages.geo.wvu.edu/~wilson/gmsys_49.pdf (accessed on 9 January 2022).
18. Elkhateeb, S.O.; Eldosouky, A.M. Detection of porphyry intrusions using analytic signal (AS), Euler Deconvolution, and Centre for Exploration Targeting (CET) Technique Porphyry Analysis at Wadi Allaqi Area, South Eastern Desert, Egypt. *Int. J. Eng. Res.* **2016**, *7*, 471–477.
19. Smith, R.S.; Thurston, J.B.; Dai, T.; MacLeod, I.N. ISPITM—The improved source parameter imaging method. *Geophys. Prospect.* **1998**, *46*, 141–151. [CrossRef]
20. Roest, W.; Verhoef, J.; Pilkington, M. Magnetic interpretation using the 3-D analytic signal. *Geophysics* **1992**, *57*, 116–125. [CrossRef]
21. Nabighian, M.N. The analytical signal of two dimensional magnetic bodies with polygon cross-section: Its proper-ties and use for automated anomaly interpretation. *Geophysics* **1972**, *37*, 507–517. [CrossRef]
22. Nabighian, M.N. Towards the three-dimensional automatic interpretation of potential field data via generalized Hilbert transforms. Fundamental relations. *Geophysics* **1984**, *53*, 957–966. [CrossRef]
23. Nwankwo, C.N.; Emujakporue, G.O.; Nwosu, L.I. Evaluation of the petroleum potentials and prospect of the Chad Basin Nigeria from heat flow and gravity data. *J. Pet. Explor. Prod. Technol.* **2011**, *2*, 1–6. [CrossRef]
24. Genik, G.J. Regional framework structure and petroleum aspects of the rift basins in Niger, Chad and Central African Republic (C.A.R.). *Tectonophysics* **1992**, *213*, 169–185. [CrossRef]
25. Genik, G.J. Petroleum geology of Cretaceous to Tertiary rift basins in Niger, Chad and the Central African Republic. *Bull. Am. Assoc. Pet. Geol.* **1993**, *77*, 1405–1434.
26. Ekwok, S.E.; Akpan, A.E.; Ebong, D.E. Enhancement and modelling of aeromagnetic data of some inland basins, southeastern Nigeria. *J. Afr. Earth Sci.* **2019**, *155*, 43–53. [CrossRef]
27. Ekwok, S.E.; Akpan, A.E.; Ebong, E.D. Assessment of crustal structures by gravity and magnetic methods in the Calabar Flank and adjoining areas of Southeastern Nigeria—A case study. *Arab. J. Geosci.* **2021**, *14*, 308. [CrossRef]
28. Ofoegbu, C.O. A model for the tectonic evolution of the Benue Trough of Nigeria. *Geol. Rundsch.* **1984**, *73*, 1007–1018. [CrossRef]
29. Abdullahi, M.; Kumar, R.; Singh, U.K. Magnetic basement depth from high-resolution aeromagnetic data of parts of lower and middle Benue Trough (Nigeria) using scaling spectral method. *J. Afr. Earth Sci.* **2018**, *150*, 337–345. [CrossRef]
30. Ola, P.S.; Adekoya, J.A.; Olabode, S.O. Source Rock Evaluation in the Lake Chad area of the Bornu Basin, Nigeria. *J. Pet. Environ. Biotechnol.* **2017**, *8*, 346.
31. Oha, I.A.; Onuoha, K.M.; Nwegbu, A.N.; Abba, A.U. Interpretation of high resolution aeromagnetic data over southern Benue Trough, southeastern Nigeria. *J. Earth Syst. Sci.* **2016**, *125*, 369–385. [CrossRef]
32. Olabode, S.O.; Adekoya, J.A.; Ola, P.S. Distribution of sedimentary formations in the Bornu Basin, Nigeria. *Pet. Explor. Dev.* **2015**, *42*, 674–682. [CrossRef]

33. Umar, T. Geologie Petroliere du Secteur Nigerian du Basin du Lac Tchad. Unpublished Ph.D. Dissertation, Universite de Pau et des Pays de l'Adour, Centre Universitaire de Recherche Scientifique Laboratoire de Geodynamique et Modelisation des Bassins Sedimentaires, Pau, France, 1999; 425p. Available online: https://www.theses.fr/1999PAUU3013 (accessed on 9 January 2022).

34. Ekwok, S.E.; Akpan, A.E.; Kudamnya, E.A. Exploratory mapping of structures controlling mineralization in South-east Nigeria using high resolution airborne magnetic data. *J. Afr. Earth Sci.* **2020**, *162*, 103700. [CrossRef]

35. Eldosouky, A.; Sehsah, H.; Elkhateeb, S.O.; Pour, A.B. Integrating aeromagnetic data and Landsat-8 imagery for detection of post-accretionary shear zones controlling hydrothermal alterations: The Allaqi-Heiani Suture zone, South Eastern Desert, Egypt. *Adv. Space Res.* **2019**, *65*, 1008–1024. [CrossRef]

36. Eldosouky, A.M.; Mohamed, H. Edge detection of aeromagnetic data as effective tools for structural imaging at Shilman area, South Eastern Desert, Egypt. *Arab. J. Geosci.* **2021**, *14*, 13. [CrossRef]

37. Elkhateeb, S.O.; Eldosouky, A.M.; Khalifa, M.O.; Aboalhassan, M. Probability of mineral occurrence in the Southeast of Aswan area, Egypt, from the analysis of aeromagnetic data. *Arab. J. Geosci.* **2021**, *14*, 1514. [CrossRef]

38. Eldosouky, A.M.; El-Qassas, R.A.; Pour, A.B.; Mohamed, H.; Sekandari, M. Integration of ASTER satellite imagery and 3D inversion of aeromagnetic data for deep mineral exploration. *Adv. Space Res.* **2021**, *68*, 3641–3662. [CrossRef]

39. Melouah, O.; Eldosouky, A.M.; Ebong, E.D. Crustal architecture, heat transfer modes and geothermal energy potentials of the Algerian Triassic provinces. *Geothermics* **2021**, *96*, 102211. [CrossRef]

40. Eldosouky, A.M.; Elkhateeb, S.O. Texture analysis of aeromagnetic data for enhancing geologic features using co-occurrence matrices in Elallaqi area, South Eastern Desert of Egypt. *NRIAG J. Astron. Geophys.* **2018**, *7*, 155–161. [CrossRef]

41. Essa, K.S.; Elhussein, M. PSO (particle swarm optimization) for interpretation of magnetic anomalies caused by sim-ple geometri-cal structures. *Pure Appl. Geophys.* **2018**, *175*, 3539–3553. [CrossRef]

42. Pallero, J.L.G.; Fernández-Martínez, J.L.; Bonvalot, S.; Fudym, O. Gravity inversion and uncertainty assessment of basement relief via Particle Swarm Optimization. *J. Appl. Geophys.* **2015**, *116*, 180–191. [CrossRef]

43. Ekinci, Y.L.; Balkaya, Ç.; Göktürkler, G.; Özyalın, Ş. Gravity data inversion for the basement relief delineation through global optimization: A case study from the Aegean Graben System, western Anatolia, Turkey. *Geophys. J. Int.* **2020**, *224*, 923–944. [CrossRef]

44. Essa, K.S.; Elhussein, M. A new approach for the interpretation of magnetic data by a 2-D dipping dike. *J. Appl. Geophys.* **2017**, *136*, 431–443. [CrossRef]

45. Okosun, E.A. Review of the geology of Bornu Basin. *J. Min. Geol.* **1995**, *31*, 113–122.

46. Nwazeapu, A.U. Hydrocarbon Exploration in Frontier Basin: The Nigeria Chad Basin Experience. In Proceedings of the 28th Annual Conference of the Nigerian Mining and Geosciences Society, Port Harcourt, Nigeria, 1–5 March 1992.

47. Avbovbo, A.A.; Ayoola, E.O.; Osahon, G.A. Depositional and structural styles in Chad Basin of north-eastem Nigeria. *AAPG Bull.* **1986**, *70*, 1787–1798.

48. Okosun, E. Cretaceous ostracod biostratigraphy from Chad Basin in Nigeria. *J. Afr. Earth Sci.* **1992**, *14*, 327–339. [CrossRef]

49. Nwankwo, C.N.; Ekine, A.S. Geothermal Gradients in the Chad Basin, Nigeria, from bottom hole temperature logs. *Int. J. Phys. Sci.* **2009**, *4*, 777–783.

50. Okosun, E.A. A preliminary assessment of the petroleum potentials from southwest Chad Basin (Nigeria). *Borno J. Geol.* **2000**, *2*, 40–50.

51. Nwachukwu, S.O. The tectonic evolution of the the southern portion of the Benue Trough, Nigeria. *Geol. Mag.* **1972**, *109*, 411–419. [CrossRef]

52. Moumouni, A.; Obaje, N.G.; Nzegbuna, A.I.; Chanda, M.S. Bulk geochemical parameters and biomarker char-acteristics of organic matter in two wells (Gaibu-1 and Kasade-1) from Bornu Basin: Implications on hydrocarbon poten-tials. *J. Pet Geoscis. Eng.* **2007**, *58*, 275–282. [CrossRef]

53. Kingston, D.R.; Dishroon, C.P.; Williams, P.A. Hydrocarbon plays and global basin classification. *AAPG Bull* **1983**, *67*, 2194–2197.

54. Hamza, H.; Hamidu, I. Hydrocarbon resource potential of the Bornu Basin Northeastern Nigerian. *Glob. J. Geol. Sci.* **2011**, *10*, 71–84.

55. Fairhead, J.; Green, C. Controls on rifting in Africa and the regional tectonic model for the Nigeria and East Niger rift basins. *J. Afr. Earth Sci.* **1989**, *8*, 231–249. [CrossRef]

56. Petters, S.W.; Ekweozor, C.M. Petroleum geology of Benue Trough and southeastern Chad Basin Nigeria. *AAPG Bull* **1982**, *66*, 1141–1149.

57. Carter, J.D.; Barber, W.; Jones, G.P. The geology of parts of Adamawa, Bauchi and Bornu provinces in north-eastem Nigeria. *Bull. Geol. Sum. Nigeria* **1963**, *30*, 109.

58. Olugbemiro, R.O.; Ligouis, B.; Abaa, S.I. The Cretaceous series in the Chad Basin, NE Nigeria source rock poten-tial and thermal maturity. *J. Pet. Geol.* **1997**, *20*, 51–58. [CrossRef]

59. Burke, K. The chad basin: An active intra-continental basin. *Tectonophysics* **1976**, *36*, 197–206. [CrossRef]

60. Reeves, C.; Reford, S.; Millingan, P. Airborne Geophysics: Old Methods, New Images. In *Proceedings of the Fourth Decennial International Conference on Mineral Exploration*; Prospectors and Developers Associaton: Toronto, ON, Canada, 1997; pp. 13–30. Available online: https://www.worldcat.org/title/geophysics-and-geochemistry-at-the-millenium-proceedings-of-exploration-97-fourth-decennial-international-conference-on-mineral-exploration/oclc/40988105 (accessed on 9 January 2022).

61. United States Geological Survey (USGS). Setting and Origin of Iron Oxide Copper- Cobalt-Gold-Rare Earth Element Deposits of Southeast Missouri. 2013. Available online: http://minerals.usgs.gov/east/semissouri/index.html (accessed on 9 January 2022).

62. Eldosouky, A.M.; Abdelkareem, M.; Elkhateeb, S.O. Integration of remote sensing and aeromagnetic data for map-ping structural features and hydrothermal alteration zones in Wadi Allaqi area, South Eastern Desert of Egypt. *J. Afr. Earth Sci.* **2017**, *130*, 28–37. [CrossRef]
63. Shebl, A.; Abdellatif, M.; Elkhateeb, S.; Csámer, Á. Multisource Data Analysis for Gold Potentiality Mapping of Atalla Area and Its Environs, Central Eastern Desert, Egypt. *Minerals* **2021**, *11*, 641. [CrossRef]
64. Ekwok, S.E.; Akpan, A.E.; Achadu, O.-I.M.; Eze, O.E. Structural and lithological interpretation of aero-geophysical data in parts of the Lower Benue Trough and Obudu Plateau, southeast Nigeria. *Adv. Space Res.* **2021**, *68*, 2841–2854. [CrossRef]
65. Ekwok, S.E.; Akpan, A.E.; Kudamnya, E.A.; Ebong, D.E. Assessment of groundwater potential using geophysical data: A case study in parts of Cross River State, south-eastern Nigeria. *Appl. Water Sci.* **2020**, *10*, 144. [CrossRef]
66. Kearey, P.; Brooks, M.; Hill, I. *An Introduction to Geophysical Exploration*, 3rd ed.; Blackwell Science Ltd. Editorial Offices: New York, NY, USA, 2002.
67. Telford, W.M.; Geldart, L.P.; Sheriff, R.E. *Applied Geophysics*, 2nd ed.; Cambridge University Press: Cambridge, UK, 1990.
68. Casto, D.W. *Calculating Depths to Shallow Magnetic Sources Using Aeromagnetic Data from the Tucson Basin*; US Department of the Interior: Washington, DC, USA, 2001. [CrossRef]
69. Essa, K.S.; Mehanee, S.A.; Soliman, K.S.; Diab, Z.E. Gravity profile interpretation using the R-parameter imaging technique with application to ore exploration. *Ore Geol. Rev.* **2020**, *126*, 103695. [CrossRef]
70. Santos, D.F.; Silva, J.B.C.; Lopes, J.F. In-depth 3D magnetic inversion of basement relief. *Geophysics* **2022**, *87*, 1–72. [CrossRef]
71. Mehanee, S.A. Simultaneous Joint Inversion of Gravity and Self-Potential Data Measured Along Profile: Theory, Numerical Examples, and a Case Study from Mineral Exploration with Cross Validation From Electromagnetic Data. *IEEE Trans. Geosci. Remote Sens.* **2021**, *60*, 1–20. [CrossRef]
72. Won, I.J.; Bevis, M. Computing the gravitational and magnetic anomalies due to a polygon: Algorithms and Fortran subroutines. *Geophysics* **1987**, *52*, 232–238. [CrossRef]
73. Talwani, M.; Hiertzler, J.R. Computation of magnetic anomalies caused by two dimensional bodes of arbitrary shape. *Geol. Sci.* **1964**, *9*, 464–480.
74. Talwani, M.; Worzel, J.L.; Landisman, M. Rapid gravity computations for two-dimensional bodies with application to the Mendocino submarine fracture zone. *J. Geophys. Res. Earth Surf.* **1959**, *64*, 49–59. [CrossRef]
75. Gunn, P. Regional magnetic and gravity responses of extensional sedimentary Basins. *AGSO J. Aust. Geol. Geophys.* **1997**, *17*, 115–131.
76. Benkhelil, M.G.; Ponsard, J.F.; Saugy, L. The Bornu-Benue Trough, the Niger Delta and its Offshore: Tectonosedimentary reconstruction during the cretaceous and tertiary from geophysical data and geology. In *Geology of Nigeria*; Kogbe, C.A., Ed.; Elizabethan Press: Lagos, Nigeria, 1975; pp. 277–309.
77. Ekwok, S.E.; Akpan, A.E.; Achadu, O.-I.M.; Thompson, C.E.; Eldosouky, A.M.; Abdelrahman, K.; Andráš, P. Towards Understanding the Source of Brine Mineralization in Southeast Nigeria: Evidence from High-Resolution Airborne Magnetic and Gravity Data. *Minerals* **2022**, *12*, 146. [CrossRef]
78. Ekwok, S.E.; Akpan, A.E.; Achadu, O.I.M.; Ulem, C.A. Implications of tectonic anomalies from potential feld data in some parts of Southeast Nigeria. *Environ. Earth Sci.* **2022**, *81*, 6. [CrossRef]
79. Mineral Resources of the Western US. The Teacher-Friendly Guide to the Earth Scientist of the Western US. 2017. Available online: http://geology.Teacherfriendlyguide.Org/index.php/mineral-w (accessed on 9 January 2022).

MDPI

Article

Towards Understanding the Source of Brine Mineralization in Southeast Nigeria: Evidence from High-Resolution Airborne Magnetic and Gravity Data

Stephen E. Ekwok [1], Anthony E. Akpan [1], Ogiji-Idaga M. Achadu [2], Cherish E. Thompson [1], Ahmed M. Eldosouky [3,*], Kamal Abdelrahman [4] and Peter Andráš [5]

[1] Applied Geophysics Programme, Department of Physics, University of Calabar, Calabar 540271, Nigeria; styvnekwok@unical.edu.ng (S.E.E.); anthonyakpan@unical.edu.ng (A.E.A.); cherishthompson@unical.edu.ng (C.E.T.)

[2] Department of Geology, University of Calabar, Calabar 540271, Nigeria; martinsachadu@unical.edu.ng

[3] Department of Geology, Suez University, Suez 43518, Egypt

[4] Department of Geology & Geophysics, College of Science, King Saud University, Riyadh 11451, Saudi Arabia; khassanein@ksu.edu.sa

[5] Faculty of Natural Sciences, Matej Bel University in Banska Bystrica, 974 01 Banska Bystrica, Slovakia; peter.andras@umb.sk

* Correspondence: ahmed.eldosouky@sci.suezuni.edu.eg; Tel.: +20-115-326-7202

Abstract: Investigation into understanding the genesis of brines in southeast Nigeria was carried out utilizing high-resolution potential field (HRPF) data. This study reveals that igneous intrusions and associated hydrothermal fluids are responsible for brine generation. The obtained result of the analytic signal revealed the locations and spatial distribution of short- and long-wavelength geologic structures associated with igneous intrusions. The low pass filtering, upward continuation, and 2D modelling procedures showed key synclinal structures which coincided well with the location of brine fields. The results showed that salt ponds are common in the neighborhood of igneous intrusions. To validate this finding, a conceptual model describing igneous-related hydrothermal circulation systems that are driven by convective cells of the hydrothermal fluid and overburden loads was generated. This model fits reasonably well into the overall stratigraphic and geologic framework of the study area.

Keywords: magnetic method; gravity method; magmatic intrusion; hydrothermal fluid; Lower Benue Trough; southeast Nigeria

Citation: Ekwok, S.E.; Akpan, A.E.; Achadu, O.-I.M.; Thompson, C.E.; Eldosouky, A.M.; Abdelrahman, K.; Andráš, P. Towards Understanding the Source of Brine Mineralization in Southeast Nigeria: Evidence from High-Resolution Airborne Magnetic and Gravity Data. *Minerals* **2022**, *12*, 146. https://doi.org/10.3390/min12020146

Academic Editor: Frank Wendler

Received: 11 December 2021
Accepted: 24 January 2022
Published: 25 January 2022

Publisher's Note: MDPI stays neutral with regard to jurisdictional claims in published maps and institutional affiliations.

1. Introduction

Since the early twentieth century, the origin of brine fields in the Benue Trough and adjacent sub-basins has been a point of contention [1,2]. Brines can be found in ponds, springs, drilled water boreholes, and hand-dug wells [3]. In terms of their genesis in the Benue Trough (BT), [3,4] summarized three possibilities proposed by previous studies. They include formation or a connate water source [5–7], evaporite or a solid source [8–10], and a hydrothermal source [8–14]. Most of the prior studies have been focused on a single occurrence or projections, and some lack relevant hydro-chemical data [3]. In general, these studies determined that the brines in the Trough came from either connate water or evaporite.

Hydro-geochemical investigation of the Middle and Lower Benue Trough reveals that brines are the result of halite dissolution and fossil seawater sources [4]. In [15,16], the authors conducted similar research that linked brines to a marine source. In [17], on the other hand, the authors suggested that connate water was the primary source of brines in the BT. Although [1] suggested an evaporite source, [4] stated that "there was no proof of evaporites in the BT". In the western Qaidam basin, meteoric water was identified as the main source of saltwater [18], while [19] identified meteoric, carbonate, and magmatic waters as the different sources of brines in the Sichuan basin.

Recent studies by [20–25] show that hydrothermal processes play a powerful role in the generation of brines within and beyond rift environments dominated by intrusive and extrusive rocks. Hydrothermal fluid igneous-related hydrothermal circulation systems with high salinity are highly mineralized, containing dissolved minerals such as zinc, lead, gold, and copper [20,22,25] The BT is the main target of geoscience research due to its mineralization, complicated tectonics, and associated geologic features [4,26,27]. Field observations, hydro-geochemical, isotopic, and geomorphic features, as well as tectonic and stratigraphic investigations, revealed the presence of brine fields, barite–lead–zinc veins, and fracture systems in the vicinity of magmatic intrusions in the BT [3,4,28–30].

Magnetic and gravity data can be used to investigate magmatism and its associated structures. Potential field datasets can be applied to delineate deep-seated faults [31–34], basement relief [35,36], intrusive granitoids [37,38] and geological structures [39]. Intrusions and geologic features are defined as the driving mechanisms for mineralization [20]. For instance, numerous geophysical techniques allow for a quick exploration of near-surface volcanics and intrusive rock adjacent to salt ponds in the BT [3]. Mineralization is frequently linked to hydrothermal alterations and structural control produced by magmatic intrusions [25,40]. High-resolution airborne magnetic and gravity measurements are now considered essential components of mineral exploration [40,41]. Magnetization and density discrepancies induced by igneous-related hydrothermal occurrences can be mapped using this information.

The aeromagnetic method is an appropriate and effective mineral assessment tool wherever there are magnetization variations between rock types [42–46]. Enhanced magnetic data can delineate fractures, faults, dykes, rock boundaries, and even regional surficial geologic contacts [47–53]. A pseudo-geological map can be built using magnetic data from which prospective mineralization zones can be inferred [54,55]. Similarly, gravity data can reveal the shape of magmatic intrusions and related geologic structures beneath the surface, map sedimentary basins, and provide vital information on basin formation mechanisms [56]. The gravity method is commonly operated as a control in seismic studies and as a tool in petroleum development [41]. It can also be used in hydrogeological, archaeological, and engineering studies [57,58]. Furthermore, in joint base metal investigations, the technique is often applied to specified targets delineated by magnetic and electromagnetic investigations [41,59,60].

This study uses gravity and magnetic data to look for anomalies connected to igneous-related hydrothermal circulation systems, brine field locations, and structures within the Ikom-Mamfe-Rift and Abakaliki Anticlinorium in the Lower Benue Trough (LBT). Researchers have discovered that areas with tectonothermal activity and hydrothermal changes frequently have different magnetic and density properties compared with nearby areas of comparatively free igneous intrusions [61–63]. Some geophysical investigations of brines and associated lead–zinc–barite (Pb–Zn–Ba) mineralization in the BT used electromagnetic, electrical resistivity, and ground gravity methods [17,53,64–67] The genesis of brine fields in southeast Nigeria is investigated using high-resolution potential field (HRPF) data. In addition, the genetic environment of these brines is explored with regard to the research area's geo-tectonic setting. Image analysis, enhancement procedures, and 2D modeling involving HRPF data are applied to delineate the hydrothermally altered zones and locate some geologic structures, possible locations of intrusions, and brines.

2. Geologic Setting of the Study Area

2.1. Location

The area of investigation is situated in southeast Nigeria (Long. 7°30′ E to 9°00′ E and Lat. 6°00′ N to 6°30′ N). The elevation above sea level map (Figure 1) of the studied region shows that the altitudes varied, starting approximately 29.4–186.5 m above sea level, with the eastern and northwestern parts having maximum heights.

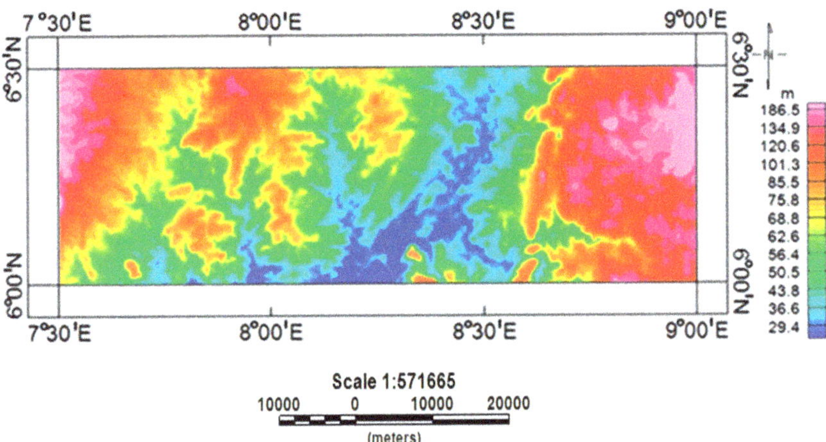

Figure 1. Elevation (above sea level) map of the study location.

2.2. Geology

The study area covers some geologic regions of the Abakaliki Anticlinorium (AA), Ikom-Mamfe Rift (IMR), and Obudu Plateau (OP) (Figure 2). The OP occupies the northeastern flank. It is edged in the west by AA and southeast by IMR. The IMR which intruded the study territory at the southeastern flank is bounded by the AA in the west, northwest, and north, and the OP in the northeast. The AA occupies southwestern, western, northwestern, and northern domains.

The Bamenda Massif is an extension of the OP, which is one of the Nigerian basement outcrops in southeast Nigeria [68,69]. The region's lithological differences include high-grade metamorphic rocks, primarily gneisses, and schists, intruded by un-metamorphosed dolerites, granites, aplites, and quartzo-feldspathic veins, and they are heavily migmatized [68]. Rocks from this area have been dated to be Eburnean, Archaean, and Pan-African in age [70]. Workers concluded that the evolutionary history of the southeast basement in Nigeria is related to the mobile Pan-African belt in central Africa, after comparing the lithologies and ages of rocks from the central African Fold Belt, northern Cameroon, and southeast Nigeria basement complexes [68,70]. The Pan-Central-African belt is the result of a continent–continent collision, with the Congo craton's northern edge acting as a passive margin, while the western Cameroon domains and Adamawa-Yade acted as active margins [2,68]. The OP is associated with migmatitic gneisses, which are classified as garnet–hornblende gneiss, garnet–sillimanite gneiss, or simply migmatite gneiss [68,71].

The IMR is an eastern extension of the Lower Benue Trough that runs into Cameroon, and terminates below the Tertiary–Recent cover of the Cameroon Volcanic Line (CVL) [72]. Igneous intrusions in the IMR, which were caused by Tertiary–Recent tectonic processes related to the CVL, resulted in extensive deformations and metamorphisms of the geologic materials in the region. As a result, the covering sedimentary sequences became severely fractured, baked, deformed, and domed [63].

The Asu River Group (ARG), which includes conglomeratic sandstones, conglomerates, mudstones, shales, calcareous, and carbonaceous rocks, is the first sedimentary group in the IMR [73]. The ARG sits on highly fractured Precambrian basement rocks [73]. The Eze-Aku Formation (EAF) was deposited during the Turonian regression period. The post-Santonian Nkporo-Afikpo Shale Formation overlies the EAF. The sandstone, mudstone, and shale strata are the foremost rock components of this formation [29].

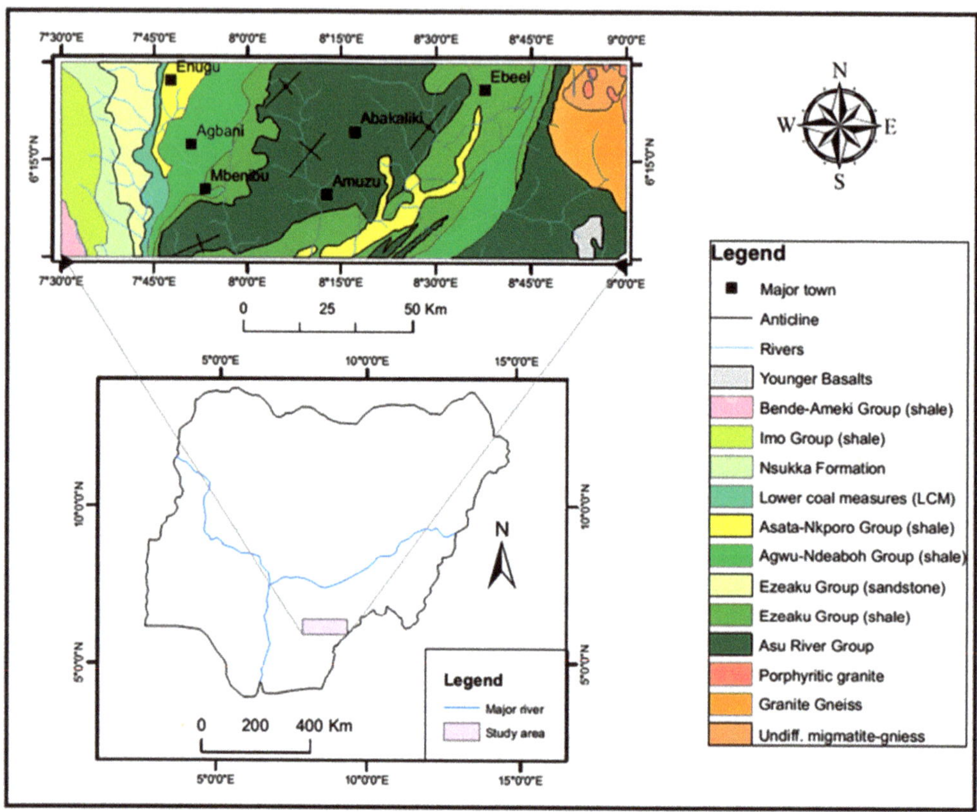

Figure 2. Geologic map of the study location.

The Campanian Nkporo Shale, Coniacian Awgu Shale, Turonian Eze-Aku Shale, and Albian Asu River Group are among the strata found in the AA [74]. The Albian ARG is made up of fissile, heavily fractured, bluish-black shales with very few sandstone strata [74]. The EAF is made up of calcareous sandstones, calcareous siltstones and shales, and thin sandy and shelly limestones [75]. Grey bluish shales, marine fossiliferous limestones, and calcareous sandstones from the Coniacian period make up the Awgu Shales [74]. The Awgu Shales are overlain by the Campanian Nkporo Shales, which are predominantly marine with some arenaceous sandstone units (Figure 3).

The folding of the overlying layers was caused by massive Santonian tectonic processes that occurred in two periods [29,74]. The Abakaliki Anticlinorium arose from the dominantly compressional nature of the forces. In [76], the authors compared the geological development of the Abakaliki province to that of a complete orogenic cycle encompassing sedimentation, magmatism, metamorphism, and compressive tectonics in thorough research on the geology of the province. According to [76], the compression that caused the large-scale folding and cleavage was directed at N 155° E. As a result of the magmatism, multiple intrusive masses were injected into the Eze-Aku and Asu River Group shales [74].

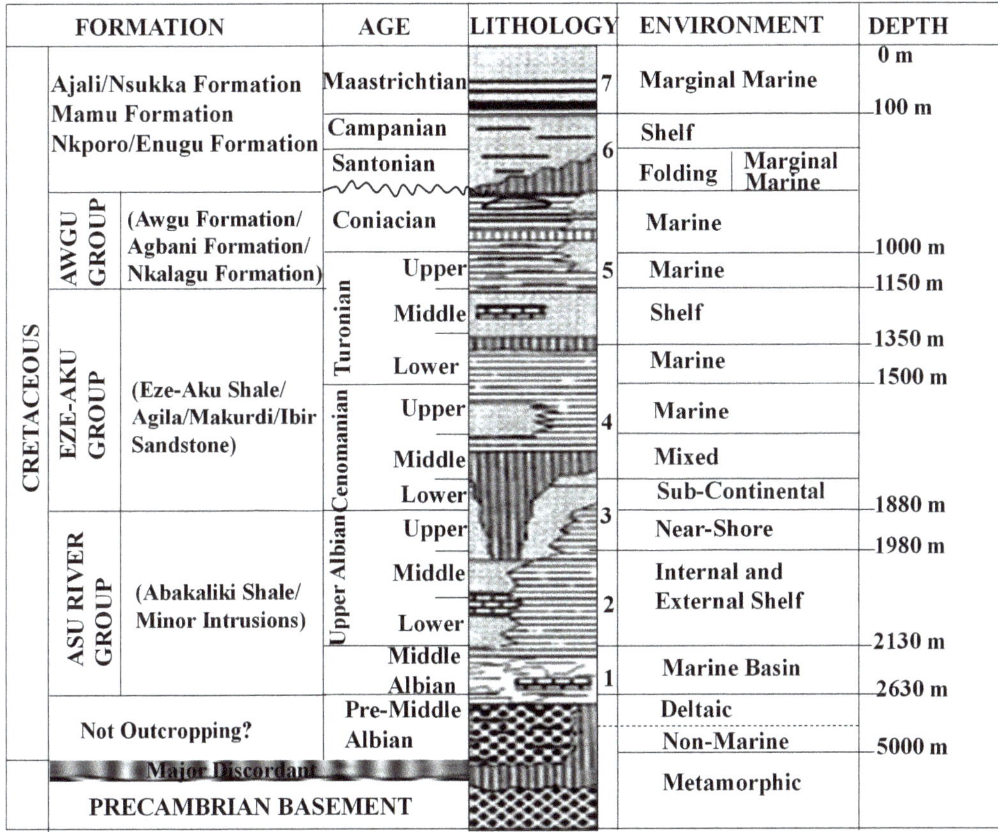

Figure 3. Lithostratigraphic sequence of the Lower Benue Trough.

3. Materials and Method

3.1. Data Acquisition

The Nigerian Geological Survey Agency (NGSA) provided the airborne gravity and magnetic data used in this investigation. Between 2005 and 2010, Fugro Airborne Surveys, Canada, measured, reduced, and compiled the high-resolution geophysical datasets, which were then submitted to the NGSA in digitized and gridded forms. The data were collected along 826,000 lines using the Flux-Adjusting Surface Data Assimilation System (FASDAS) with tie-line spacing, flight-line spacing, and terrain clearance of 0.5 km, 0.1 km, and 0.08–0.1 km, respectively.

The tenth (10th) generation of the International Geomagnetic Reference Field (IGRF-version 4.0) and International Gravity Standardization Net 1971 (IGSN71) algorithms were used to subtract regional fields from observed HRPF data. The IGSN71 and IGRF, both of which are widely used and approved, provide consistency in potential field exploration techniques [69]. The employed HRPF data in this study were reduced and processed to total magnetic intensity (TMI) and Bouguer gravity gridded (BG) maps (Figure 4).

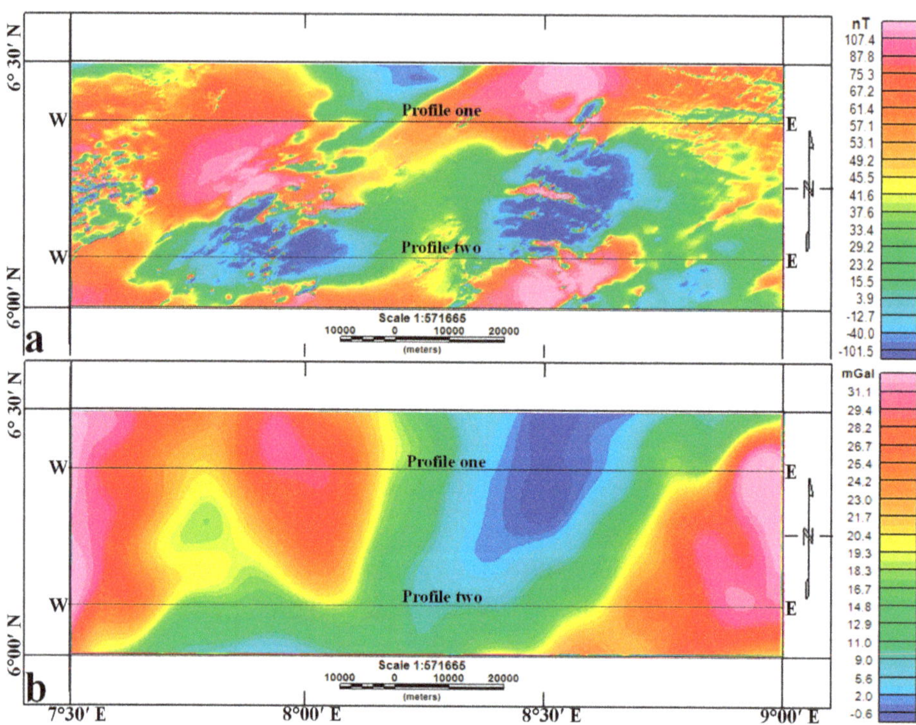

Figure 4. (**a**) Total magnetic intensity and (**b**) Bouguer gravity maps (with profiles 1 and 2 used for modelling).

3.2. Methodology

Local datum transformation and projection technique were used to record the potential field data in the World Geodetic System 84 (WGS-84) and Universal Transverse Mercator coordinate system at zone 32 of the northern hemisphere (UTM-32N). Using the Oasis montaj "add grid" menu, the gridded data were loaded into the toggle project explorer platform. The data were loaded into the MAG-MAP, source parameter imaging (SPI), Euler deconvolution, and GM-SYS tools, which generated control files for the various enhancement and modelling methods.

Filtering algorithms deliberately enhance anomalies from a certain set of geologic sources relative to anomalies due to other geologic sources and they are used in potential field data enhancements [77]. The main igneous intrusions in the research area were delineated using typical potential field (PF) data enhancement methods such as the analytic signal (*ASig*) [78–81], low pass filter [82], and upward continuation (UPWC) method [53,54]. In addition, 2D forward modelling was operated to map igneous intrusions, estimate sediment thicknesses, and define basement topography.

The *ASig* filter [83,84] generates maximum responses over magnetic/gravity anomalies and detects the edges of magnetic/gravity source bodies. This filter is commonly applied at low-magnetic latitudes due to the in-built complication associated with the RTP technique. The most important advantage of the analytical signal is that, in 2D cases, it does not depend on the direction of magnetization, but this is incorrect in 3D case [78]. In [78], the authors demonstrated that the *ASig* amplitude can be obtained from the three orthogonal derivatives of the potential field as:

$$\left| ASig_{(x,y)} \right| = \sqrt{\left(\frac{\partial A}{\partial x}\right)^2 + \left(\frac{\partial A}{\partial y}\right)^2 + \left(\frac{\partial A}{\partial z}\right)^2} \tag{1}$$

where, A is the measured PF.

The low pass filter is based on the method proposed by [82], and the filter cut-off wavelength is the only variable parameter [85]. Although it passes low-frequency signals, and weakens signals with frequencies above the cut-off frequency, low-pass filtering is used to eliminate unwanted short-wavelength anomalies [86]. The rectangular low-pass filter in a 1D Fourier transform is provided by:

$$L(k) = \begin{cases} 1, k \le k_c \\ 0, k > k_c \end{cases} \tag{2}$$

where $L(k)$ is the amplitude spectrum of the transfer function of the rectangular filter and k_c is the cut-off frequency.

In analyzing regional magnetic/gravity structures originating from deep-seated PF sources, UPWC is used. The upward continuation filter transforms the observed magnetic/gravity field on a surface to a higher level. In comparison to deep causative sources, this enhancement reduces the effect of near-surface bodies [53]. The wavenumber domain approach for UPWC [41] is expressed as follows:

$$F(\omega) = e^{-h\omega} \tag{3}$$

where, h is the height of continuation. This procedure decreases progressively with increasing wavenumber, reducing the higher wavenumbers more severely, thus generating a map in which more regional anomalies predominate [87]. The 2D modelling method involves developing a geologic hypothetical model and computing magnetic/gravity responses, applying [88–90]. Every crustal block has a certain susceptibility and/or density value. The anomaly across the entire profile is the sum of all the crustal block contributions. The interpretation of PF anomalies is based on identifying the probable structures, locations, and physical properties of the geologic features that created the anomaly. Two profiles in the east–west direction were obtained from the TMI and BG data. The PFs for the sedimentary top and basal basement structural cross-section were estimated iteratively until acceptable matches between the observed and synthetic curves were achieved. Using programs found in Oasis montaj version 7.0.1 (OL), potential field data enhancement, automatic depth estimation, and 2D forward modelling operations were carried out (2008). The inverse problem normally associated with potential field data is ill-posed thus making the solution unstable and non-unique [27,91]. A consistent solution for such a problem can be acquired by having sufficient geologic knowledge and the use of improved techniques in data corrections, enhancements and interpretations [92].

Tight folds, cleaves, and igneous intrusions such as basic sills and sub-volcanic intrusions characterize the Cretaceous depositions in the study area [93]. In the Middle and Lower Benue Trough, [3] documented shallow volcanics and intrusions in the vicinity of salt ponds. PF methods can be operated to investigate magmatic intrusions, mineralization, and related geologic structures [8,20,43]. Enhanced PF data showed variable gravity and magnetic disparities from numerous causative sources caused by short- and long-wavelength anomalies [31]. According to [27], mineralization is typically associated with locations characterized by complex geologic formations, created by tectonic processes. Low-frequency structures on the improved maps were greatly highlighted to demarcate areas marked by long-wavelength anomalies. To map the main tectonics and depocenters in the research area, these anomalies were mapped using *ASig*, low pass, and UPWC filters.

4. Results

The highest and lowest magnetic (pink = 0.0891 nT/m and blue = 0.0026 nT/m) and gravity (pink = 0.00215 mGal/m and blue = 0.00040 mGal/m) intensities can be easily identified (Figure 5). The eastern section of the investigated area (Figure 5) is dominated by basement biotite–gneiss and granite rocks of OP, where Kakube, Iso-Bendegheg, and Odumekpang fall. This region is characterized by basement rocks. They are believed to be the principal origin of high magnetization and density observed in the OP. Edor, Obubra, and Agbaragba regions that correspond with IMR have high magnetization and density believed to be caused by the Tertiary–Recent igneous intrusions related to the CVL [63]. Sparsely distributed high, moderate, and low potential field signatures occupy the remaining parts of the investigated area. This region is part of the AA, which is characterized by a post-depositional tectonic event [27]. Low gravity and magnetic strengths are also present in isolated locations denoted by various blue colors (indicating depocenters).

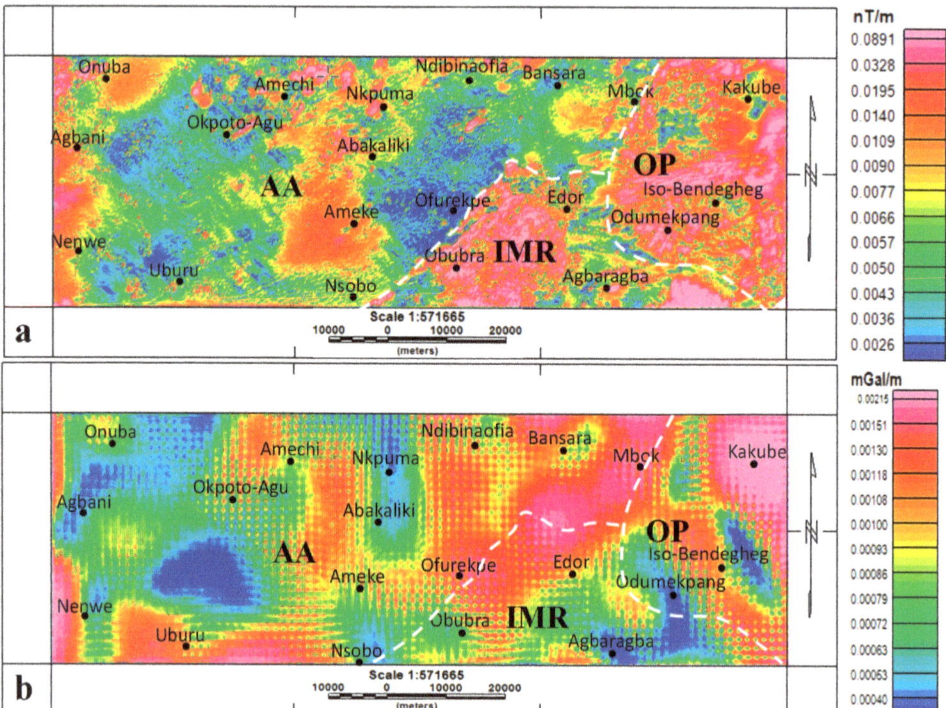

Figure 5. Analytic signal maps of (**a**) total magnetic intensity and (**b**) Bouguer gravity data. AA: Abakaliki Anticlinorium, OP: Obudu Plateau, IMR: Ikom-Mamfe Rift.

The main synclinal structures (defined by a blue color) were delineated using the low pass [82] (Figure 6) and UPWC (Figure 7) [78] filters, which roughly overlap with the zones of Obubra (southern), Uburu (southwest) and Okpoma (northeast) areas that fall under the Ikom-Mamfe Rift, Afikpo and Ogoja Synclines, respectively [58,94]. These structures sandwiched the main Santonian [93] and Tertiary–Recent [63] tectonic intrusions (red-pink color) of the AA and IMR, respectively. Some geophysical studies have reported the coexistence of intrusions and troughs in the BT [74,95]. The OP eastern part, which is dominated by magnetite, is characterized by intrusions and outcrops of igneous basements [9].

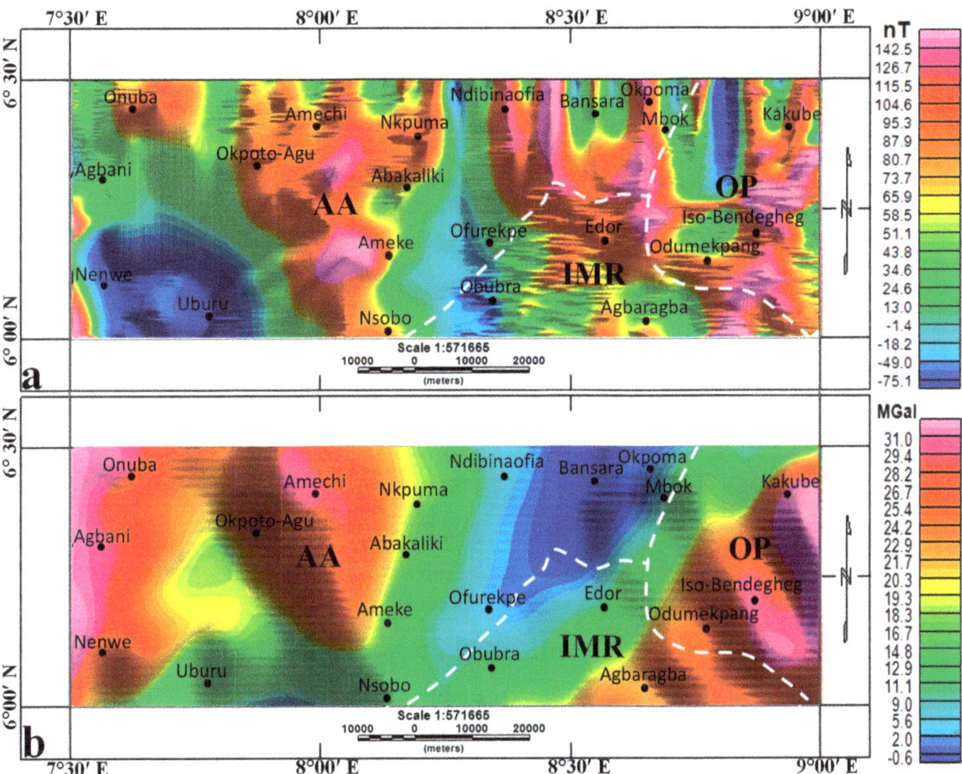

Figure 6. Low pass maps of (**a**) total magnetic intensity and (**b**) Bouguer gravity data. AA: Abakaliki Anticlinorium, OP: Obudu Plateau, IMR: Ikom-Mamfe Rift.

To view the lithologic units of the underlying basement based on their susceptibilities and densities, the high susceptibility (0.00057 in cgs units) and density (2.91 g/cc) values are due to basic rocks, whereas lower susceptibility (0.0003 in cgs units) and density (2.81 g/cc) values are due to acidic rocks. The models (Figures 8 and 9) reveal that the Cretaceous deposits are severely fractured, folded, and baked [26,57]. Again, normal fault blocks associated with the intrusions were delineated by the models. Profile 1 (Figure 8) that cuts across part of AA and ends at OP (Figure 4) is characterized by two prominent igneous intrusions related to the Santonian AA [58,93]. The location of the igneous intrusion at the western end of profile 2 (Figure 9) seems to match closely with the position of the intrusion in Figure 8. This indicates the elongated nature of the anticlinal structure of the AA that trends in the N–S direction (Figures 6 and 7). Towards the eastern part of profile 2 (Figure 9), the intrusion is outcropped in Agbaragba. Within this area and neighborhood, there are several reported extrusive rocks such as syenites, basalts, and trachytes [96,97] related to the CVL [63]. Because of the proliferation of intrusions [93] and basement rocks in outcrop sections in the OP [68,98], the investigated area is characterized by depths generally <3000 m. This depth estimate coincides relatively well with previous findings in the area [27,31,32,35,37].

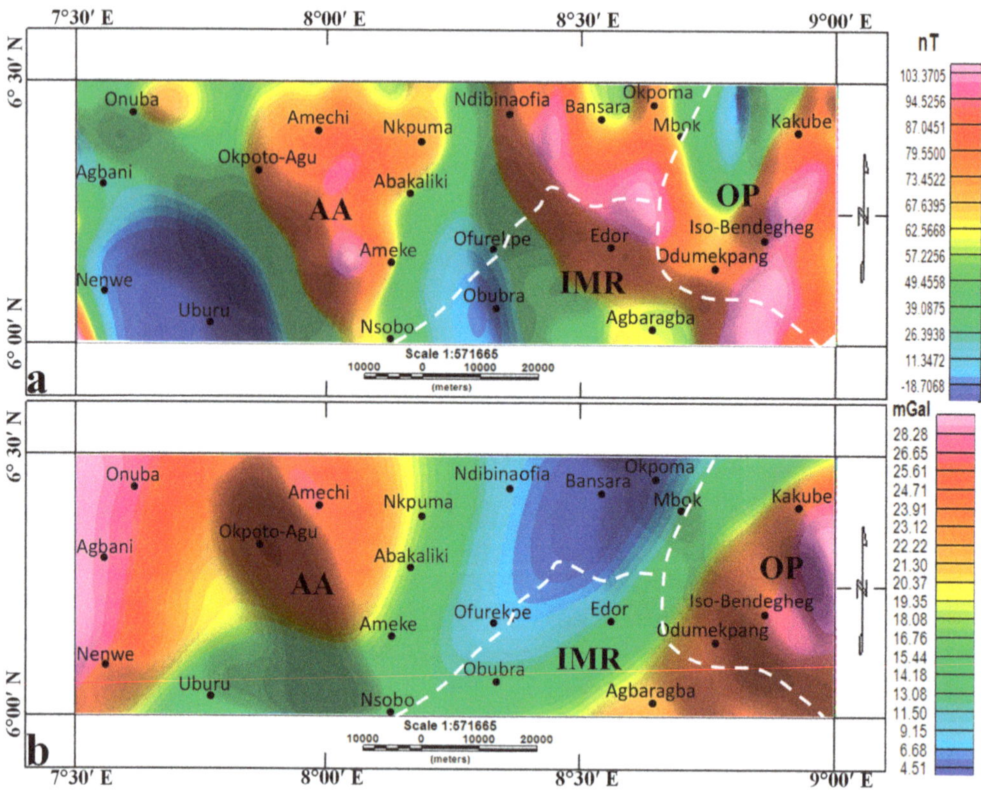

Figure 7. Upward, continued to 3000 m, maps of (**a**) total magnetic intensity and (**b**) Bouguer gravity data. AA: Abakaliki Anticlinorium, OP: Obudu Plateau, IMR: Ikom-Mamfe Rift.

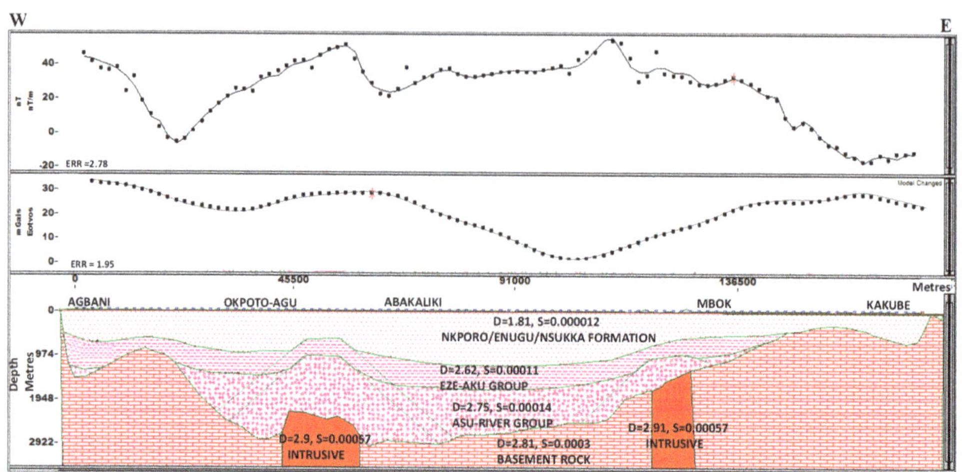

Figure 8. 2D joint magnetic and gravity model obtained from profile 1.

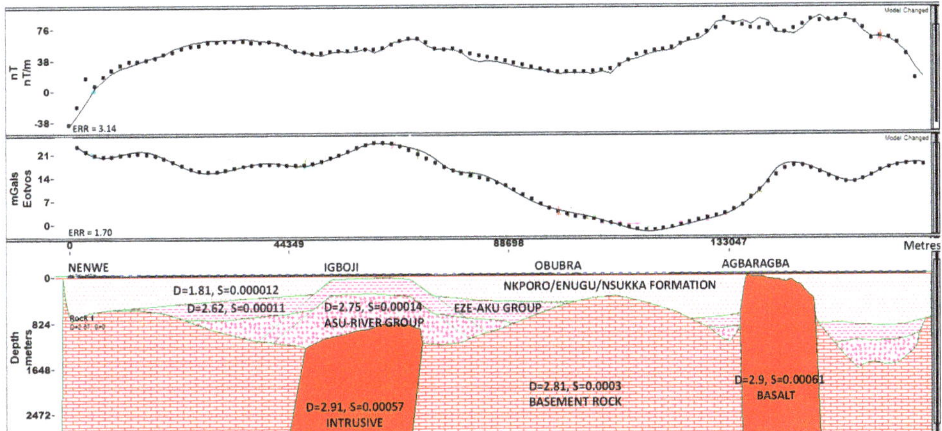

Figure 9. 2D joint magnetic and gravity model obtained from profile 2.

5. Discussion of Results

The LBT is relatively well known because of geoscience investigations for natural resources. Hydrocarbon, lead, zinc, barite, and brine are commonly explored in southeast Nigeria [3,63,93]. The occurrence of brines and genesis in the LBT have been investigated by several geoscientists [1,3,4,15–17,65,66,99] with contrasting reports. In [17,65,66], the authors stated from their various geophysical studies that brine fields are characterized by multi-layered saline zones in the subsurface. In [4,15], the authors suggested that these brines are products of the dissolution of halite and palaeo/fossil seawater. In [16], the authors also opined marine rather than a continental source for these brines. While [1] suggested evaporite minerals and chalcophile element sources, Ref. [3] suggested that brines originated from connate water.

Modern investigations by [10,20–25,100–103] connected the origin of brines in a rift environment, such as LBT, to igneous intrusions. Previously, sodium chloride and hydrochloride acid were detected by [104] in the emanations of volcanoes. In addition, Refs. [105,106] discussed the volcanic origin of salt. In [3,26,100], the authors reported the occurrence of brine in the neighborhood of igneous intrusions in the BT. Figures 6 and 7 indicated the locations of the Afikpo syncline (Uburu and the surrounding area), Obubra (in the c), and the Ogoja syncline (Okpoma and the adjoining region), represented by a blue color. These areas are depocenters bordering intrusions (represented by red-pink color) associated with the Santonian AA [74] and Tertiary–Recent intrusions of the IMR [7]. These synclinal structures, that are dominated by argillite, coincide with the sites of the major brine fields in the investigation area [1,4,16,17,66,99]. Intrusions delineated by Figures 8 and 9 occur in the neighborhood of the brine fields. This finding validates the results of Figures 7 and 8. Igneous intrusion systems in rifted basins are usually distinguished by networks of interrelated, laterally, and vertically wide-ranging complexes of dykes and sills that transgress basin stratigraphy [42]. These intrusions are associated with normal fault blocks, faults, uplifts, and folds. The faults and other openings serve as a pathway for the hydrothermal fluid's migration upward. The widespread occurrence of igneous intrusions in the investigated area signifies a critical geological risk in hydrocarbon surveys [101].

Furthermore, intrusions are commonly connected to hydrothermal fluid [21,40,100] that is as salty or even saltier than seawater, and may have some traces of dissolved minerals such as lead, zinc, copper, and gold. In the LBT, there are massive occurrences of lead–zinc–barite coexistence with salt ponds [3,28,31]. The presence of salt in the water halts the metallic minerals from precipitating out of the brine, as the chlorides in the salt preferentially bond with the metals [22,100]. Nevertheless, since the brine is hot, the minerals dissolve more easily [40]. As magma cools, it frees its super-heated, mineral-enriched water (metalliferous brines) into adjoining rock [21,40]. As they travel long distances laterally, they experience modifications in pressure and/or composition [100]. The hydrothermal fluid becomes diluted once they come into contact with meteoric water (groundwater). Fluid flow in the fractured upper crust is usually driven by hydraulic gradients, which may perhaps result from a number of different possible causes and imbalances, including thermal and chemical disequilibrium, topography, overburden loads [100], and attendant convective cells in the hydrothermal system. They drive the brines up through hydraulic boundary faults/fractures via porous and permeable strata until oozing takes place at the surface (Figure 10). This conceptual model was created based on the geologic, stratigraphic (Figure 3) and tectonic history of the study area. Additionally, geodynamic information on the robust thermally driven convection process [105], related heat, and movement of hydrothermal fluid [22], served as a control in the design of the model. The metals are precipitated out when the brines rise and cool [22]. The cool brines are then trapped by argillite close to the Earth's surface and occur as salt ponds such as those present at Uburu, Okposi, Okpoma, etc. in the studied area. The absence of hot springs in the studied area may be due to a lack of deep-reaching vertical circulation systems [107,108].

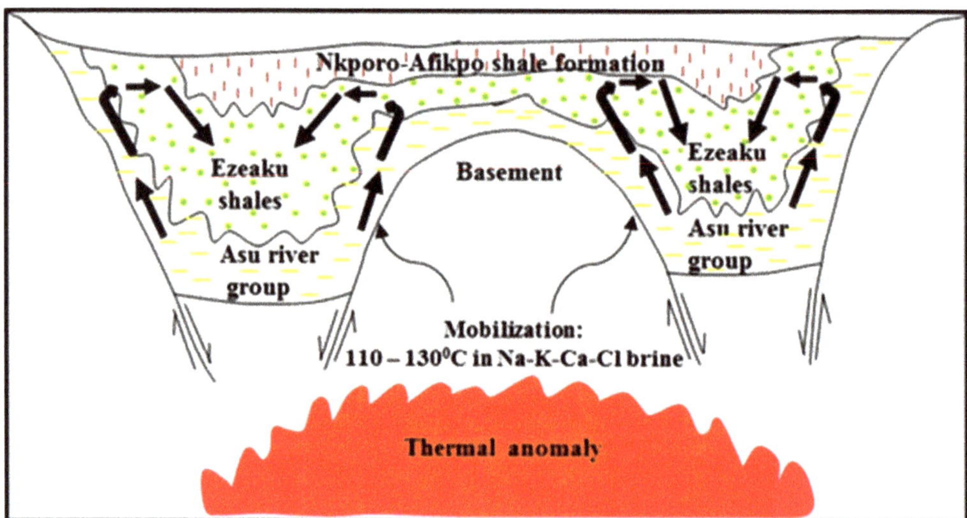

Figure 10. Conceptual model of brine source in the Lower Benue Trough.

Moreover, ascent channels for hydrothermal fluids are very conducting structures allowing high flow rates so that hot brines can reach the surface environment at high temperatures [108]. The conceptual model (Figure 10) of the hydrothermal system of the LBT agrees with the Darcy flow law [109], describing fluid flow through porous media. This defines the ability of a fluid to flow through a porous media such as rock. It relies on the fact that the amount of flow between two points is directly related to the difference in pressure between the points, the distance between the points, and the interconnectivity of flow pathways in the rock between the points. The measurement of interconnectivity is called permeability.

The Darcy law is expressed as:

$$\vec{q} = -[K]\frac{1}{g}\nabla P \tag{4}$$

where, $\vec{q}$ is the specific discharge vector per cross sectional area (m^3 s^{-1} m^{-2}) with the components q_x, q_z, q_z, (K) (m s^{-1}) the tensor of hydraulic conductivity, K (kg m^{-3}) the density of the fluid, g (m s^{-2}) the acceleration due to gravity, and ∇P (kg m^{-2} s^{-2} or Pa m^{-1}) the vector of the pressure gradient.

6. Conclusions

Investigation connecting igneous intrusions and associated hydrothermal fluids as a brine source in a rifted environment such as BT was carried out in southeastern Nigeria. To map these intrusions and determine their spatial distributions within the study area, HRPF data were used. The *ASig* results showed the spatial distribution of the short- and long-wavelength geologic structures. The main igneous bodies and bordering synclines, which coincide with the sites of Uburu, Obubra, and Okpoma salt ponds, were delineated by the low pass, UPWC, and 2D GM-SYS results. In general, the low-frequency maps and 2D models showed brine field sites in the neighborhood of magnetic intrusions, which are commonly associated with metalliferous brines. To fit these interpretations into the overall stratigraphic and geologic settings of the area, a conceptual model of the brine source was generated. This model is more akin to a hydrothermal system that is driven by associated convective cells of the hydrothermal fluid and overburden loads. Furthermore, studies involving geochemical analyses of the brines and adjoining igneous rocks should be carried out to compare their elemental composition, which will be the subject of future research.

Author Contributions: Conceptualization, A.M.E.; Data curation, S.E.E.; Formal analysis, O.-I.M.A.; Funding acquisition, K.A. Investigation, A.E.A., C.E.T. and P.A.; Methodology, S.E.E.; Supervision, A.M.E.; Validation, O.-I.M.A.; Writing—original draft, A.E.A., O.-I.M.A., C.E.T. and K.A.; Writing—review & editing, A.M.E. All authors have read and agreed to the published version of the manuscript.

Funding: This research was funded by Researchers Supporting Project number (RSP-2022/351), King Saud University, Riyadh, Saudi Arabia.

Acknowledgments: Deep thanks and gratitude to the Researchers Supporting Project number (RSP-2022/351), King Saud University, Riyadh, Saudi Arabia for funding this research article.

Conflicts of Interest: The authors declare no conflict of interest.

References

1. Eseme, E.; Agyingi, C.M.; Foba-Tendo, J. Geochemistry and genesis of brine emanations from Cretaceous strata of the Mamfe Basin, Cameroon. *J. Afr. Earth Sci.* **2002**, *35*, 467–476. [CrossRef]
2. Toteu, S.F.; Penaye, J.; Djomani, Y.P. Geodynamic evolution of the Pan-African belt in Central Africa with special reference to Cameroon. *Can. J. Earth Sci.* **2004**, *41*, 73–85. [CrossRef]
3. Uma, K.O. The brine fields of the Benue Trough, Nigeria: A comparative study of geomorphic, tectonic and hydrochemical properties. *J. Afr. Earth Sci.* **1998**, *26*, 261–275. [CrossRef]
4. Tijani, M.N. Evolution of saline waters and brines in the Benue-Trough, Nigeria. *Appl. Geochem.* **2004**, *19*, 1355–1365. [CrossRef]
5. Offodile, M.E. The geology of the Middle Benue, Nigeria. Ph.D. Thesis, University of Upsala, Uppsala, Sweden, 1976; 166p.
6. Tattam, C.H. Preliminary report on the salt industry in Nigena. *Geol. Surv. Niger. Rep.* **1943**, *778*.
7. Uzuakpunwa, A.B. The geochemistry and origin of the evaporite deposits in the southern half of the Benue Trough. *Earth Evol. Sci.* **1981**, *2*, 136–139.
8. Egboka, B.C.E.; Uma, K.O. Hydrochemistry, contaminant transport and their tectonic effects in the Okposi-Uburu salt lake area, Imo State, Nigeria. *Hydrol. Sci. J.* **1986**, *31*, 205–221. [CrossRef]
9. Ford, S.O. The economic mrneral resources of the Benue Trough. *Earth Evol. Sci.* **1980**, *1*, 154–163.
10. Urajaka, S.O. Salt water resources of Fast Central State of Nigeria. *J. Min. Geol.* **1972**, *7*, 35–41.
11. Akande, S.O.; Horn, E.E.; Reutel, C. Mineralogy, fluid inclusion and genesis of the Arufu and Akwana Pb Zn F mineralizahon, Middle Benue Trough, Nigeria. *J. Afr. Earth Sci.* **1988**, *7*, 167–180. [CrossRef]
12. Farrington, I.L. A preliminary description of the Nigerian Lead-Zinc field. *Econ. Geol.* **1952**, *7*, 483–608. [CrossRef]
13. McConnel, R.B. Notes on the Lead-Zinc deposits of Nigeria and Cretaceous stratigraphy of the Benue and Cross River Valleys. Geological Survey of Nigeria Report. No. 752. *Niger. Geol. Surv. Rep.* 1949; *unpublished*.
14. Olade, M.A. Evolution of Nrgeria's Benue Trough of Nigeria (aulacogen): A tectonic model. *Geol. Mag.* **1975**, *112*, 575–583. [CrossRef]

15. Tijani, M.N.; Loehnert, E.P.; Uma, K.O. Origin of saline groundwaters in the Ogoja area, Lower Benue Trough, Nigeria. *J. Afr. Earth Sci.* **1996**, *23*, 237–252. [CrossRef]

16. Ushie, F.; Eminue, O.; Nwankwoala, H. Occurrence of Brines (NaCl) and their effect on the Groundwater of Okpoma and Environs, Southeastern Nigeria. *Int. J. Adv. Sci. Tech. Res.* **2014**, *4*, 2249–9954.

17. Umar, N.D.; Idris, I.G.; Abdullahi, A.I. Hydrogeophysical investigation of the aquifers of brine field of awe and environs, Central Benue Trough, Nigeria. *Int. J. Sci. Eng. Res.* **2018**, *9*, 1852–1868.

18. Fan, Q.; Ma, H.; Lai, Z.; Tan, H.; Li, T. Origin and evolution of oilfield brines from Tertiary strata in western Qaidam Basin: Constraints from 87Sr/86Sr, δD, δ18O, δ34S and water chemistry. *Chin. J. Geochem.* **2010**, *29*, 446–454. [CrossRef]

19. Xun, Z.; Cijun, L.; Xiumin, J.; Qiang, D.; Lihong, T. Origin of subsurface brines in the Sichuan basin. *Ground Water* **1996**, *35*, 53–58. [CrossRef]

20. Blundy, J.; Mavrogenes, J.; Tattitch, B.; Sparks, S.; Gilmer, A. Generation of porphyry copper deposits by gas–brine reaction in volcanic arcs. *Nat. Geosci.* **2015**, *8*, 235–240. [CrossRef]

21. Dill, H.G.; Botz, R.; Berner, Z.; Hamad, A.B.A. The origin of pre- and synrift, hypogene Fe-P mineralization during the Cenozoic along the Dead Sea Transform Fault, Northwest Jordan. *Econ. Geol.* **2010**, *105*, 1301–1319. [CrossRef]

22. Mineral Resources of the Western US. The Teacher-Friendly Guide to the Earth Scientist of the Western US. 2017. Available online: http://geology.Teacherfriendlyguide.Org/index.php/mineral-w (accessed on 1 December 2021).

23. Risacher, F.; Alonsob, H.; Salazar, C. The origin of brines and salts in Chilean salars: A hydrochemical review. *Earth-Sci. Rev.* **2003**, *63*, 249–293. [CrossRef]

24. Sharma, R.; Srivastava, P.K. Hydrothermal fluids of magmatic origin. In *Modelling of Magmatic and Allied Processes*; Society of Earth Scientists Series; Springer: Berling/Heidelberg, Germany, 2014. [CrossRef]

25. Xu, K.; Yu, B.; Gong, H.; Ruan, Z.; Pan, Y.; Ren, Y. Carbonate reservoirs modified by magmatic intrusions in the Bachu area, Tarim Basin, NW China. *Geosci. Front.* **2015**, *6*, 779–790. [CrossRef]

26. Ekwok, S.E.; Akpan AEKudamnya, E.A.; Ebong, D.E. Assessment of groundwater potential using geophysical data: A case study in parts of Cross River State, south-eastern Nigeria. *Appl. Water Sci.* **2020**, *10*, 144. [CrossRef]

27. Ekwok, S.E.; Akpan, A.E.; Ebong, D.E. Enhancement and modelling of aeromagnetic data of some inland basins, southeastern Nigeria. *J. Afr. Earth Sci.* **2019**, *155*, 43–53. [CrossRef]

28. Akande, S.O.; Muecke, A. Co-existing copper sulphides and sulphosalts in the Abakaliki Pb–Zn deposits, lower Benue-Trough (Nigeria) and their genetic significance. *Miner. Pet.* **1993**, *47*, 183–192. [CrossRef]

29. Nwachukwu, S.O. The tectonic evolution of the southern portion of the Benue Trough, Nigeria. *Geol. Mag.* **1972**, *109*, 411–419. [CrossRef]

30. Olade, M.A.; Morton, R.D. Origin of lead–zinc mineralization in the southern Benue-Trough, Nigeria: Fluid inclusion and trace element studies. *Mineral. Dep.* **1985**, *20*, 76–80. [CrossRef]

31. Ekwok, S.E.; Akpan, A.E.; Kudamnya, E.A. Exploratory mapping of structures controlling mineralization in Southeast Nigeria using high resolution airborne magnetic data. *J. Afr. Earth Sci.* **2020**, *162*, 103700. [CrossRef]

32. Ekwok, S.E.; Akpan, A.E.; Achadu, O.I.M.; Eze, O.E. Structural and lithological interpretation of aero-geophysical data in parts of the Lower Benue Trough and Obudu Plateau, Southeast Nigeria. *Adv. Space Res.* **2021**, *68*, 2841–2854. [CrossRef]

33. Ekinci, Y.L.; Balkaya, Ç.; Göktürkler, G. Parameter estimations from gravity and magnetic anomalies due to deep-seated faults: Differential evolution versus particle swarm optimization. *Turk. J. Earth Sci.* **2019**, *28*, 860–881.

34. Mehanee, S. A new scheme for gravity data interpretation by a faulted 2-D horizontal thin block: Theory, numerical examples and real data investigation. In *IEEE Transactions on Geoscience and Remote Sensing*; IEEE: Piscataway, NJ, USA, 2022. [CrossRef]

35. Ekwok, S.E.; Akpan, A.E.; Ebong, E.D. Assessment of crustal structures by gravity and magnetic methods in the Calabar Flank and adjoining areas of Southeastern Nigeria-a case study. *Arab. J. Geosci.* **2021**, *14*, 1–10. [CrossRef]

36. Ekinci, Y.L.; Balkaya, Ç.; Göktürkler, G.; Özyalın, Ş. Gravity data inversion for the basement relief delineation through global optimization: A case study from the Aegean Graben System, western Anatolia, Turkey. *Geophys. J. Int.* **2021**, *224*, 923–944. [CrossRef]

37. Ekwok, S.E.; Akpan, A.E.; Achadu, O.I.M.; Ulem, C.A. Implications of tectonic anomalies from potential feld data in some parts of Southeast Nigeria. *Environ. Earth Sci.* **2022**, *81*, 6. [CrossRef]

38. Balkaya, Ç.; Ekinci, Y.L.; Göktürkler, G.; Turan, S. 3D non-linear inversion of magnetic anomalies caused by prismatic bodies using differential evolution algorithm. *J. Appl. Geophys.* **2017**, *136*, 372–386. [CrossRef]

39. Balkaya, Ç.; Göktürkler, G.; Erhan, Z.; Ekinci, Y.L. Exploration for a cave by magnetic and electrical resistivity surveys: Ayvacık Sinkhole example, Bozdağ, İzmir (western Turkey). *Geophysics* **2012**, *77*, 135–146. [CrossRef]

40. United States Geological Survey (USGS). Setting and Origin of Iron Oxide-Copper-Cobalt-Gold-Rare Earth Element Deposits of Southeast Missouri. 2013. Available online: http://minerals.usgs.gov/east/semissouri/index.html (accessed on 23 January 2022).

41. Telford, W.M.; Geldart, L.P.; Sheriff, R.E. *Applied Geophysics*, 2nd ed.; Cambridge University Press: Cambridge, UK, 1990.

42. Eldosouky, A.M.; Sehsah, H.; Elkhateeb, S.O.; Pour, A.B. Integrating aeromagnetic data and Landsat-8 imagery for detection of post-accretionary shear zones controlling hydrothermal alterations: The Allaqi-Heiani Suture zone, South Eastern Desert, Egypt. *Adv. Space Res.* **2020**, *65*, 1008–1024. [CrossRef]

43. Eldosouky, A.M.; Elkhateeb, S.O. Texture analysis of aeromagnetic data for enhancing geologic features using co-occurrence matrices in Elallaqi area, South Eastern Desert of Egypt. *NRIAG J. Astron. Geophys.* **2018**, *7*, 155–161. [CrossRef]

44. Eldosouky, A.M.; Abdelkareem, M.; Elkhateeb, S.O. Integration of remote sensing and aeromagnetic data for mapping structural features and hydrothermal alteration zones in Wadi Allaqi area, South Eastern Desert of Egypt. *J. Afr. Earth Sci.* **2017**, *130*, 28–37. [CrossRef]

45. Eldosouky, A.M.; Mohamed, H. Edge detection of aeromagnetic data as effective tools for structural imaging at Shilman area, South Eastern Desert, Egypt. *Arab. J. Geosci.* **2021**, *14*, 13. [CrossRef]

46. Elkhateeb, S.O.; Eldosouky, A.M.; Khalifa, M.O.; Aboalhassan, M. Probability of mineral occurrence in the Southeast of Aswan area, Egypt, from the analysis of aeromagnetic data. *Arab. J. Geosci.* **2021**, *14*, 1514. [CrossRef]

47. Pham, L.T.; Eldosouky, A.M.; Melouah, O.; Abdelrahman, K.; Alzahrani, H.; Oliveira, S.P.; András, P. Mapping subsurface structural lineaments using the edge filters of gravity data. *J. King Saud Univ.-Sci.* **2021**, *33*, 101594. [CrossRef]

48. Pham, L.T.; Nguyen, D.A.; Eldosouky, A.M.; Abdelrahman, K.; Vu, T.V.; Al-Otaibi, N.; Ibrahim, E.; Kharbish, S. Subsurface structural mapping from high-resolution gravity data using advanced processing methods. *J. King Saud Univ. Sci.* **2021**, *33*, 101488. [CrossRef]

49. Pham, L.T.; Oksum, E.; Do, T.D.; Nguyen, D.V.; Eldosouky, A.M. On the performance of phase-based filters for enhancing lateral boundaries of magnetic and gravity sources: A case study of the Seattle uplift. *Arab. J. Geosci.* **2021**, *14*, 129. [CrossRef]

50. Pham, L.T.; Eldosouky, A.M.; Oksum, E.; Saada, S.A. A new high resolution filter for source edge detection of potential data. *Geocarto Int.* **2020**, 1–18. [CrossRef]

51. Saada, A.S.; Eldosouky, A.M.; Kamal, A.; Al-Otaibi, N.; Ibrahim, E.; Ibrahim, A. New insights into the contribution of gravity data for mapping the lithospheric architecture. *J. King Saud Univ.-Sci.* **2021**, *33*, 101400. [CrossRef]

52. Saada, A.S.; Mickus, K.; Eldosouky, A.M.; Ibrahim, A. Insights on the tectonic styles of the Red Sea rift using gravity and magnetic data. *Mar. Pet. Geol.* **2021**, *133*, 105253. [CrossRef]

53. Sehsah, H.; Eldosouky, A.M. Neoproterozoic hybrid forearc—MOR ophiolite belts in the northern Arabian-Nubian Shield: No evidence for back-arc tectonic setting. *Int. Geol. Rev.* **2020**, 1–13. [CrossRef]

54. Elkhateeb, S.O.; Eldosouky, A.M. Detection of porphyry intrusions using analytic signal (AS), Euler Deconvolution, and Centre for Exploration Targeting (CET) Technique Porphyry Analysis at Wadi Allaqi Area, South Eastern Desert, Egypt. *Int. J. Eng. Res.* **2016**, *7*, 471–477.

55. Essa, K.S.; Mehanee, S.; Soliman, K.; Diab, Z.E. Gravity profile interpretation using the R-parameter imaging technique with application to ore exploration. *Ore Geol. Rev.* **2020**, *126*, 103695. [CrossRef]

56. Kearey, P.; Brooks, M.; Hill, I. *An Introduction to Geophysical Exploration*, 3rd ed.; Blackwell Science Ltd Editorial Offices: New York, NY, USA, 2002.

57. Abu El-Magd, S.A.; Eldosouky, A.M. An improved approach for predicting the groundwater potentiality in the low desert lands; El-Marashda area, Northwest Qena City, Egypt. *J. Afr. Earth Sci.* **2021**, *179*, 104200. [CrossRef]

58. Ekwok, S.E.; Akpan, A.E.; Ebong, E.D.; Eze, O.E. Assessment of depth to magnetic sources using high resolution aeromagnetic data of some parts of the Lower Benue Trough and adjoining areas, Southeast Nigeria. *Adv. Space Res.* **2021**, *67*, 2104–2119. [CrossRef]

59. Mehanee, S. Simultaneous joint inversion of residual gravity and self-potential data measured along profile: Theory, numerical examples and a case study from mineral exploration with cross validation from electromagnetic data. *IEEE Trans. Geosci. Remote Sens.* **2022**, *60*, 1–20. [CrossRef]

60. Cherkashov, G.; Poroshina, I.; Stepanova, T.; Ivanov, V.; Bel'tenev, V.; Lazareva, L.; Rozhdestvenskaya, I.; Samovarov, M.; Shilov, V.; Glasby, G.P.; et al. Seafloor Massive Sulfides from the Northern Equatorial Mid-Atlantic Ridge: New Discoveries and Perspectives. *Mar. Georesources Geotechnol.* **2010**, *28*, 222–239. [CrossRef]

61. Gunn, P.J.; Dentith, M.C. Magnetic responses associated with mineral deposits. *AGSO J. Aust. Geol. Geophys.* **1997**, *17*, 145–158.

62. Correia, A.; Jones, F.W. On the existence of a geothermal anomaly in southern Portugal. *Tectonophysics* **1997**, *271*, 123–134. [CrossRef]

63. Akpan, A.E.; Ebong, D.E.; Ekwok, S.E.; Joseph, S. Geophysical and Geological Studies of the Spread and Industrial Quality of Okurike Barite Deposit. *Am. J. Environ. Sci.* **2014**, *10*, 566–574. [CrossRef]

64. Mbah, V.O.; Onwuemesi, A.G.; Aniwetalu, E.U. Exploration of lead-zinc (Pb-Zn) mineralization using very low frequency electromanetic (VLF-EM) in Ishiagu, Ebonyi State. *J. Geol. Geophys.* **2015**, *4*, 1–7.

65. Mbipom, E.W.; Okon-Umoren, O.E.; Umoh, J.U. Geophysical investigations of salt ponds in Okpoma area south-eastern Nigeria. *J. Min. Geol. Niger.* **1990**, *26*, 285–290.

66. Okoyeh, E.I.; Akpan, A.E.; Egboka, B.C.E., Okolo, M.C.; Okeke, H.C. Geophysical delineation of subsurface fracture associated with Okposi-Uburu Salt Lake Southeastern, Nigeria. *Int. Res. J. Environ. Sci.* **2015**, *4*, 1–6.

67. Ugbor, D.O.; Okeke, F.N. Geophysical investigation in the Lower Benue trough of Nigeria using gravity method. *Int. J. Phys. Sci.* **2010**, *5*, 1757–1769.

68. Agbi, I.; Ekwueme, B.N. Preliminary review of the geology of the hornblende biotite gneisses of Obudu Plateau Southeastern Nigeria. *Glob. J. Geol. Sci.* **2018**, *17*, 75–83. [CrossRef]

69. Dada, S.S. Crust-forming ages and Proterozoic crustal evolution in Nigeria: A reappraisal of current interpretations. *Precambrian Res.* **1998**, *87*, 65–74. [CrossRef]

70. Ukwang, E.E.; Ekwueme, B.N.; Kröner, A. Single zircon evaporation ages: Evidence for the Mesoproterozoic crust in S.E. Nigerian basement complex. *Chin. J. Geochem.* **2012**, *31*, 48–54. [CrossRef]
71. Haruna, I.V. Review of the Basement Geology and Mineral Belts of Nigeria. *J. Appl. Geol. Geophys.* **2017**, *5*, 37–45.
72. Fairhead, J.D.; Okereke, C.S.; Nnange, J.M. Crustal Structure of the Mamfe basin, West Africa, based on gravity data. *Tectonophysics* **1991**, *186*, 351–358. [CrossRef]
73. Dumort, J.C. *Carte Géologique de Reconnaissance et Note Explicative sur la Feuille Douala-Ouest (1:500000)*; Direction des Mines et de la Géologie du Cameroun: Yaoundé, Cameroun, 1968.
74. Ofoegbu, C.O.; Onuoha, K.M. Analysis of magnetic data over the Abakaliki Anticlinorium of the Lower Benue Trough, Nigeria. *Mar. Pet. Geol.* **1991**, *8*, 174–183. [CrossRef]
75. Reyment, R.A. *Aspects of the Geology of Nigeria*; Ibadan University Press: Ibadan, Nigeria, 1965.
76. Benkhelil, J. Structure et evolution geodynamique du Basin intracontinental de la Benoue (Nigeria). *Bull. Cent. Rech. Explor. Prod. Elf-Aquitaine* **1988**, *12*, 29–128.
77. Milligan, P.; Gunn, P. Enhancement and presentation of airborne geophysical data. *AGSO J. Aust. Geol. Geophys.* **1997**, *17*, 63–75.
78. Roest, W.R.; Verhoef, J.; Pilkington, M. Magnetic interpretation using the 3-D analytic signal. *Geophysics* **1992**, *57*, 116–125. [CrossRef]
79. Essa, K.S.; Mehanee, S.; Elhussein, M. Magnetic data profiles interpretation for mineralized buried structures identification applying the variance analysis method. *Pure Appl. Geophys.* **2020**, *178*, 973–993. [CrossRef]
80. Klingele, E.E.; Marson, I.; Kahle, H.G. Automatic interpretation of gravity gradiometric data in two dimensions: Vertical gradients. *Geophys. Prospect.* **1991**, *39*, 407–434. [CrossRef]
81. Mehanee, S.; Essa, K.S.; Diab, Z.E. Magnetic data interpretation using a new R-parameter imaging method with application to mineral exploration. *Nat. Resour. Res.* **2020**, *30*, 77–95. [CrossRef]
82. Fraser, D.C.; Fuller, B.D.; Ward, S.H. Some numerical techniques for application in mining exploration. *Geophysics* **1966**, *31*, 1066–1077. [CrossRef]
83. Nabighian, M.N. The analytical signal of two dimensional magnetic bodies with polygon cross-section: Its properties and use for automated anomaly interpretation. *Geophysics* **1972**, *37*, 507–517. [CrossRef]
84. Nabighian, M.N. Towards the three-dimensional automatic interpretation of potential field data via generalized Hilbert transforms: Fundamental relations. *Geophysics* **1984**, *53*, 957–966. [CrossRef]
85. Zahra, H.S.; Oweis, H.T. Application of high-pass filtering techniques on gravity and magnetic data of the eastern Qattara Depression area, Western Desert, Egypt. *NRIAG J. Astron. Geophys.* **2016**, *5*, 106–123. [CrossRef]
86. Allen, R.L.; Mills, D.W. *Signal Analysis, Time, Frequency, Scale, and Structure*; IEEE Press: Piscataway, NJ, USA, 2004.
87. Reeves, C.; Reford, S.; Millingan, P. Airborne geophysics: Old methods, new images. In *Proceedings of the Fourth Decennial. International Conference on Mineral Exploration*; Gubins, A., Ed.; Prospectors and Developers Association of Canada: Toronto, ON, Canada, 1997; pp. 13–30.
88. Talwani, M.; Hiertzler, J.R. Computation of magnetic anomalies caused by two dimensional bodes of arbitrary shape. *Geol. Sci.* **1964**, *9*, 464–480.
89. Talwani, M.; Worzel, J.L.; Landisman, M. Rapid gravity computations for 2 dimensional bodies with application to the Mendocino submarine fracture zone. *J. Geophys. Res.* **1959**, *64*, 49–59. [CrossRef]
90. Won, I.J.; Bevis, M. Computing the gravitational and magnetic anomalies due to a polygon: Algorithms and Fortran subroutines. *Geophysics* **1987**, *52*, 232–238. [CrossRef]
91. Essa, K.S.; Elhussein, M. A new approach for the interpretation of magnetic data by a 2-D dipping dike. *J. Appl. Geophys.* **2017**, *136*, 431–443. [CrossRef]
92. Essa, K.S.; Elhussein, M. PSO (particle swarm optimization) for interpretation of magnetic anomalies caused by simple geometrical structures. *Pure Appl. Geophys.* **2018**, *175*, 3539–3553. [CrossRef]
93. Benkhelil, J. Cretaceous deformation, magmatism, and metamorphism in the Lower Benue Trough, Nigeria. *Geol. J.* **1987**, *22*, 467–493. [CrossRef]
94. Kogbe, C.A. *Geology of Nigeria: A Review*; Elizabethan Publishing Co.: Lagos, Nigeria, 1976; pp. 436–468.
95. Ofoegbu, C.O.; Mohan, N.L. Interpretation of aeromagnetic anomalies over part of southeastern Nigeria using three-dimensional Hilbert transformation. *Pure Appl. Geophys.* **1990**, *134*, 13–29. [CrossRef]
96. Abolo, M.G. Geology and petroleum potential of the Mamfe basin, Cameroon, central Africa. *Afr. Geosci. Rev.* **2008**, *12*, 65–77.
97. Ajonina, H.N.; Ajibola, O.A.; Bassey, C.E. The Mamfe Basin, SE Nigeria and SW Cameroon: A review of the basin filling model and tectonic evolution. *J. Geosci. Soc. Cameroon* **2002**, *1*, 24–25.
98. Arinze, J.I.; Emedo, O.C.; Ngwaka, A.C. Analysis of aeromagnetic anomalies and structural lineaments for mineral and hydrocarbon exploration in Ikom and its environs southeastern Nigeria. *J. Afr. Earth Sci.* **2019**, *151*, 274–285.
99. Lar, U.A.; Sallau, A.K. Trace element geochemistry of the Keana brines field, Middle Benue Trough. *Environ. Geochem. Health* **2005**, *27*, 331–339. [CrossRef]
100. Barton, M.D.; Johnson, D.A. Alternative brine sources for Fe-Oxide (-Cu-Au) systems: Implications for hydrothermal alteration and metals. In *Hydrothermal Iron Oxide Copper-Gold & Related Deposits: A Global Perspective*; Porter, T.M., Ed.; Australian Mineral Foundation: Adelaide, Australia, 2000; pp. 43–60.

101. Holford, S.P.; Schofield, N.; Jackson, C.A.L.; Magee, C.; Green, P.F.; Duddy, I.R. Impacts of igneous intrusions on source and reservoir potential in prospective sedimentary basins along the Western Australian continental margin. In Proceedings of the West Australian Basins Symposium, Perth, WA, Australia, 18–21 August 2013; pp. 1–12.
102. Neupane, G.; Wendt, D.S. Assessment of Mineral Resources in Geothermal Brines in the US. In Proceedings of the 42nd Workshop on Geothermal Reservoir Engineering, Stanford, CA, USA, 13–15 February 2017; pp. 1–18.
103. Vehling, F.; Hasenclever, J.; Rüpke, L. Brine formation and mobilization in submarine hydrothermal systems: Insights from a novel multiphase hydrothermal flow model in the system H_2O–NaCl. *Transp. Porous Media* **2021**, *136*, 65–102. [CrossRef]
104. Daubeny, C. *A Description of Active and Extinct Volcanoes*; Cambridge University Press: Cambridge, UK, 1826; pp. 168–172.
105. Ballaert, W. Origin of Salt Deposits. *Proc. Brit. Assoc. Adv. Sci.* **1852**, *2*, 100.
106. Darwin, C. *The Structure and Distribution of Coral Reefs. Being the First Part of the Geology of the Voyage of the Beagle, under the Command of Capt. Fitzroy, R.N. during the Years 1832 to 1836*; Smith Elder and Co. Geological Observations: London, UK, 1842; Volume 3, p. 235.
107. Roy, A.B. *Fundamentals of Geology*; Narosa Publishing House Pvt. Ltd.: Oxford, UK, 2010; p. 149.
108. Stober, I.; Bucher, K. Hydraulic conductivity of fractured upper crust: Insights from hydraulic tests in boreholes and fluid-rock interaction in crystalline basement rocks. *Geofluids* **2015**, *15*, 161–178. [CrossRef]
109. Singh, H.; Cai, J. Permeability of fractured shale and two-phase relative permeability in fractures. In *Encyclopedia of Electrochemical Power Sources*; Elsevier: Amsterdam, The Netherlands, 2019.

Article

Geochemical and Hydrothermal Alteration Patterns of the Abrisham-Rud Porphyry Copper District, Semnan Province, Iran

Timofey Timkin [1,*], Mahnaz Abedini [2], Mansour Ziaii [2] and Mohammad Reza Ghasemi [2]

[1] School of Earth Sciences & Engineering, Tomsk Polytechnic University, 634050 Tomsk, Russia
[2] Faculty of Mining, Petroleum & Geophysics Engineering, Shahrood University of Technology, Shahrood 3619995161, Iran; abedini.m@shahroodut.ac.ir (M.A.); mziaii@shahroodut.ac.ir (M.Z.); mr.ghasemi@shahroodut.ac.ir (M.R.G.)
* Correspondence: timkin@tpu.ru; Tel.: +7-9069-567-150

Abstract: In this study, the zonality method has been used to separate geochemical anomalies and to calculate erosional levels in the regional scale for porphyry-Cu deposit, Abrisham-Rud (Semnan province, East of Iran). In geochemical maps of multiplicative haloes, the co-existence of both the supra-ore elements and sub-ore elements local maxima implied blind mineralization in the northwest of the study area. Moreover, considering the calculated zonality indices and two previously presented geochemical models, E and NW of the study have been introduced as ZDM and BM, respectively. For comparison, the geological layer has been created by combining rock units, faults, and alterations utilizing the K-nearest neighbor (KNN) algorithm. The rock units and faults have been identified from the geological map; moreover, alterations have been detected by using remote sensing and ASTER images. In the geological layer map related to E of the study area, many parts have been detected as high potential areas; in addition, both geochemical and geological layer maps only confirmed each other at the south of this area and suggested this part as high potential mineralization. Therefore, high potential areas in the geological layer map could be related to the mineralization or not. Due to the incapability of the geological layer in identifying erosional levels, mineralogy investigation could be used to recognize this level; however, because of the high cost, mineralogy is not recommended for application on a regional scale. The findings demonstrated that the zonality method has successfully distinguished geochemical anomalies including BM and ZDM without dependent on alteration and was able to predict erosional levels. Therefore, this method is more powerful than the geological layer.

Keywords: zonality method; remote sensing; vertical zonality index; geological layer; alteration; K-nearest neighbor; porphyry-Cu deposit

Citation: Timkin, T.; Abedini, M.; Ziaii, M.; Ghasemi, M.R. Geochemical and Hydrothermal Alteration Patterns of the Abrisham-Rud Porphyry Copper District, Semnan Province, Iran. *Minerals* **2022**, *12*, 103. https://doi.org/10.3390/min12010103

Academic Editors: Amin Beiranvand Pour, Omeid Rahmani and Mohammad Parsa

Received: 15 October 2021
Accepted: 27 December 2021
Published: 16 January 2022

Publisher's Note: MDPI stays neutral with regard to jurisdictional claims in published maps and institutional affiliations.

1. Introduction

The utilization of geochemical methods for ore deposit exploration dates back to 1930. Fersman (1939) carried out the first survey of such an exploration [1]. Since then, further studies on the theory and application of geochemical exploration methods have been carried out, and these techniques have been increasingly modified and improved. Mining geochemistry is a branch of applied geochemistry, which is based on the utilization of geochemical methods that helps increase the ore reserves of known mines by assessing the ore potential of deep horizons. In other words, local and mine scale exploration models for anomaly recognition are created and developed by using mining geochemistry. Recent experiences in the application of mining geochemistry illustrate its efficiency in the discovery of blind and zone-dispersed mineralization (BM and ZDM) within areas of active and abandoned mines. Due to increasing ore reserves and mining income, this trend in geochemical exploration is very important [2]. The recognition of various alteration zones is a qualitative method that cannot help geochemists in separating BM from ZDM at a local scale. In mining geochemistry, the alteration has no basic role in separating anomalies at a local scale [3,4]. Optimal drilling points were determined by using mining geochemistry as a quantitative method without being time consuming and inducing high costs. In the past decades, several models and methods based on geochemistry have been developed for predicting geochemical anomalies as well as the locations of hidden orebodies [2,5–7]. Most of these models and methods are focused on the identification of geochemical anomalies reflecting the presence of hidden orebodies and the prediction of horizons of erosional surfaces [8]. In active mines, vertical geochemical zonality is the most important feature of primary halos because of the relation to the direction of the ore-bearing fluid [9–11]. Beus and Grigorian (1977) used vertical geochemical zonality to predict hidden mineralization at the mine scale [12]. Grigorian (1985, 1992) presented a zonality model to identify BM from ZDM [5,6]. Since then, the zonality method has been used in many studies [2–4,13–21]. Solovov (1987) used metallometric methods for the identification of geochemical anomaly (IGA) and the quantitative evaluation of ore reserves [22]. Baranov (1987) introduced a model in which horizons of the erosional surface were computed for geochemical associations [23]. Solovov (1990) introduced different relations to predict hidden orebodies by using metallometric exploration [24]. Liu and Peng (2004) presented a model to predict hidden orebodies by the synthesis of geological, geophysical, and geochemical information based on a dynamic approach [25].

In most mineral exploration methods (e.g., porphyry-Cu), a mineral potential map is obtained by using one layer or a combination of different layers [26–31], which includes field geological surveys, geochemical surveys, field geophysical surveys, and remote sensing [32–34]. This map consists of shallow to the deep layers, which poses a problem when these layers are not associated with mineralization. Each of these layers has a value for mapping the areas with the potential of mineralization, and a number of these layers are generally surficial and cannot be useful for identifying BM. Hydrothermally altered rocks result from chemical attacks of pre-existing rocks by hydrothermal fluids. The spatial distribution of hydrothermally altered rocks is a key to locating the main outflow zones of hydrothermal systems, which may result in the recognition of mineral deposits [35]. Minerals associated with alteration can be detected by remote sensing. These tasks are achieved by using the analysis of the spectral signatures recorded in the visible-near infrared (VNIR), short wave infrared (SWIR), and thermal infrared (TIR) regions of the electromagnetic spectrum with this spectral signature constituting the key mineral identification criterion [36]. Furthermore, the mineral deposits are spatially and genetically associated with the various types of geological structures including faults or fractures [37]. Faults and fractures, which transport magmatic, meteoric, and metal carrying hydrothermal fluids, subsequently deposit metals [38]. Zarasvandi et al. (2005); Sillitoe (2010); Mirzaie et al. (2015); Habibkhah et al. (2020); and Yumul Jr. et al. (2020) have investigated the importance of the role of faults/fractures in porphyry-Cu [38–42]. The strong advance in remote sensing allows exploiting a variety of sources and methods in the characterization of lineaments [43]. Remote sensing is a valuable technical resource for mineral exploration when it is properly employed [44–53].

In this study, the results of the zonality method were compared with the geological layer including rock units, faults, and alterations. For this goal, a part of the 1:100,000 scale map of Abrisham-Rud (Semnan province, East of Iran) was examined. This area is a part of the Troud Range in the Khorasan porphyry tract, which few studies on porphyry mineralization have been conducted [54]. Orojnia (2003) studied the lithology and provenance of Eocene volcanic rocks in this area, which suggested identifying the economic potential of Cu [55]. Mars (2014) suggested the potential of hydrothermal alteration in the Torud Range that could be associated with an unidentified porphyry system [56]. Thus, study data are concentrated in the province of Abrisham-Rud, where it has high propects for porphyry copper mineralization.

2. Geological Setting of the Study Area

The study area is a part of the 1:100,000 scale map of Abrisham-Rud (Semnan province, East of Iran), which is located in the north of the Central Iran zone [57], and it is a part of Troud Range in Khorasan porphyry tract (Figure 1a). Khorasan porphyry tracts are delimited by permissive units of island arc setting of Late Cretaceous to Early Miocene age [58]. This tract includes four main ranges: Taknar-Kashmar, Kuh Mish, Sabzevar, and Torud [54]. Igneous units of this tract are shown in Figure 1a, along with the location of known porphyry-related mineral occurrences and other geologic features mentioned in this section. In the Torud Range (west of Khorasan porphyry tract), middle Eocene volcano-sedimentary rocks are overlain by Eocene–Oligocene calc-alkaline and alkaline volcanic rocks, which are interlayered with shallow marine, lacustrine, and subaerial sedimentary successions. These successions are intruded by basic tholeiitic dikes and calc-alkaline quartz monzodioritic to granodioritic stocks [59,60]. The aluminum content of hornblendes indicates shallow emplacement depths [61]. In this range, Chah Shirin and Chah Mussa have been introduced as porphyry/porphyry-related deposits [58].

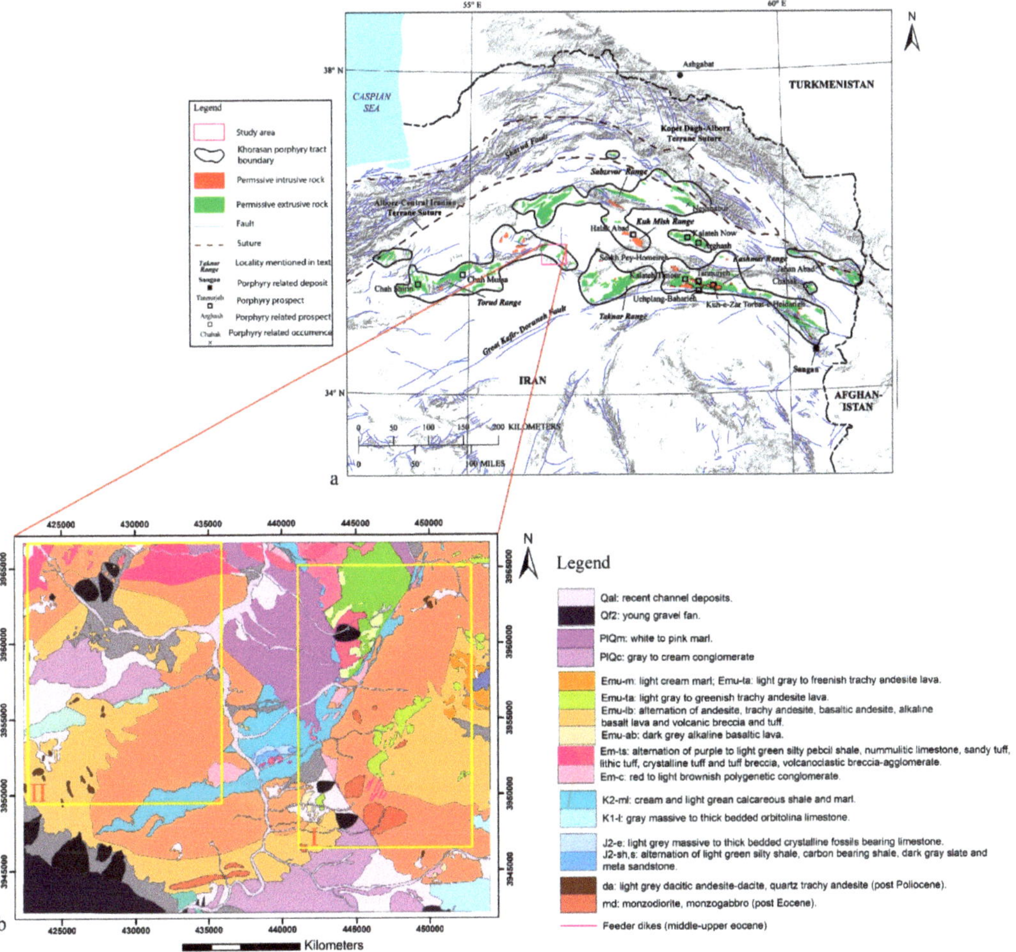

Figure 1. (**a**) Location of the study area in Troud Range (modified after Reference [54]); (**b**) lithological map of the study area (modified after Reference [62]).

The study area has a dry climate, mountainous topography, and poor vegetation cover. The oldest geological units are metamorphic rocks for which its bedrock was metamorphosed in the Late Triassic. The latest units are associated with a quaternary that consists of gravel fan, terrace, clay and salt deposits, and channel deposits (Figure 1b). Most volcanic activity occurred in semi-arid settings in the form of lava. The middle-upper Eocene rock units are the most extended in the study area and mostly include volcanic breccia-agglomerates and tuffs, intermediate lavas, basic and acidic rocks, and pyroclastic-sedimentary rocks. These units have outcrops and are sometimes dispersed or indistinguishable, and there are many faults in them. The volcanic breccia–agglomerate units that sometimes had intermediate lavas observed in red-brown and sometimes had dusty colors (argillic alteration), as well as the presence of green minerals (chlorite, epidote) in the volcanic fragments of this breccia, indicate lava explosive eruption in a semi-arid and shallow setting with the effect of water on alterations [55].

3. Material

3.1. Geological Data

The 1:100,000 scale map of Abrisham-Rud was purchased from the Geological Survey of Iran (GSI). As shown in Figure 1b, a part of this geological map was selected.

3.2. Geochemical Data

The geochemical database of the 1:100,000 scale map of Abrisham-Rud was also purchased from GSI. The 364 samples have been surveyed in a systematic network with intervals of 1400×1400 m^2 (Figure 2), and these soil samples have been analyzed using the ICP-MS method. The concentrations of Cu, Mo, Pb, and Zn were considered for this study.

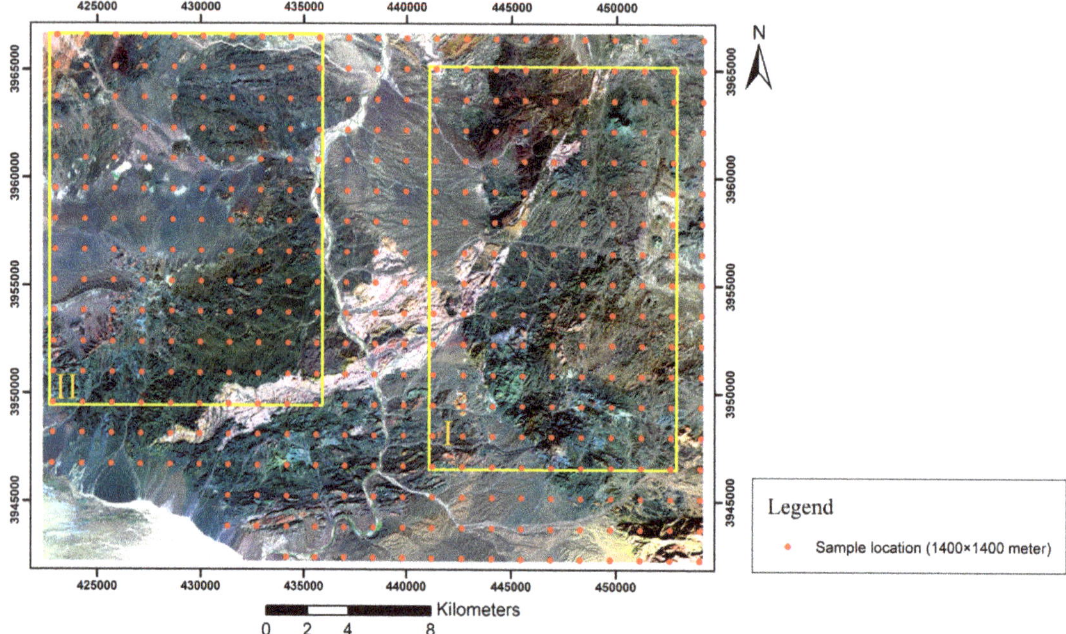

Figure 2. Geochemical sampling network in the study area (the false-color composites of Sentinel-2 MSI; band 12 in red, band 8 in green, and band 2 in blue).

3.3. Remote Sensing Data

In this study, Sentinel-2 MSI-Level 1C and ASTER-Level 1T images were downloaded from the USGS Earth Explorer website (earthexplorer.usgs.gov, accessed on 22 December 2021; Sentinel 2 File: T40SDE_20181118T07015, 20181118-Date of Acquisition (YYYYMMDD); Aster File: AST_L1T_003032820050712 48_20150508204820_5306, 03282005-Date of Acquisition (MMDDYYYY), 20150508-Date of Processing (YYYYMMDD)).

3.3.1. Sentinel-2 MSI Data

The Sentinel-2 multi-spectral instrument (MSI) satellite carried a high-resolution multispectral imager with 13 bands spanning VNIR through SWIR regions. Sentinel-2 MSI includes 4 spectral bands (bands 2, 3, 4 and 8) at 10 m, 6 bands (bands 5, 6, 7, 8a, 11 and 12) at 20 m, and 3 bands (bands 1, 9, and 10) at 60 m. Sentinel-2 MSI measures reflected radiation in ten bands between 0.433 and 0.955 μm (VNIR) and three bands between 1.36 and 2.28 μm (SWIR) [63–65]. In this study, a Sentinel-2 MSI-Level 1C image was used, which is produced from the Sentinel-2 MSI-Level 1B product by radiometric and geometric corrections.

3.3.2. ASTER Data

Advanced Spaceborne Thermal Emission and Reflection Radiometer (ASTER) sensors are one of the multi-spectral sensors that have been installed on the Terra satellite. ASTER measures reflected radiation in three bands between 0.52 and 0.86 μm (VNIR) and in six bands between 1.60 and 2.43 μm (SWIR) and five bands of emitted radiation in the 8.125 μm to 11.65 μm (TIR) region with 15 m, 30 m, and 90 m resolution, respectively [66–68]. In this study, an ASTER-Level 1T image was utilized. ASTER-Level 1T data contain calibrated at-sensor radiance, which corresponds with ASTER-Level 1B that has been geometrically corrected and rotated to the north up UTM projection.

3.3.3. Data Preparation

Atmospheric correction was used to minimize influences of atmospheric factors in multispectral data. The Internal Average Relative Reflection (IARR) method was applied to Sentinel-2 and ASTER data. The IARR technique for mineral mapping requires no prior knowledge of geological features [69]. The average radiance for each band of the image was calculated; therefore, an average spectrum was created, and this average spectrum was divided into actual radiance for each band of each pixel to create an image of apparent reflectance. It has been suggested as the best method for arid areas with no vegetation cover [70].

4. Methodology

4.1. Zonality Method

A zonation of a geochemical halo has a spatial nature and vectorial context that can be defined by the three parameters of dimension (space), direction, and element concentration [4]. Recognition of zonality of geochemical halos associated with BM can be achieved using four cases of complementary analyses [2]: (1) analysis of element associations representing supra-ore and sub-ore halos of mineral deposits; (2) analysis of a single component, implying false anomaly; (3) analysis of mean values of indicator elements outside significant geochemical anomalies to eliminate background noise in data analysis; and (4) mapping of multiplicative geochemical anomalies.

One of the most important indices in porphyry-Cu deposits is the ratio of Pb and Zn to Cu and Mo, which is often defined as a zonality index [2,12,71]. This index represents different exhumation levels of mineral deposits [6]. Input variables can be subdivided into the following: supra-ore, upper-ore, ore, lower-ore, and sub-ore [24]. Ziaii et al. (2012) and Ziaii et al. (2009) showed that these groups provide the necessary information to separate BM from ZDM in porphyry-Cu mineralization [2,3].

4.1.1. Anomaly Separation

For mapping the multiplicative index of supra-ore (Pb × Zn) and sub-ore (Cu × Mo) elements and calculating the zonality index, the threshold value is calculated by using Equation (1) for each element:

$$C_A = C_x \, \varepsilon^t, \tag{1}$$

where C_A is the anomaly value, and $C_x = C_o$ is the geometric mean of the elements contained within the background area, which is calculated by using Equation (2):

$$\widetilde{C_x} = \operatorname{ant} \log\left(\frac{1}{N}\sum_{i=1}^{N} \log C_i\right), \tag{2}$$

$$\varepsilon = \operatorname{ant} \log\left(\sqrt{\frac{\sum_{i=1}^{N} \left(\log C_i - \log \widetilde{C_x}\right)^2}{N-1}}\right), \tag{3}$$

where C_i is the element concentration of samples, N is the number of samples, and ε is generally called a standard factor, which is calculated from Equation (3). Due to the above consideration as to the selection of the value of t, the lowest anomalous content, when trying to detect weak anomalies determined from one sampling point, in geochemical prospecting it is taken equal to $C_{Al} \geq C_o \varepsilon^3$. This relationship corresponds to a "three standard deviations" criterion extensively used in many engineering disciplines to determine quantities falling outside the probable values of a random anomaly distributed quantity. Thus, weak anomalies formed by a sequence of adjacent sampling points with increased pathfinder elements below C_{Al} can be detected, and it is conventional to lower the threshold value according to the criterion $C_{Am} \geq C_o \varepsilon^{3/\sqrt{m}}$, where m = 2, 3, 4 . . . 9 is the number points that may be joined, which can show a common anomaly in the geochemical map [22].

4.1.2. Erosional Surface

In order to predict the erosional level, Solovov (1987) suggested using areal productivity, and Beus and Grigorian (1977) suggested using the coefficient of mineralization to eliminate the syngenetic parameters of the halos, which increase anomaly detection [12,22]. It should be noted that both the areal productivity and mineralization coefficients are used to calculate the vertical geochemical zonality index.

In the systematic sampling network, linear productivity is calculated according to Equation (4):

$$M = \Delta x\left(\sum_{x=1}^{n} C_x - nC_o\right), \tag{4}$$

where M is the linear productivity, Δx is the distance between the samples in each profile, C_x is the values greater than the anomaly concentration, and n is the number of anomal samples. If the values of M_i are preliminarily estimated in each of m profiles across the anomaly, P is determined according to Equation (5) [22]:

$$P = 2l\left(\sum_{i=1}^{m} M_i\right), \tag{5}$$

where 2l is the distance between profiles. Therefore, the zonality index introduced by Solovov (1987) is calculated for zones I and II by using Equation (6) [22].

$$K_S = \frac{P(Pb) \times P(Zn)}{P(Cu) \times P(Mo)}, \tag{6}$$

The zonality index introduced by Beus and Grigorian (1977) is calculated for each zone by using Equation (7) [12]:

$$K_G = \frac{\eta(\alpha)_{Pb} \times \overline{CA}_{Pb} \times \eta(\alpha)_{Zn} \times \overline{CA}_{Zn}}{\eta(\alpha)_{Cu} \times \overline{CA}_{Cu} \times \eta(\alpha)_{Mo} \times \overline{CA}_{Mo}}, \tag{7}$$

where CA is the arithmetic mean of element contents, and $\eta(\alpha)$ is the coefficient of mineralization calculated for each element by using Equation (8):

$$\eta(\alpha) = \frac{\eta_A(\alpha)_{ore}}{\eta_A(\alpha)},$$
(8)

where $\eta(\alpha)_{ore}$ is the number of anomalous samples, and $\eta_A(\alpha)$ is the total number of samples in each zone [12].

4.2. Remote Sensing

4.2.1. Lineaments Extraction

O'Leary et al. (1976) defined the term "lineaments" as a simple or composite linear feature of a surface for which its parts are aligned in a rectilinear or slightly curvilinear relationship and differs from the pattern of adjacent features and probably reflects some subsurface phenomena [72]. Faults, fractures, and large crush zones are formed by extension or compression processes and are considerable and fundamental factors on ore mineral deposition. Areas with concentrations or intersections of these structures could be suitable for the penetration of magma, ore-forming solutions, and, afterward, mineralization [73]. In other words, a detailed geological study imperatively means acquiring knowledge of present structural information, principally the lineaments [74]. The significance of lineaments is also manifested by their localization often close to several mineralogical deposits [75].

Several studies have been based on Sentinel-2 MSI for the detection of lineaments [76–79]. According to Bentahar et al. (2020), Sentinel-2 MSI allows extracting more lineaments and extraction of the smallest structural lineaments [79].

Lineament extraction methods can be conducted by using manual photointerpretation by an expert, semi-automatic detection using computer vision techniques, and automatic methods. Automatic methods have resulted in a more efficient lineament extraction process [43,80–83]. The main steps of lineaments extraction are mentioned below:

- Applying principal component analysis (PCA) and choosing PC1 to recognize lines;
- Filter operations using Directional filter with azimuths of $0°$, $45°$, $90°$, and $135°$;
- Automatic lineaments extraction using LINE module in the PCI Geomatica software;
- Merging lineaments obtained from azimuths of $0°$, $45°$, $90°$, and $135°$;
- Lineament mapping.

Principal Component Analysis

PCA (Pearson, 1901) is a statistical method that has the advantage of compressing information contained in initial bands into new bands called principal components (PCs) [84–86]. This method has been commonly used in lithological mapping and lineament extraction [79]. The PCA method can reduce redundancy in different bands, which can obtain aimed dimension reduction [87], isolation of noise, and enhancement of the targeted information in the image [88]. Each PC can reflect the maximum information of the original variable, and the information contained therein is not repeated.

Filter Operations

Filter operations were used to emphasize or de-emphasize spatial frequency in the image. This frequency can be attributed to the presence of the lineaments in the area. In other words, the filtering operator can sharpen the boundary that exists between adjacent units. A directional filter is a first-order derivative edge enhancement filter that selectively enhances image features possessing specific direction components (gradients) [89]. Directional filters are used strictly for the structural analysis. These filters improve the perception of lineaments, causing an optical effect of shade worn on the image as if it was illuminated by light grazing [78].

PCI Geomatica Software

The automatic extraction of the lineaments was carried out by algorithm LINE EXTRACTION of the PCI Geomatica software [80], which is a widely used module for automatic lineament extraction. The LINE module of PCI Geomatica software extracts linear features from an image and records polylines in vector segments by two main steps, namely edge detection and line detection, and using six parameters [43]:

Edge detection

- RADI (filter radius) (in pixels): The radius of the filter that is used in contours detection. Values between 3 and 8 are recommended in order to avoid introducing noise;
- GTHR (Edge Gradient Threshold): The value of the gradient to be taken as the threshold in contour detection (between 0 and 255). Values between 10 and 70 are acceptable;
- Line detection;
- LTHR (Curve Length Threshold) (in pixels): The minimum length of a curve to be taken as the lineament (a value of 10 is suitable);
- FTHR (Line Fitting Threshold) (In pixels): The tolerance allowed in the curve fitting (results of the previous parameter) to form a polyline. Values between 2 and 5 are recommended;
- ATHR (Angular Difference Threshold) (In degree): Defines the angle not to be exceeded between two polylines to be linked. Values between 3 and 20 are suitable;
- DTHR (Linking Distance Threshold) (In pixels): The maximum distance between two polylines to be linked. Values between 10 and 45 are acceptable.

4.2.2. Iron Mineralization and Alteration Detection

Lowell and Guilbert (1970) described the San Manuel-Kalamazoo deposit and compared the results with 27 other porphyry-Cu deposits. According to this model, four alteration zones were introduced, which are often used for porphyry-Cu exploration. As shown in Figure 3, the zones in this model from the center to the outside are potassic, phyllic, argillic, and propylitic zones [90,91].

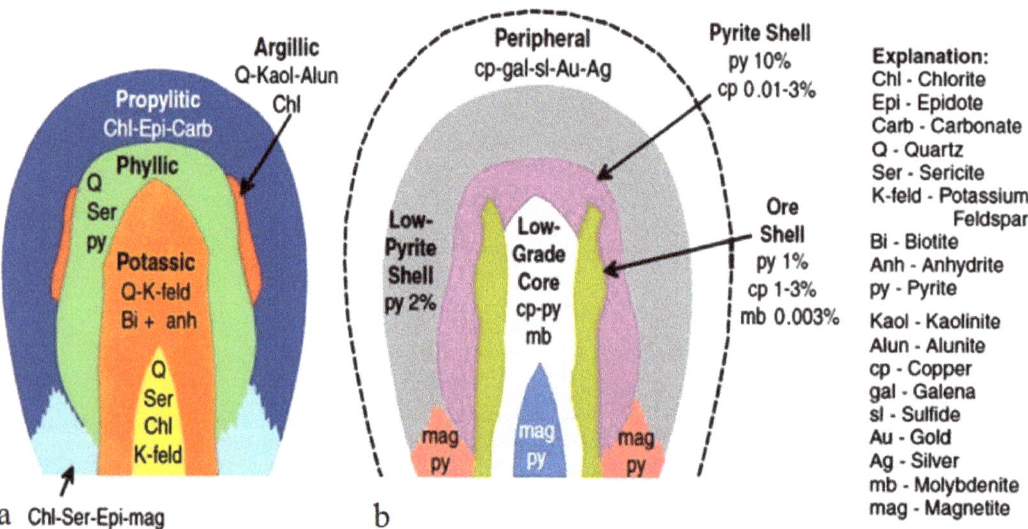

Figure 3. Alteration zones associated with porphyry-Cu deposit (modified after Reference [90]): (**a**) schematic cross section of alteration zones; (**b**) schematic cross section of ores associated with each alteration zone.

After presenting the Lowell–Guilbert model, some porphyry-Cu deposits were recognized to be associated with intrusive rocks posessing low silica. Hollister (1975) introduced this model and called it the diorite model, although the host pluton may be syenite, monzonite, and diorite [92].

Diorite's model differs from the Lowell–Guilbert model. In the Diorite model, sulfur concentrations were relatively low in mineralizing fluids. As a result, not all the iron oxides in the host rocks were converted to pyrite, and there are many iron remains in chlorites and biotites. Excess iron tends to occur as magnetite, which may be present in all alteration zones. Therefore, phyllic and argillic alteration zones are usually absent so that the potassic zone is surrounded by the propylitic zone [93,94].

Band Ratio

Band ratios are a very useful method for highlighting certain features or materials that cannot be seen in raw bands [95]. This method was applied to the Sentinel-2 MSI image to detect iron mineralization.

Color Composite

Colors provide more visual and conceptual information of the image. The combination of three black and white images creates a new image that can provide a better interpretation of surface features [96]. This method was used in the ASTER image for better visual interpretation of the alteration areas.

Logical Operator Algorithm

Mars and Rowan (2006) developed two logical operator algorithms based on ASTER-defined band ratios for regional mappings of argillic and phyllic-altered rocks in the Zagros magmatic arc, Iran [97]. Mars (2013) used thermal images in-band ratios to map hydrothermal alterations [98]. The logical operator algorithm presented by Mars (2013) can be the best suited for hydrothermal alteration associated with porphyry-Cu mineralization on a regional scale. The logical operator algorithm performs a series of band ratios for each pixel. Each logical operator determines a true (one) or false (zero) value for each ratio by comparing the band ratio to a predetermined range of threshold values. All of the ratios in the algorithm have to be true for a value of one to be assigned to the byte image; otherwise, a zero value is produced. Thus, a byte image consisting of zeros and ones is produced with each algorithm [97]. Due to the geological settings in the study area, the logical operator algorithm was applied to the Aster image to map alterations.

4.2.3. Generation of The Geological Layer

The geological layer is created by combining rock units, faults, and alterations using the K-nearest neighbor (KNN) algorithm. The 1:100,000 scale map of Abrisham-Rud has complex and multistage geological settings, and some units may not have outcrop, whereas this scale does not pose a problem for this study. The layer of rock units was created by using the high value for intrusive rock and intermediate to base units, as well as the low value for other units. In the study area, faults are dense and intersect in different directions. Fault layers were created by buffers at 100 to 500 at 100 m intervals. The alteration layers for each argillic, phyllic, and propylitic (epidote-chlorite) alteration were created by buffers at 100 to 300 at 100 m intervals around the alteration.

K-Nearest Neighbor Algorithm

KNN is a non-parametrically supervised algorithm designed to solve regression and classification problems [99]. KNN is the fundamental and the simplest classification technique when there is little or no prior knowledge about the distribution of data [100]. This algorithm is quite successful when a large training data set [101] and many geological studies are provided [102–106].

KNN classifies objects based on the closest training examples in the feature space [107,108]. In order to classify or predict a new case, KNN relies on finding similar cases in training data. These cases are classified by voting for neighbor classes [109]. The optimal choice of the number of neighbors "K" depends on the metrics used for classification and regression purposes [109]. Thus, KNN algorithm predicts the target class through three steps [101,110]: (1) preparing the dataset consists of training, test, and feature; (2) measuring the distances between each test data and all training data depending on the weight values of each individual; and (3) finding "K" the neighbors nearest to the test data from training data based on distance and weight measurements.

The most common and simple distance metrics are Euclidean, Manhattan, and Minkowski. The Minkowski distance is generally a more complete form of distance metrics and is calculated based on Equation (9):

$$\text{Minkowski Distance} = \sqrt[\lambda]{\sum_{i=1}^{k} |x_i - y_i|^{\lambda}}, \tag{9}$$

where x and y are points to calculate the distance, k is the number of neighbors, and λ is the order of the Minkowski distance, which contains values greater than zero. Thus, where $\lambda = 2$, the Minkowski distance is equivalent to the Manhattan distance, and where $\lambda = 1$, it is equivalent to the Euclidean distance. The Manhattan distance is usually preferred over the more common Euclidean distance when there is high dimensionality in the data set [111].

In this study, the Manhattan distance was used as the nearest neighbor classifier, and weights were calculated on Equation (10) based on the distance from the target to predict in a neighborhood:

$$\text{weighted Manhattan Distance} = \sum_{i=1}^{k} W_i |x_i - y_i|, \tag{10}$$

where W is the weights for each nearest neighbor, $0 < W_i < 1$ and $\sum_{i=1}^{k} Ww_i = 1$.

5. Result and Discussion

5.1. Zonality Method

Table 1 shows the calculation of background and threshold values, as well as the Clark values for supra-ore (Pb and Zn) and sub-ore (Cu and Mo) elements in the study area.

Table 1. Background, threshold, and Clarke values for supra-ore (Pb and Zn) and sub-ore (Cu and Mo) elements in the study area.

Values	Pb (ppm)	Zn (ppm)	Cu (ppm)	Mo (ppm)
Background	13.5	62	43.3	0.58
Threshold	19.6	90.6	78.1	0.88
Clarke (Beus and Grigorian, 1977) [12]	12	75	40	1.1

Figure 4 shows the anomaly map of multiplicative geochemical halos of the supra-ore and sub-ore (Cu × Mo) elements. As shown in Figure 4, zones I and II have been considered to predict erosional levels, and they create the geological layer. As shown in Figure 4, zone I mostly implied sub-ore (Cu and Mo) elements, and zone II included both supra-ore (Pb and Zn) and sub-ore (Cu and Mo) elements. The co-existence of both supra-ore and sub-ore elements' local maxima implies blind mineralization [20].

The geochemical and geometrical similarity of genetically similar orebodies, the uniformity of ore and halo, and the tentative nature of geological and economic boundaries are all crucial for considering mining geochemical models. In addition, the vertical geochemical zonality index and their spatial associations with particular geological and geochemical factors are important aspects of mineral distributions for exploration and insight into ore geometry. The vertical geochemical zonality index could be used to estimate the erosional level of porphyry-Cu deposits [2,3]. In order to identify the erosional level in zones I and II, the presented models by Ziaii (1996) and Ziaii et al. (2009) were used [2,67].

Ziaii (1996) introduced the vertical zonality model for porphyry-Cu deposits using areal productivity and the zonality index (Equation (6)) based on porphyry-Cu deposits in Kazakhstan, Bulgaria, Armenia, and Iran (Figure 5a). The vertical variations in three zonality indices associated with porphyry-Cu deposits in areas of the same landscape-geochemical conditions in different countries are shown in Figure 5a. Values of each zonality index decrease downward uniformly despite considerable differences in local geological settings of individual porphyry-Cu deposits, suggesting the existence of uniform vertical zonality in primary halos of porphyry-Cu deposits [2,6,71]. Therefore, vertical variations in the indices allow the distinction of mineralization levels and their primary halos (supra ore, upper-ore, ore, lower-ore, and sub-ore) [6,22,24].

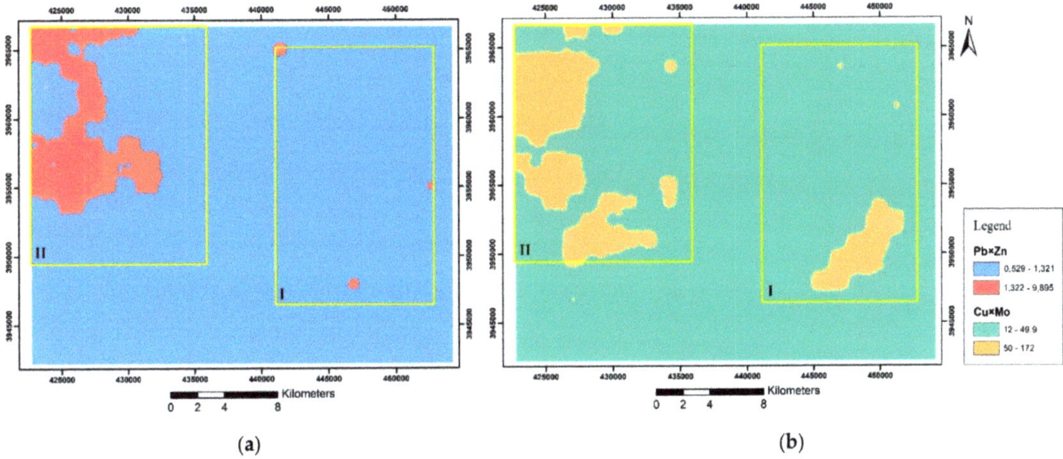

Figure 4. Geochemical maps of (**a**) supra-ore and (**b**) sub-ore elements, zones I and II, in the study area.

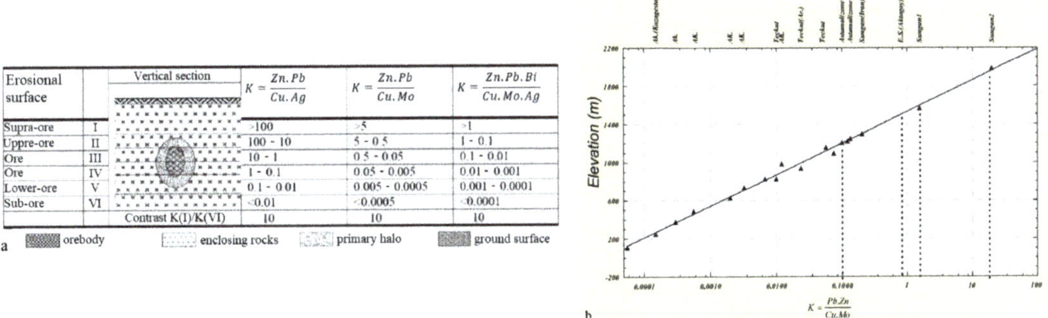

Figure 5. (**a**) Vertical geochemical zonality models for porphyry-Cu deposits based on typical standard porphyry-Cu deposits in Kazakhstan, Bulgaria, Armenia, and Iran (modified after Reference [71]); (**b**) geochemical model for porphyry-Cu deposit based on the porphyry Cu deposits database comprising Aktogy (Kazakhstan), Asarel (Bulgaria), Tekhut (Armenia), and Sungun (Iran) (modified after Reference [2]).

Moreover, it can be deduced from Figure 5a that similar values of the zonality index imply similar depths of mineralization and primary halos within an ore field. Thus, primary halos of mineral deposits at different depths are characterized by specific values of the zonality index. The practical exploration significance of the zonality index is for the recognition of erosional surfaces representing vertical levels of geochemical anomalies. Concerning the present erosional level, high values of the zonality index imply the presence of sub-cropping to BM, whereas low values of the index imply outcropping or already eroded deposits [14].

The values of areal productivity, mineralization coefficient, and zonality indices were introduced by Solovov (K_S) and Grigorian (K_G) in two zones (Table 2). The values of (Ks) in zones I and II are equal to 0.18 and 26.57, respectively. Considering the presented model by Ziaii (1996) [71], zones I and II are ZDM and BM, respectively (Figure 5a).

Table 2. Areal productivity, mineralization coefficient, and zonality index introduced by Solovov and zonality index introduced by Grigorian for Cu, Mo, Pb, and Zn elements in zones I and II.

Zone	Elements	P, m²%	$\eta(\alpha)$	K_S	K_G
Zone I	Cu	1,430,818,851	19.62		
	Mo	6,626,021	0.0735		
	Pb	12,649,676	0.3564	0.18	0.71
	Zn	133,323,564	2.856		
Zone II	Cu	1,094,297,325	8.6		
	Mo	32,266,712	0.275		
	Pb	552,971,529	4.2	26.57	28.23
	Zn	1,696,662,825	16		

Ziaii et al. (2009) introduced the geochemical model for porphyry-Cu deposits in Aktogy (Kazakhstan), Asarel (Bulgaria), Tekhut (Armenia), and Sungun (Iran) using mineralization coefficient and zonality index (Figure 5b) [2]. This plot shows the depth of mineralization versus the zonality index (KG). Despite considerable differences in geological settings, the linear relationship suggests the existence of a quantitatively uniform vertical geochemical zonality in the structure of primary halos of the deposits.

In this study, the values of (KG) in zones I and II are equal to 0.71 and 28.23, respectively (Table 2). According to the presented model by Ziaii et al. (2009), erosional levels in zone I and II are nearly similar to Astamal and Songun 2 areas, respectively (Figure 5b). Based on the previous studies, Sungun 2 and Astamal areas have been recognized as BM and ZDM, respectively [2].

Therefore, the results of the introduced models by Ziaii (1996) and Ziaii et al. (2009) confirm each other in identifying the erosional level in each zone [2,71].

5.2. Remote Sensing

Lineaments Extraction

The procedure of lineament extraction was accomplished in this manner: the PCA image of six Sentinel-2 MSI bands (bands 2, 3, 4, 8, 11, and 12), as shown in Figure 6a. PC1 explains the largest amount of eigenvalue among six bands. PC1 with the loading of the same signs represents overall brightness in all bands [14], and it shows that the albedo is related largely to the topographic features [76] (Figure 6b). Then, a directional filter was applied using 3×3 kernels in four directions with azimuths of $0°$, $45°$, $90°$, and $135°$. By using these azimuths, this filter visually enhances edges striking N-S, NE-SW, E-W, and NW-SE, respectively.

The sum of the directional filter kernel arrays is zero. The result is that areas with uniform pixel values are zeroed in the output image, while those with variable pixel values are presented as bright edges. In PCI Geomatica software, differences in the values of six parameters of edge and line detection indicate the differences of opinion among researchers. The values proposed by Adiri et al. (2017) were used [43]. Finally, the lineaments obtained from azimuths of $0°$, $45°$, $90°$, and $135°$ were merged, and the repetitive segments and non-geological lineaments (river, road, etc.) were deleted.

Lineaments and lineaments density maps in two zones were demonstrated in Figure 7. Lineament density was used to find the correlation between the concentration of lineaments and the distribution of existing faults in the study area (Figure 7). In two zones, this comparison proved that the fault is well related to lineament density in most areas. These areas were generally recognized in the middle-upper Eocene units and intrusive rocks.

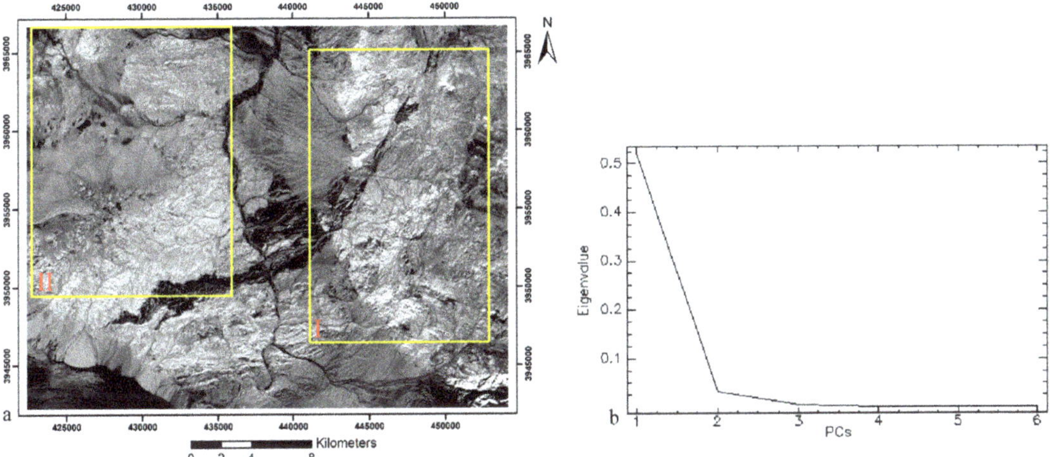

Figure 6. (**a**) PC1 image in the study area; (**b**) eigenvector obtained from PCA of six Sentinel-2 MSI bands (bands 2, 3, 4, 8, 11 and 12).

Lineament orientation allows identifying the most frequent directions of lineaments, and they can be compared with directions related to the existing faults [43]. As shown in Figure 8, in zones I and II, the directions of the lineaments correspond to existing faults.

Lineaments may be formed, for example, by structural alignment, geomorphologic consequences, structural weaknesses, faults, valleys, rivers, the boundaries between the different lithological units, vegetation cover, and artificial objects (road, bridge, etc.) [43]. In this study, due to the importance of faults in porphyry-Cu mineralization [6], the faults obtained from the geological map were used to create a geological layer.

5.3. Iron Mineralization and Alteration Detection

The results of the band ratio applied to Sentinel-2 MSI were shown in Figure 9. According to the rock units in the studied zones and Porphyry-Cu alterations zones, iron mineralization is dispersed and dense in two zones (Figure 9), which is well identified in areas containing volcanic and intrusive rocks. In the south of zone I, which includes intrusive rock consisting of monzodiorite and monzogabbro, iron mineralization is well recognized.

In ASTER, false-color composites of SWIR bands were used for better visual interpretation of the alteration areas. Empirical combinations have shown that an image with a false-color composite (band 4 in red, band 6 in green, and band 8 in blue) is the most suitable color composite for identifying alteration areas in porphyry-Cu deposits. As shown in Figure 10, areas with the propylitic alteration are shown in green to dark green based on the alteration intensity, and areas with the argillic and phyllic alteration are shown in white and pink to red. This is due to the high reflectivity of alunite, kaolinite, and muscovite minerals in band 4 compared to bands 6 and 8.

Geological settings in the study area have made it difficult to identify some alterations, especially in zone II. Rocks containing hydrous quartz, chalcedony, opal, and amorphous silica (hydrothermal silica-rich rocks); calcite-dolomite and epidote-chlorite (propylitic); alunite-pyrophyllite-kaolinite (argillic); and sericite-muscovite (phyllic) were mapped using

ASTER and logical operator algorithms (Figure 11). It is observed that Mars (2013) used images that have to differ in correction levels from the images used in this paper. Mars (2013) used "and (b4 gt 260)" to remove the black pixel [94], but in the images that have been used in this paper, the pixel has no value higher than 260. Thus, "and" in the algorithm causes the result to be zero (Table 3).

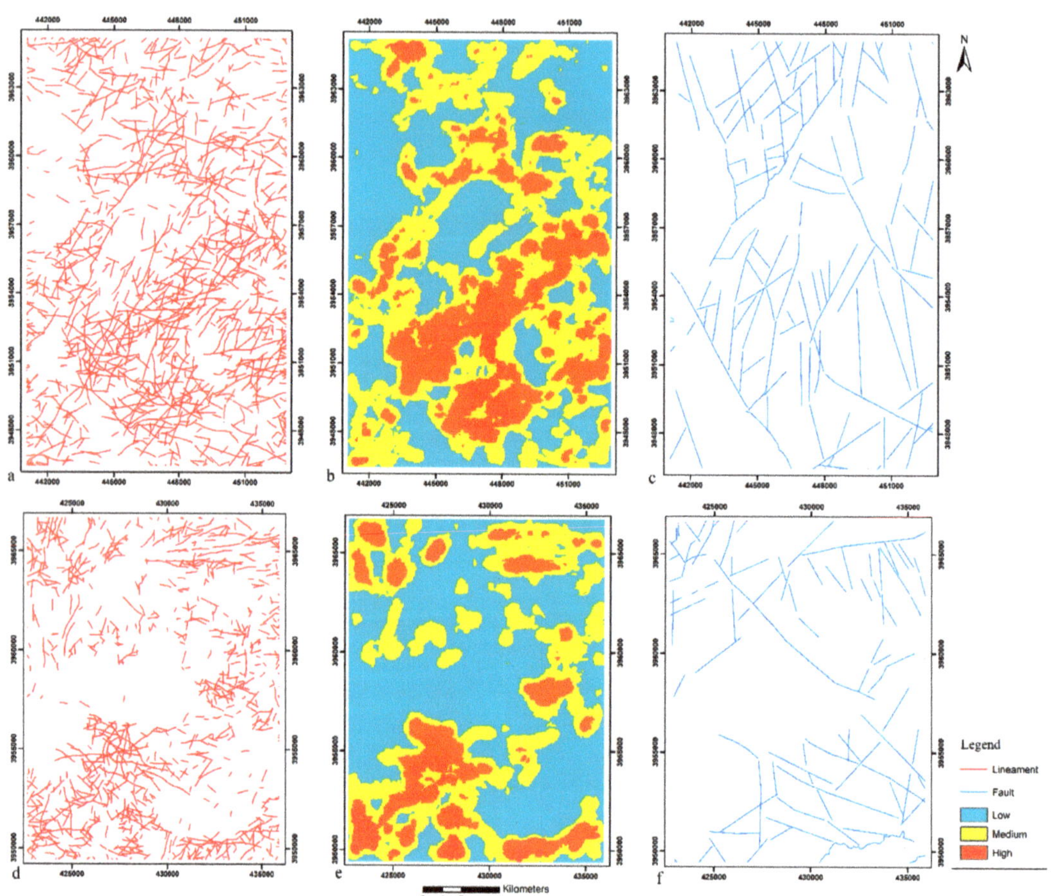

Figure 7. Maps of lineaments: (**a**) zone I and (**d**) zone II; lineaments density for (**b**) zone I and (**e**) zone II; existing faults for (**c**) zone I and (**f**) zone II.

Although not all alterations are associated with ore bodies and not all ore bodies are accompanied by alteration, the presence of altered rocks is a valuable indicator of possible deposits [112]. Kaolinite is mostly related to weathering feldspars, and epidote can be related to regional metamorphism. Kaolinite and epidote anomalies can have genesis related to deposit when they have a close relationship with muscovite anomalies [65].

Due to intrusive rocks, there is a possibility of the diorite model in the south of zone I. In this area, propylitic alteration (epidote–chlorite) was identified less than other parts of zone I, and argillic and phyllic alterations are well-identified around these intrusive rocks. It should be noted that all intrusive rocks have not shown alterations, which could be due to erosion or geological settings. Furthermore, alterations have been identified in other areas of zone I, which could imply mineralization areas or presence of minerals associated with alterations. In zone II, propylitic alteration (epidote–chlorite) was not identified, but argillic and phyllic alterations were detected in part of the studied area. Therefore, because of the absence of the alterations, it is not reasonable to create a geological layer in this zone.

5.4. Geological Layer

Rock units, faults, and alterations layers were combined by using the KNN algorithm and the geological layer, as shown in Figure 12. In this procedure, faults and alterations have an important role in mapping high potential areas. The density and intersection of faults and the extent of alterations represent these areas. Considering the geological layer, the detected areas as high potential could be related to mineralization or not. These areas can be compared to anomalous areas obtained by using the zonality method. As a result, more parts of the geological layer map were detected as high potential in comparison to the zonality method. However, the results of both methods confirm each other in the south of zone I. The geological layer and mineralogical investigation cannot identify erosional levels. In addition, mineralogy is not economical for application on a regional scale. In the study area, it is not recommended.

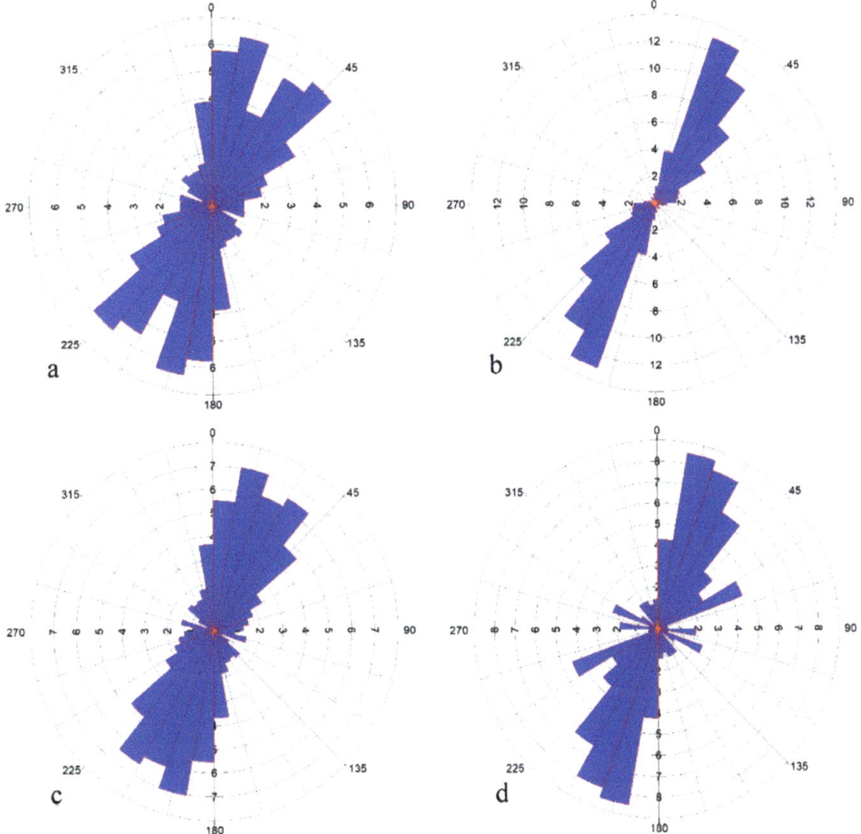

Figure 8. Orientations of lineaments of (**a**) zone I and (**c**) zone II compared to the faults of (**b**) zone I and (**d**) zone II.

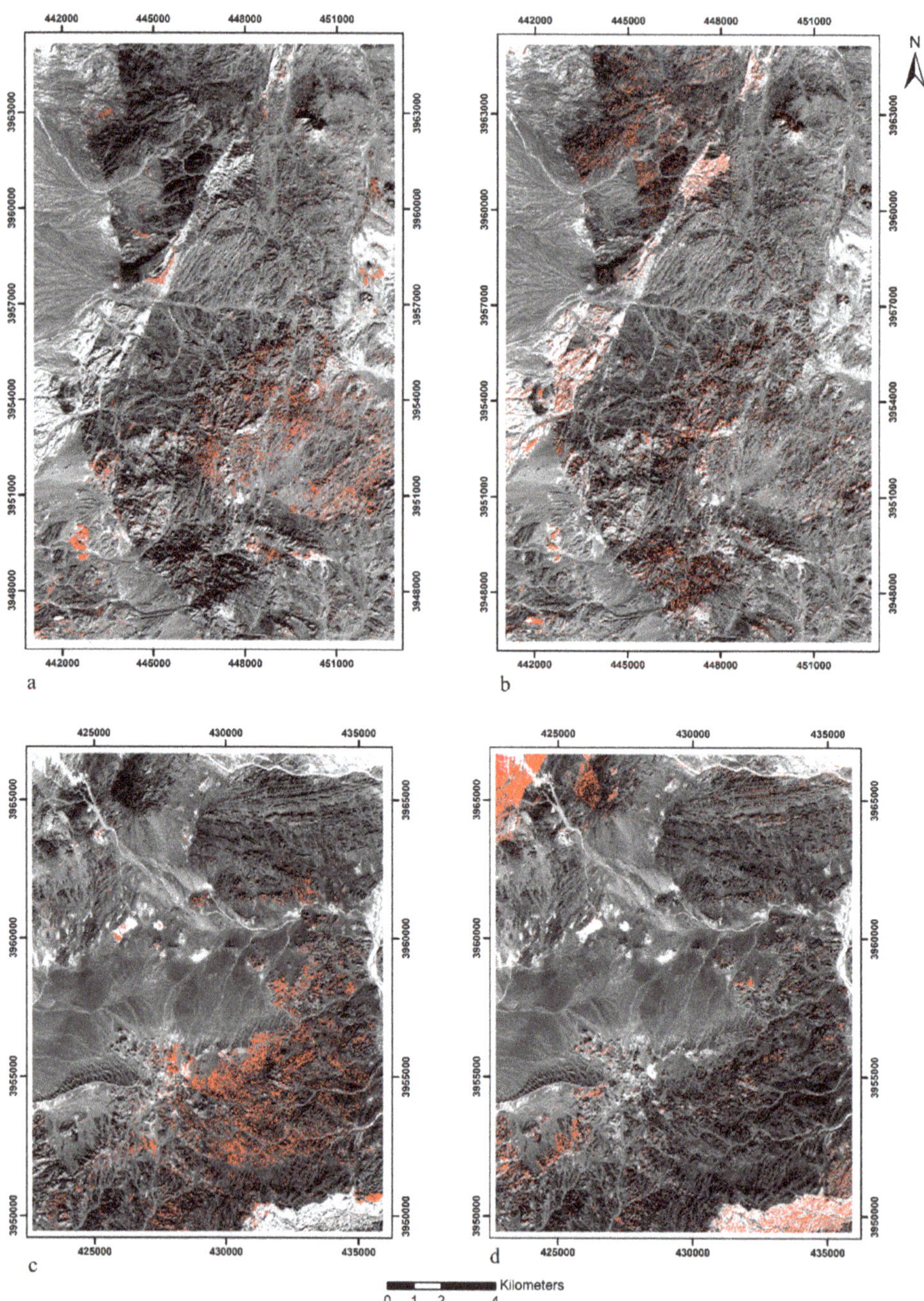

Figure 9. Band ratios of Sentinel-2 MSI image for identifying ferric iron (band 4/band 3): (**a**) zone I and (**c**) zone II; ferrous iron (band 12/band 8) + (band 3/band 4) in (**b**) zone I and (**d**) zone II on band 8.

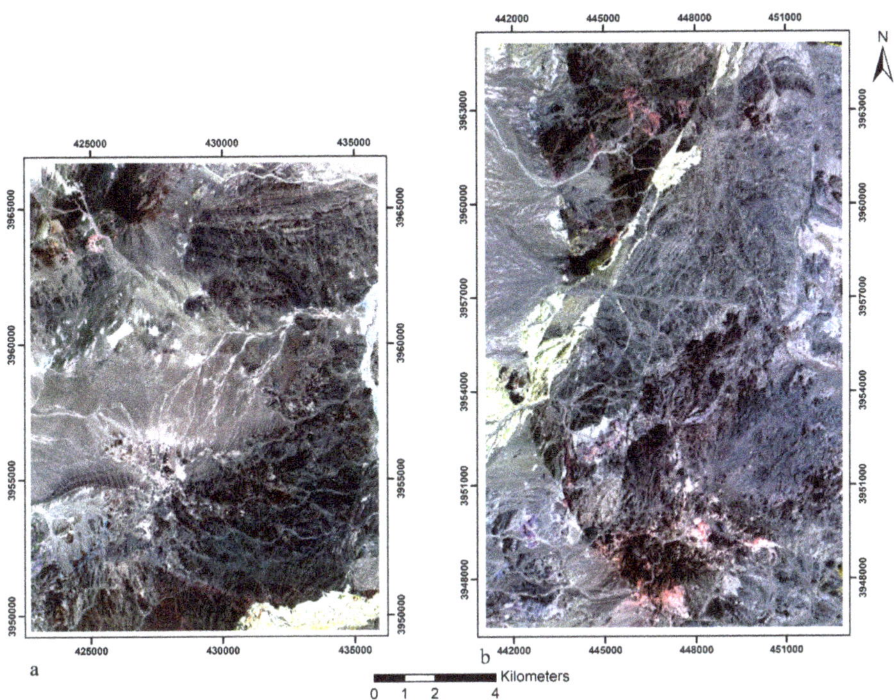

Figure 10. False-color composites of ASTER (band 4 in red, band 6 in green, and band 8 in blue) in (**a**) zone II and (**b**) zone I.

Table 3. The logical operator algorithms used with ASTER to map hydrothermally altered rocks in zones I and II (modified after Reference [98]). (b: band; float: floating point; le: less than or equal to; lt: less than; ge: greater than or equal to; gt: greater than.)

Zone	Hydrothermal Alteration	Algorithm
Zone I	Hydrothermal silica-rich (hydrous silica, chalcedony, opal) Propylitic (carbonate) Propylitic (epidote–chlorite) Argillic (alunite, kaolinite) Phyllic (sericite–muscovite)	((float(b3)/b2) le 1.06) and ((float(b4)/b7) ge 1.06) and ((float(b13)/b12) ge 1.016) and ((float(b12)/b11) lt 1.08) ((float(b3)/b2) le 1.06) and ((float(b6)/b8) gt 1.04) and (b5 gt b6) and (b7 gt b8) and (b9 gt b8) and ((float(b13)/b14) gt 1.005) ((float(b3)/b2) le 1.06) and ((float(b6)/b8) gt 1.04) and ((float(b5)/(float(b4)/b6)) gt 0.513) and (b5 gt b6) and (b6 gt b7) and (b7 gt b8) and (b9 gt b8) and ((float(b13)/b14) le 1.005) ((float(b3)/b2) le 1.06) and ((float(b4)/b6) gt 1.06) and ((float(b5)/b6) le 1.04) and ((float(b7)/b6) ge 1.04) ((float(b3)/b2) le 1.06) and ((float(b4)/b6) gt 1.06) and ((float(b5)/b6) gt 1.04) and ((float(b7)/b6) ge 1.04)
Zone II	Hydrothermal silica-rich (hydrous silica, chalcedony, opal) Propylitic (carbonate) Propylitic (epidote–chlorite) Argillic (alunite, kaolinite) Phyllic (sericite–muscovite)	((float(b3)/b2) le 0.66) and ((float(b4)/b7) ge 1.03) and ((float(b13)/b12) ge 1.156) and ((float(b12)/b11) lt 1.065) ((float(b3)/b2) le 0.66) and ((float(b6)/b8) gt 1.066) and (b5 gt b6) and (b7 gt b8) and (b9 gt b8) and ((float(b13)/b14) gt 0.91) ((float(b3)/b2) le 0.66) and ((float(b6)/b8) gt 1.066) and ((float(b5)/(float(b4)/b6)) gt 0.5) and (b5 gt b6) and (b6 gt b7) and (b7 gt b8) and (b9 gt b8) and ((float(b13)/b14) le 0.91) ((float(b3)/b2) le 0.66) and ((float(b4)/b6) gt 0.97) and ((float(b5)/b6) le 1.04) and ((float(b7)/b6) ge 1.04) ((float(b3)/b2) le 0.66) and ((float(b4)/b6) gt 0.97) and ((float(b5)/b6) gt 1.04) and ((float(b7)/b6) ge 1.04)

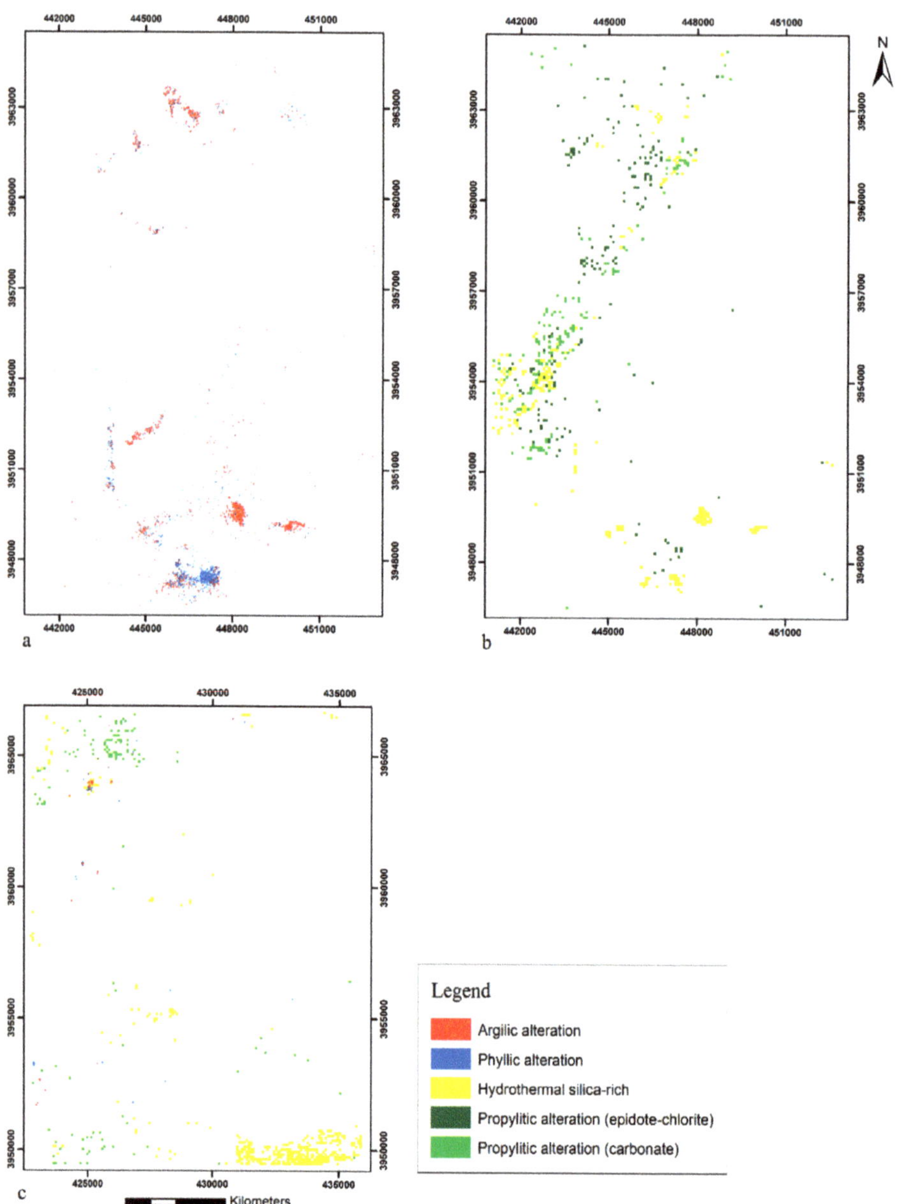

Figure 11. Hydrothermal alteration obtained from ASTER using logical operator algorithms. (**a**) Argillic and phyllic alteration in zone I; (**b**) hydrothermal silica-rich, propylitic alteration (carbonate), and propylitic alteration (epidote-chlorite) in zone I; (**c**) argillic, phyllic, and propylitic (carbonate) alteration and hydrothermal silica-rich in zone II.

Unlike the geological layer, the application of the zonality method in mineral prospecting allows further interpretation about whether delineated desirable areas are attractive for the exploration of ZDM or BM deposits.

This comparison demonstrates that the zonality method for detecting anomalous areas is more powerful than the geological layer.

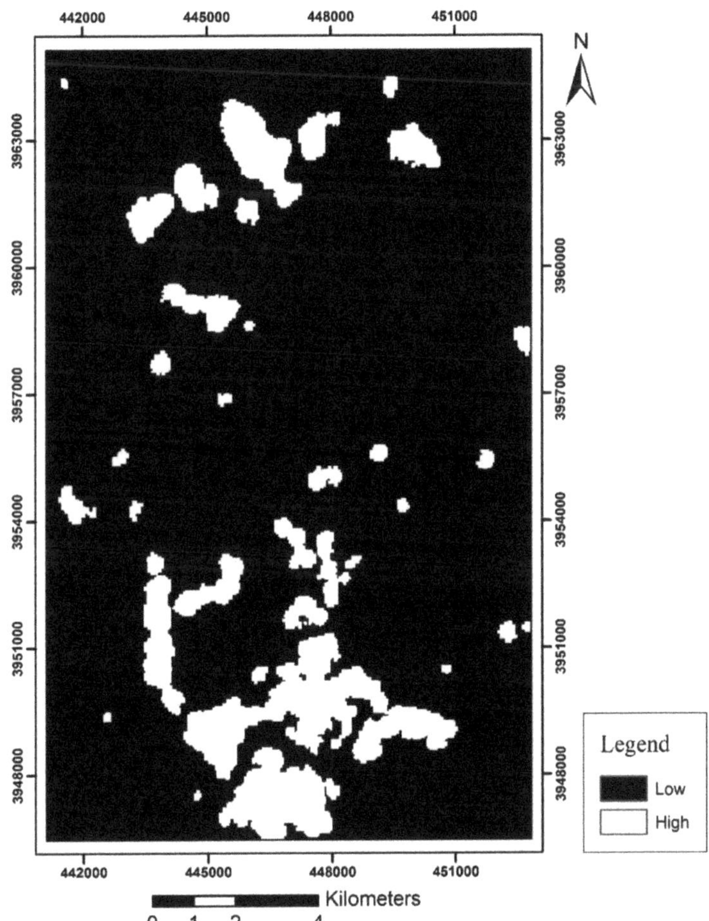

Figure 12. Geological layer map obtained by combining rock units, faults, and alterations layers by using the KNN algorithm in zone I.

6. Conclusions

The traditional zonality method has been used in the exploration of porphyry-Cu deposits for many years and is an effective method for the distinction between sub-ore and supra-ore halos, prediction of the erosional level of mineralization, and exploration of blind mineral deposits.

Utilizing the zonality method, the geochemical maps of multiplicative haloes were mapped. In the east of the study area, multiplicative haloes of sub-ore elements (Cu and Mo) were observed, and these represent zone dispersed mineralization. In the northwest, both multiplicative supra-ore (Pb and Zn) and sub-ore (Cu and Mo) element haloes existed, and these imply blind mineralization. Thus, zones I and II, which are located in the east and northwest of the study area, were selected for calculating erosional levels and for creating the geological map. Zonality indices introduced by Solovov (1987) (Ks) and Beus and Grigorian (1977) (KG) were calculated in zones I and II [12,22]. The (Ks) values were equal to 0.18 and 26.57 in zones I and II, respectively; moreover, (KG) values in zones I and II were equal to 0.71 and 28.23, respectively. Then, the presented models by Ziaii (1996) and Ziaii et al. (2009) were considered for identifying erosional levels in these zones [2,67]. Due to these models, zones I and II were recognized as ZDM and BM, respectively. Therefore, the zonality method was successfully applied in the identification of anomalous areas, separate BM from ZDM, and predicted erosional levels.

The results of the zonality method were compared to the geological layer, which was created by rock units, faults, and alterations by using the KNN algorithm. Each of these layers plays an important role in prospecting and exploring mineral deposits. Thus, high potential areas can be identified by combining these layers. For zones I and II, rock units and faults were identified from the geological map, and the alterations were detected using ASTER images and logical operator algorithms. It was observed that the alterations layers had a significant contribution in constructing the geological layer. The alteration zones of porphyry-Cu deposits include propylitic (chlorite and epidote), argillic (alunite and kaolinite), and phyllic (sericite and muscovite), and they are important for identifying possible areas associated with porphyry-Cu systems. These alterations were detected in zone I, especially around the intrusive rocks in the S of this zone. In zone II, only argillic and phyllic alterations were identified in part of the studied area. Due to the lack of alteration in zone II, the geological layer was created only in zone I. Comparing the results of the two methods showed that more parts of the geological layer map were highlighted as having high potential. These high potential areas could be related to mineralization or not; in other words, the geological layer cannot separate BM from ZDM. However, the results of both methods correspond to each other in the south of zone I. In other, the geological layer is unable to recognize erosional levels. Therefore, mineralogy investigation is required, which is not recommended to apply on a regional scale because of its high cost. It could be concluded that the geological layer, which is based on alteration, cannot help geochemists in separating BM from ZDM and in predicting erosional levels.

Author Contributions: M.Z. and M.R.G. conceived and designed the study and wrote the initial draft of the manuscript. M.R.G. and M.A. collected the samples and conducted sample preparation for analytical work. M.R.G. and M.Z. analyzed the data and conducted processing graphic and analytical work. T.T. revised and corrected the final version. All authors have read and agreed to the published version of the manuscript.

Funding: This research received no external funding.

Data Availability Statement: All information is collected by the researchers of this article and is not accessible anywhere.

Acknowledgments: This research was funded by the School of Mining, Petroleum and Geophysics Engineering at the Shahrood University of Technology. Laboratory research was carried out within the framework of the Tomsk Polytechnic University Development Program. Authors are thankful to anonymous reviewers and the editor for constructive criticisms and useful suggestions on the paper.

Conflicts of Interest: The authors declare no conflict of interest.

References

1. Fersman, A.E. *Geochemical and Mineralogical Methods of Prospecting for Mineral Deposits*; Academy of Science: Moscow, Russia, 1939. (In Russian)
2. Ziaii, M.; Pouyan, A.A.; Ziaei, M. Neuro-fuzzy modelling in mining geochemistry: Identification of geochemical anomalies. *J. Geochem. Explor.* **2009**, *100*, 25–36. [CrossRef]
3. Ziaii, M.; Ardejani, F.D.; Ziaei, M.; Soleymani, A.A. Neuro-fuzzy modeling based genetic algorithms for identification of geochemical anomalies in mining geochemistry. *Appl. Geochem.* **2012**, *27*, 663–676. [CrossRef]
4. Hamedani, M.L.; Plimer, I.R.; Xu, C. Orebody Modelling for Exploration: The Western Mineralisation, Broken Hill, NSW. *Nat. Resour. Res.* **2012**, *21*, 325–345. [CrossRef]
5. Grigorian, S.V. *Secondary Lithochemical Haloes in Prospecting for Hidden Mineralization*; Nedra Publishing House: Moscow, Russia, 1985. (In Russian)
6. Grigorian, S.V. *Mining Geochemistry*; Nedra Publishing House: Moscow, Russia, 1992. (In Russian)

7.   Ziaii, M.; Abedi, A.; Ziaei, M. Prediction of hidden ore bodies by new integrated computational model in marginal Lut region in east of Iran. *Proc. Explor.* **2007**, *7*, 957–961.

8.   Ziaii, M.; Pouyan, A.A.; Ziaei, M. Geochemical anomaly recognition using fuzzy C-means cluster analysis. *Wseas Trans. Syst.* **2006**, *5*, 2424–2429.

9.   Li, H.; Wang, Z.; Li, F. Ideal models of superimposed primary halos in hydrothermal gold deposits. *Geochem. Explor.* **1995**, *55*, 329–336. [CrossRef]

10. Li, H.; Zhang, G.Y.; Yu, B. *Tectonic Primary Halo Model and the Prospecting Effect During Deep Buried Ore Prospecting in Gold Deposits*; Geological Publishing House: Beijing, China, 2006.

11. Yongqing, C.; Pengda, Z. Zonation in primary halos and geochemical prospecting pattern for the Guilaizhuang gold deposit, eastern China. *Nonrenew. Resour.* **1998**, *7*, 37–44. [CrossRef]

12. Beus, A.A.; Grigorian, S.V. *Geochemical Exploration Methods for Mineral Deposits*; Applied Publishing Ltd.: Wilmette, IL, USA, 1977.

13. Yongqing, C.; Jingning, H.; Zhen, L. Geochemical characteristics and zonation of primary halos of pulang porphyry copper deposit, Northwestern Yunnan Province, Southwestern China. *J. China Univ. Geosci.* **2008**, *19*, 371–377. [CrossRef]

14. Ziaii, M.; Carranza, E.J.M.; Ziaei, M. Application of geochemical zonality coefficients in mineral prospectivity mapping. *Comput. Geosci.* **2011**, *37*, 1935–1945. [CrossRef]

15. Harraz, H.Z.; Hamdy, M.M. Zonation of primary haloes of Atud auriferous quartz vein deposit, Central Eastern Desert of Egypt: A potential exploration model targeting for hidden mesothermal gold deposits. *J. Afr. Earth Sci.* **2015**, *101*, 1–18. [CrossRef]

16. Safari, S.; Ziaii, M.; Ghoorchi, M. Integration of singularity and zonality methods for prospectivity map of blind mineralization. *Int. J. Min. Geo Eng.* **2016**, *50*, 189–194.

17. Imamalipour, A.; Mousavi, R. Vertical geochemical zonation in the Masjed Daghi porphyry copper-gold deposit, northwestern Iran: Implications for exploration of blind mineral deposits. *Geochem. Explor. Environ. Anal.* **2018**, *18*, 120–131. [CrossRef]

18. Imamalipour, A.; Karimlou, M.; Hajalilo, B. Geochemical zonality coefficients in the primary halo of Tavreh mercury prospect, northwestern Iran: Implications for exploration of listwaenitic type mercury deposits. *Geochem. Explor. Environ. Anal.* **2019**, *19*, 347–357. [CrossRef]

19. Safari, S.; Ziaii, M.; Ghoorchi, M.; Sadeghi, M. Application of concentration gradient coefficients in mining geochemistry: A comparison of copper mineralization in Iran and Canada. *J. Min. Environ.* **2018**, *9*, 277–292.

20. Ziaii, M.; Safari, S.; Timkin, T.; Voroshilov, V.; Yakich, T. Identification of geochemical anomalies of the porphyry–Cu deposits using concentration gradient modelling: A case study, Jebal-Barez area, Iran. *J. Geochem. Explor.* **2019**, *199*, 16–30. [CrossRef]

21. Safari, S.; Ziaii, M. Evaluation of geochemical anomalies in kerver deposit. *Iran. J. Min. Eng. IRJME* **2019**, *14*, 76–91. (In Persian)

22. Solovov, A.P. *Geochemical Prospecting for Mineral Deposits*; Mir: Moscow, Russia, 1987. (In Russian)

23. Baranov, E.V. *Endogenetic Halos Associated with Massive Sulphide Deposits*; Nedra Publishing House: Moscow, Russia, 1987. (In Russian)

24. Solovov, A.P.; Arkhipov, A.Y.; Bugrov, V.A. *Guidebook on Geochemical Exploration of Mineral Resources*; Nedra Publishing House: Moscow, Russia, 1990. (In Russian)

25. Liu, L.M.; Peng, S.L. Prediction of hidden ore bodies by synthesis of geological, geophysical and geochemical information based on dynamic model in Fenghuangshan ore field, Tongling district, China. *J. Geochem. Explor.* **2004**, *81*, 81–98. [CrossRef]

26. Asadi, H.H.; Sansoleimani, A.; Fatehi, M.; Carranza, E.J.M. An AHP–TOPSIS Predictive Model for District-Scale Mapping of Porphyry Cu–Au Potential: A Case Study from Salafchegan Area (Central Iran). *Nat. Resour. Res.* **2016**, *25*, 417–429. [CrossRef]

27. Yousefi, M.; Carranza, E.J.M. Union score and fuzzy logic mineral prospectivity mapping using discretized and continuous spatial evidence values. *J. Afr. Earth Sci.* **2017**, *128*, 47–60. [CrossRef]

28. Shabankareh, M.; Hezarkhani, A. Application of support vector machines for copper potential mapping in Kerman region, Iran. *J. Afr. Earth Sci.* **2017**, *128*, 116–126. [CrossRef]

29. Pazand, K.; Hezarkhani, A. Predictive Cu porphyry potential mapping using fuzzy modelling in Ahar–Arasbaran zone, Iran. *Geol. Ecol. Landsc.* **2018**, *2*, 229–239. [CrossRef]

30. Mars, J.C.; Robinson, G.R., Jr.; Hammarstrom, J.M.; Zürcher, L.; Whitney, H.; Solano, F.; Gettings, M.; Ludington, S. Porphyry copper potential of the United States southern basin and range using ASTER data integrated with geochemical and geologic datasets to assess potential near-surface deposits in well-explored permissive tracts. *Econ. Geol.* **2019**, *114*, 1095–1121. [CrossRef]

31. Ford, A. Practical implementation of random forest-based mineral potential mapping for porphyry Cu–Au mineralization in the Eastern Lachlan Orogen, NSW, Australia. *Nat. Resour. Res.* **2020**, *29*, 267–283. [CrossRef]

32. Voroshilov, V.G. Anomalous structures of geochemical fields of hydrothermal gold deposits: Formation mechanism, methods of geometrization, typical models, and forecasting of ore mineralization. *Geol. Ore Depos.* **2009**, *51*, 1–16. [CrossRef]

33. Shirazy, A.; Hezarkhani, A.; Timkin, T.; Shirazi, A. Investigation of Magneto-/Radio-Metric Behavior in Order to Identify an Estimator Model Using K-Means Clustering and Artificial Neural Network (ANN) (Iron Ore Deposit, Yazd, IRAN). *Minerals* **2021**, *11*, 1304. [CrossRef]

34. Shirazy, A.; Ziaii, M.; Hezarkhani, A.; Timkin, T.V.; Voroshilov, V.G. Geochemical behavior investigation based on K-means and artificial neural network prediction for titanium and zinc, Kivi region, Iran. *Bull. Tomsk Polytech. Univ. Geo Assets Eng.* **2021**, *332*, 113–125.

35. Carranza, E.J.M. Geologically Constrained Mineral Potential Mapping: Examples from the Philippines. Ph.D. Thesis, Technische Universiteit Delft, Delft, The Netherlands, 2002.

36. Adiri, Z.; Lhissou, R.; El Harti, A.; Jellouli, A.; Chakouri, M. Recent advances in the use of public domain satellite imagery for mineral exploration: A review of Landsat-8 and Sentinel-2 applications. *Ore Geol. Rev.* **2020**, *117*, 103332. [CrossRef]
37. Cox, D.P.; Singer, D.A. Mineral deposit models. *U.S. Geol. Surv. Bull.* **1693**, *1986*, 393.
38. Yumul, G.P., Jr.; Dimalanta, C.B.; Gabo-Ratio, J.A.S.; Armada, L.T.; Queaño, K.L.; Jabagat, K.D. Mineralization parameters and exploration targeting for gold—Copper deposits in the Baguio (Luzon) and Pacific Cordillera (Mindanao) Mineral Districts, Philippines: A review. *J. Asian Earth Sci.* **2020**, *191*, 104232. [CrossRef]
39. Zarasvandi, A.; Liaghat, S.; Zentilli, M. Geology of the Darreh-Zerreshk and Ali-Abad porphyry copper deposits, central Iran. *Int. Geol. Rev.* **2005**, *47*, 620–646. [CrossRef]
40. Sillitoe, R.H. Porphyry Copper Systems. *Econ. Geol.* **2010**, *105*, 3–41. [CrossRef]
41. Mirzaie, A.; Bafti, S.S.; Derakhshani, R. Fault control on Cu mineralization in the Kerman porphyry copper belt, SE Iran: A fractal analysis. *Ore Geol. Rev.* **2015**, *71*, 237–247. [CrossRef]
42. Habibkhah, N.; Hassani, H.; Maghsoudi, A.; Honarmand, M. Application of numerical techniques to the recognition of structural controls on porphyry Cu mineralization: A case study of Dehaj area, Central Iran. *Geosystem Eng.* **2020**, *23*, 159–167. [CrossRef]
43. Adiri, Z.; El Harti, A.; Jellouli, A.; Lhissou, R.; Maacha, L.; Azmi, M.; Zouhair, M.; Bachaoui, E.M. Comparison of Landsat-8, ASTER and Sentinel 1 satellite remote sensing data in automatic lineaments extraction: A case study of Sidi Flah-Bouskour inlier, Moroccan Anti Atlas. *Adv. Space Res.* **2017**, *60*, 2355–2367. [CrossRef]
44. Azizi, H.; Tarverdi, M.A.; Akbarpour, A. Extraction of hydrothermal alterations from ASTER SWIR data from east Zanjan, northern Iran. *Adv. Space Res.* **2010**, *46*, 99–109. [CrossRef]
45. Mielke, C.; Boesche, N.K.; Rogass, C.; Kaufmann, H.; Gauert, C.; De Wit, M. Spaceborne mine waste mineralogy monitoring in South Africa, applications for modern push-broom missions: Hyperion/OLI and EnMAP/Sentinel-2. *Remote Sens.* **2014**, *6*, 6790–6816. [CrossRef]
46. Van der Werff, H.; Van der Meer, F. Sentinel-2A MSI and Landsat 8 OLI provide data continuity for geological remote sensing. *Remote Sens.* **2016**, *8*, 883. [CrossRef]
47. Pour, A.B.; Hashim, M.; Park, Y.; Hong, J.K. Mapping alteration mineral zones and lithological units in Antarctic regions using spectral bands of ASTER remote sensing data. *Geocarto Int.* **2018**, *33*, 1281–1306. [CrossRef]
48. Zhang, N.; Zhou, K. Identification of hydrothermal alteration zones of the Baogutu porphyry copper deposits in northwest China using ASTER data. *J. Appl. Remote Sens.* **2017**, *11*, 015016. [CrossRef]
49. Safari, M.; Maghsoudi, A.; Pour, A.B. Application of Landsat-8 and ASTER satellite remote sensing data for porphyry copper exploration: A case study from Shahr-e-Babak, Kerman, south of Iran. *Geocarto Int.* **2018**, *33*, 1186–1201. [CrossRef]
50. Testa, F.J.; Villanueva, C.; Cooke, D.R.; Zhang, L. Lithological and hydrothermal alteration mapping of epithermal, porphyry and tourmaline breccia districts in the Argentine Andes using ASTER imagery. *Remote Sens.* **2018**, *10*, 203. [CrossRef]
51. Noori, L.; Pour, A.B.; Askari, G.; Taghipour, N.; Pradhan, B.; Lee, C.W.; Honarmand, M. Comparison of different algorithms to map hydrothermal alteration zones using ASTER remote sensing data for polymetallic vein-type ore exploration: Toroud–Chahshirin magmatic belt (TCMB), North Iran. *Remote Sens.* **2019**, *11*, 495. [CrossRef]
52. Adiri, Z.; El Harti, A.; Jellouli, A.; Maacha, L.; Azmi, M.; Zouhair, M.; Bachaoui, E.M. Mapping copper mineralization using EO-1 Hyperion data fusion with Landsat 8 OLI and Sentinel-2A in Moroccan Anti-Atlas. *Geocarto Int.* **2020**, *35*, 781–800. [CrossRef]
53. Shirazy, A.; Ziaii, M.; Hezarkhani, A.; Timkin, T. Geostatistical and Remote Sensing Studies to Identify High Metallogenic Potential Regions in the Kivi Area of Iran. *Minerals* **2020**, *10*, 869. [CrossRef]
54. Zürcher, L.; Bookstrom, A.A.; Hammarstrom, J.M.; Mars, J.C.; Ludington, S.D.; Zientek, M.L.; Dunlap, P.; Wallis, J.C. Tectono-magmatic evolution of porphyry belts in the central Tethys region of Turkey, the Caucasus, Iran, western Pakistan, and southern Afghanistan. *Ore Geol. Rev.* **2019**, *111*, 102849. [CrossRef]
55. Orojnia, P. Lithology and provenance of Eocene volcanic rocks in 1:100000 scale map sheet of Abrisham–Roud. Master's Thesis, Geological Survey and Mineral Explorations of Iran, Tehran, Iran, 2003; p. 140. (In Persian).
56. Mars, J.C. *Regional Mapping of Hydrothermally Altered Igneous Rocks along the Urumieh-Dokhtar, Chagai, and Alborz Belts of Western Asia Using Advanced Spaceborne Thermal Emission and Reflection Radiometer (ASTER) Data and Interactive Data Language (IDL) Logical Operators: A Tool for Porphyry Copper Exploration and Assessment: Chapter O in Global Mineral Resource Assessment*; Scientific Investigations Report 2010-5090-O; U.S. Geological Survey: Reston, VA, USA, 2014; p. 36.
57. Nabavi, M.H. *An introduction to geology of Iran, Geological Survey and Mineral Explorations of Iran, Tehran*; Geological Survey of Iran: Tehran, Iran, 1976. (In Persian)
58. Zürcher, L.; Bookstrom, A.A.; Hammarstrom, J.M.; Mars, J.C.; Ludington, S.; Zientek, M.L.; Dunlap, P.; Wallis, J.C.; Drew, L.J.; Sutphin, D.M.; et al. *Porphyry copper assessment of the Tethys region of western and southern Asia. Chapter V in Global mineral resource assessment*; Scientific Investigations Report 2010-5090-V; U.S. Geological Survey: Reston, VA, USA, 2015; p. 232.
59. Samani, B. Distribution, setting and metallogenesis of copper deposits in Iran. In *Porphyry and Hydrothermal Copper and Gold Deposits. A Global Perspective*; PGC Publishing: Adelaide, SA, Australia, 1998; pp. 135–158.
60. Shamanian, G.H.; Hedenquist, J.W.; Hattori, K.H.; Hassanzadeh, J. The Gandy and Abolhassani epithermal prospects in the Alborz magmatic Arc, Semnan province, Northern Iran. *Econ. Geol.* **2004**, *99*, 691–712. [CrossRef]
61. Ghorbani, G.; Vosoughi Abedini, M.; Ghasemi, H.A. Geothermobarometry of granitoids from Torud –Chah Shirin area (south Damghan). *Iran. Iran. J. Crystallogr. Mineral.* **2005**, *13*, 95–106. (In Persian)

62. Navab Motlagh, A. *1:100000 Scale Map Sheet of Abrisham–Roud*; Geological Survey and Mineral Explorations of Iran: Tehran, Iran, 2004.
63. Drusch, M.; Del Bello, U.; Carlier, S.; Colin, O.; Fernandez, V.; Gascon, F.; Hoersch, B.; Isola, C.; Laberinti, P.; Martimort, P.; et al. Sentinel-2: ESA's optical high-resolution mission for GMES operational services. *Remote Sens. Environ.* **2012**, *120*, 25–36. [CrossRef]
64. Van der Meer, F.D.; Van der Werff, H.M.A.; Van Ruitenbeek, F.J.A. Potential of ESA's Sentinel-2 for geological applications. *Remote Sens. Environ.* **2014**, *148*, 124–133. [CrossRef]
65. Hu, B.; Xu, Y.; Wan, B.; Wu, X.; Yi, G. Hydrothermally altered mineral mapping using synthetic application of Sentinel-2A MSI, ASTER and Hyperion data in the Duolong area, Tibetan Plateau, China. *Ore Geol. Rev.* **2018**, *101*, 384–397. [CrossRef]
66. Fujisada, H. Design and performance of ASTER instrument. In Proceedings of the International Society for Optics and Photonics, Paris, France, 15 December 1995; Fujisada, H., Sweeting, M.N., Eds.; Scientific Research Publishing: Wuhan, China, 1995; pp. 16–25.
67. Yamaguchi, Y.I.; Fujisada, H.; Kahle, A.B.; Tsu, H.; Kato, M.; Watanabe, H.; Sato, I.; Kudoh, M. ASTER instrument performance, operation status, and application to Earth sciences. In *Proceedings of IEEE 2001 International Geoscience and Remote Sensing Symposium (Cat. No. 01CH37217)*, Sydney, NSW, Australia, 9–13 July 2001; IEEE: Piscatway, NJ, USA, 2001; Volume 3, pp. 1215–1216.
68. Rowan, L.C.; Mars, J.C.; Simpson, C.J. Lithologic mapping of the Mordor, NT, Australia ultramafic complex by using the Advanced Spaceborne Thermal Emission and Reflection Radiometer (ASTER). *Remote Sens. Environ.* **2005**, *99*, 105–126. [CrossRef]
69. Ben–Dor, E.; Kruse, F.A.; Lefkoff, A.B.; Banin, A. Comparison of three calibration techniques for utilization of GER 63-channel aircraft scanner data of Makhtesh Ramon, Negev, Israel. *Int. J. Rock Mech. Min. Sci. Geomech. Abstr.* **1994**, *60*, 1339–1354.
70. Kruse, F.A. Use of airborne imaging spectrometer data to map minerals associated with hydrothermally altered rocks in the northern grapevine mountains, Nevada, and California. *Remote Sens. Environ.* **1988**, *24*, 31–51. [CrossRef]
71. Ziaii, M. Lithogeochemical Exploration Methods for Porphyry Copper Deposit in Sungun, NW Iran. Master's Thesis, Moscow State University (MSU), Moscow, Russia, 1996. Unpublished (In Russian).
72. O'leary, D.W.; Friedman, J.D.; Pohn, H.A. Lineament, linear, lineation: Some proposed new standards for old terms. *Geol. Soc. Am. Bull.* **1976**, *87*, 1463–1469. [CrossRef]
73. Tosdal, R.M.; Richards, J.P. Magmatic and structural controls on the development of porphyry Cu ± Mo ± Au deposits. *Struct. Control. Ore Genesis. Soc. Econ. Geol.* **2001**, *14*, 157–181.
74. Pour, A.B.; Hashim, M. Geolgical structure mapping of the bentong–raub suture zone, peninsular Malaysia using palsar remote sensing data. *ISPRS Ann. Photogramm. Remote Sens. Spat. Inf. Sci.* **2015**, *2*, 89. [CrossRef]
75. Meshkani, S.A.; Mehrabi, B.; Yaghubpur, A.; Sadeghi, M. Recognition of the regional lineaments of Iran: Using geospatial data and their implications for exploration of metallic ore deposits. *Ore Geol. Rev.* **2013**, *55*, 48–63. [CrossRef]
76. Zoheir, B.; El-Wahed, M.A.; Pour, A.B.; Abdelnasser, A. Orogenic gold in Transpression and Transtension Zones: Field and remote sensing studies of the barramiya–mueilha sector, Egypt. *Remote Sens.* **2019**, *11*, 2122. [CrossRef]
77. Javhar, A.; Chen, X.; Bao, A.; Jamshed, A.; Yunus, M.; Jovid, A.; Latipa, T. Comparison of multi-resolution optical Landsat-8, Sentinel-2 and radar Sentinel-1 data for automatic lineament extraction: A case study of Alichur area, SE Pamir. *Remote Sens.* **2019**, *11*, 778. [CrossRef]
78. Tamani, F.; Hadji, R.; Hamad, A.; Hamed, Y. Integrating remotely sensed and GIS data for the detailed geological mapping in semi-arid regions: Case of Youks les Bains area, Tebessa province, NE Algeria. *Geotech. Geol. Eng.* **2019**, *37*, 2903–2913. [CrossRef]
79. Bentahar, I.; Raji, M.; Mhamdi, H.S. Fracture network mapping using Landsat-8 OLI, Sentinel-2A, ASTER, and ASTER-GDEM data, in the Rich area (Central High Atlas, Morocco). *Arab. J. Geosci.* **2020**, *13*, 1–19. [CrossRef]
80. Hashim, M.; Ahmad, S.; Johari, M.A.M.; Pour, A.B. Automatic lineament extraction in a heavily vegetated region using Landsat Enhanced Thematic Mapper (ETM+) imagery. *Adv. Space Res.* **2013**, *51*, 874–890. [CrossRef]
81. Farahbakhsh, E.; Chandra, R.; Olierook, H.K.; Scalzo, R.; Clark, C.; Reddy, S.M.; Müller, R.D. Computer vision-based framework for extracting tectonic lineaments from optical remote sensing data. *Int. J. Remote Sens.* **2020**, *41*, 1760–1787. [CrossRef]
82. Sedrette, S.; Rebai, N. Assessment approach for the automatic lineaments extraction results using multisource data and GIS environment: Case study in Nefza region in North-West of Tunisia. In *Mapping and Spatial Analysis of Socio-Economic and Environmental Indicators for Sustainable Development*; Springer: Berlin/Heidelberg, Germany, 2020; pp. 63–69.
83. Aretouyap, Z.; Billa, L.; Jones, M.; Richter, G. Geospatial and statistical interpretation of lineaments: Salinity intrusion in the Kribi-Campo coastland of Cameroon. *Adv. Space Res.* **2020**, *66*, 844–853. [CrossRef]
84. Pearson, K. LIII. On lines and planes of closest fit to systems of points in space. *Lond. Edinb. Dublin Philos. Mag. J. Sci.* **1901**, *2*, 559–572. [CrossRef]
85. Gabr, S.; Ghulam, A.; Kusky, T. Detecting areas of high-potential gold mineralization using ASTER data. *Ore Geol. Rev.* **2010**, *38*, 59–69. [CrossRef]
86. Adiri, Z.; El Harti, A.; Jellouli, A.; Maacha, L.; Bachaoui, E.M. Lithological mapping using Landsat 8 OLI and Terra ASTER multispectral data in the Bas Drâa inlier, Moroccan Anti Atlas. *J. Appl. Remote Sens.* **2016**, *10*, 016005. [CrossRef]
87. Aouragh, H.; Essahlaoui, A.; Ouali, A.; Hmaidi, A.E.; Kamel, S. Lineaments frequencies from Landsat ETM+ of the Middle Atlas Plateau (Morocco). *Res. J. Earth Sci.* **2012**, *4*, 23–29.
88. Amer, R.; Kusky, T.; El Mezayen, A. Remote sensing detection of gold related alteration zones in Um Rus area, Central Eastern Desert of Egypt. *Adv. Space Res.* **2012**, *49*, 121–134. [CrossRef]

89. Zhang, L.; Wu, X. An edge-guided image interpolation algorithm via directional filtering and data fusion. *IEEE Trans. Image Process.* **2006**, *15*, 2226–2238. [CrossRef]

90. Lowell, J.D.; Guilbert, J.M. Lateral and vertical alteration-mineralization zoning in porphyry ore deposits. *Econ. Geol.* **1970**, *65*, 373–408. [CrossRef]

91. John, D.A.; Ayuso, R.A.; Barton, M.D.; Blakely, R.J.; Bodnar, R.J.; Dilles, J.H.; Gray, F.; Graybeal, F.T.; Mars, J.C.; McPhee, D.K.; et al. *Porphyry Copper Deposit Model: Chapter B of Mineral Deposit Models for Resource Assessment*; U.S. Geological Survey Science Investigation Report, 2010-5070-B; U.S. Geological Survey: Reston, VA, USA, 2010; p. 169.

92. Hollister, V.F. An appraisal of the nature and source of porphyry copper deposits. *Miner. Sci. Eng.* **1975**, *7*, 225–233.

93. Hutchison, C.S. *Economic Deposits and Their Tectonic Setting*; Macmillan Higher Education: London, UK, 1983.

94. Evans, A.M. *Ore Geology and Industrial Minerals: An Introduction*; Blackwell Science Publications: Hoboken, NJ, USA, 1993.

95. Rowan, L.C.; Goetz, A.F.H.; Ashley, R.P. Discrimination of hydrothermally altered and unaltered rocks in visible and near infrared multispectral images. *Geophysics* **1977**, *42*, 522–535. [CrossRef]

96. Jun, L.; Songwei, C.; Duanyou, L.; Bin, W.; Shuo, L.; Liming, Z. Research on false color image composite and enhancement methods based on ratio images. *Int. Arch. Photogramm. Remote Sens. Spat. Inf. Sci.* **2008**, *37*, 1151–1154.

97. Mars, J.C.; Rowan, L.C. Regional mapping of phyllic- and argillic-altered rocks in the Zagros magmatic arc, Iran, using advanced spaceborne thermal emission and reflection radiometer (ASTER) data and logical operator algorithms. *Geosphere* **2006**, *2*, 161–186. [CrossRef]

98. Mars, J.C. *Hydrothermal Alteration Maps of the Central and Southern Basin and Range Province of the United States Compiled from Advanced Spaceborne Thermal Emission and Reflection Radiometer (ASTER) Data*; U.S. Geological Survey, Open-File Report 2013–1139; U.S. Geological Survey: Reston, VA, USA, 2013; p. 5.

99. Altman, N.S. An introduction to kernel and nearest-neighbor nonparametric regression. *Am. Stat.* **1992**, *46*, 175–185.

100. Devroye, L.; Gyorfi, L.; Krzyzak, A.; Lugosi, G. On the strong universal consistency of nearest neighbor regression function estimates. *Ann. Stat.* **1994**, *22*, 1371–1385. [CrossRef]

101. Mitchell, T.M. *Machine Learning*; McGraw-Hill Education: Burr Ridge, IL, USA, 1997.

102. Abedi, M.; Norouzi, G.H. Integration of various geophysical data with geological and geochemical data to determine additional drilling for copper exploration. *J. Appl. Geophys.* **2012**, *83*, 35–45. [CrossRef]

103. Zaremotlagh, S.; Hezarkhani, A.; Sadeghi, M. Detecting homogenous clusters using whole-rock chemical compositions and REE patterns: A graph-based geochemical approach. *J. Geochem. Explor.* **2016**, *170*, 94–106. [CrossRef]

104. Ghannadpour, S.S.; Hezarkhani, A.; Roodpeyma, T. Combination of separation methods and data mining techniques for prediction of anomalous areas in Susanvar, Central Iran. *J. Afr. Earth Sci.* **2017**, *134*, 516–525. [CrossRef]

105. Golmohammadi, A.; Jafarpour, B. Reducing uncertainty in conceptual prior models of complex geologic systems via integration of flow response data. *Comput. Geosci.* **2020**, *24*, 161–180. [CrossRef]

106. Shahabi, H.; Shirzadi, A.; Ghaderi, K.; Omidvar, E.; Al-Ansari, N.; Clague, J.J.; Geertsema, M.; Khosravi, K.; Amini, A.; Bahrami, S.; et al. Flood detection and susceptibility mapping using Sentinel-1 Remote Sensing Data and a Machine Learning Approach: Hybrid intelligence of bagging ensemble based on k-nearest neighbor classifier. *Remote Sens.* **2020**, *12*, 266. [CrossRef]

107. Cover, T.; Hart, P. Nearest neighbor pattern classification. *IEEE Trans. Inf. Theory* **1967**, *13*, 21–27. [CrossRef]

108. Devroye, L. On the asymptotic probability of error in nonparametric discrimination. *Ann. Stat.* **1981**, *9*, 1320–1327. [CrossRef]

109. Kumar, Y.; Janardan, R.; Gupta, P. Efficient algorithms for reverse proximity query problems. In *Proceedings of the 16th ACM SIGSPATIAL International Conference on Advances in Geographic Information Systems, Irvine, CA, USA, 5–7 November 2008*; Association for Computing Machinery: New York, NY, USA; pp. 1–10.

110. Ramamohanarao, K.; Fan, H. Patterns based classifiers. *World Wide Web* **2007**, *10*, 71–83. [CrossRef]

111. Aggarwal, C.C.; Hinneburg, A.; Keim, D.A. On the surprising behavior of distance metrics in high dimensional space. In Proceedings of the International Conference on Database Theory; Springer: Berlin/Heidelberg, Germany, 2001; pp. 420–434.

112. Sabins, F.F. Remote sensing for mineral exploration. *Ore Geol. Rev.* **1999**, *14*, 157–183. [CrossRef]

minerals

Article

Application of Dirichlet Process and Support Vector Machine Techniques for Mapping Alteration Zones Associated with Porphyry Copper Deposit Using ASTER Remote Sensing Imagery

Mastoureh Yousefi [1], Seyed Hassan Tabatabaei [1,*], Reyhaneh Rikhtehgaran [2], Amin Beiranvand Pour [3,*] and Biswajeet Pradhan [4,5,6]

1   Department of Mining Engineering, Isfahan University of Technology, Isfahan 84156-83111, Iran; m.usefi@mi.iut.ac.ir
2   Department of Mathematical Sciences, Isfahan University of Technology, Isfahan 84156-83111, Iran; r_rikhtehgaran@cc.iut.ac.ir
3   Institute of Oceanography and Environment (INOS), University Malaysia Terengganu (UMT), Kuala Nerus 21030, Terengganu, Malaysia
4   Centre for Advanced Modelling and Geospatial Information Systems (CAMGIS), Faculty of Engineering & IT, University of Technology Sydney, Sydney 2007, Australia; Biswajeet.pradhan@uts.edu.au
5   Center of Excellence for Climate Change Research, King Abdulaziz University, P.O. Box 80234, Jeddah 21589, Saudi Arabia
6   Earth Observation Centre, Institute of Climate Change, University Kebangsaan Malaysia, Bangi 43600, Selangor, Malaysia
*   Correspondence: tabatabaei@cc.iut.ac.ir (S.H.T.); beiranvand.pour@umt.edu.my (A.B.P.); Tel.: +98-3133915104 (S.H.T.); +60-9-6683824 (A.B.P.); Fax: +60-9-6692166 (A.B.P.)

Citation: Yousefi, M.; Tabatabaei, S.H.; Rikhtehgaran, R.; Pour, A.B.; Pradhan, B. Application of Dirichlet Process and Support Vector Machine Techniques for Mapping Alteration Zones Associated with Porphyry Copper Deposit Using ASTER Remote Sensing Imagery. *Minerals* **2021**, *11*, 1235. https://doi.org/10.3390/min11111235

Academic Editor: Paul Alexandre

Received: 2 September 2021
Accepted: 4 November 2021
Published: 6 November 2021

Publisher's Note: MDPI stays neutral with regard to jurisdictional claims in published maps and institutional affiliations.

Abstract: The application of machine learning (ML) algorithms for processing remote sensing data is momentous, particularly for mapping hydrothermal alteration zones associated with porphyry copper deposits. The unsupervised Dirichlet Process (DP) and the supervised Support Vector Machine (SVM) techniques can be executed for mapping hydrothermal alteration zones associated with porphyry copper deposits. The main objective of this investigation is to practice an algorithm that can accurately model the best training data as input for supervised methods such as SVM. For this purpose, the Zefreh porphyry copper deposit located in the Urumieh-Dokhtar Magmatic Arc (UDMA) of central Iran was selected and used as training data. Initially, using ASTER data, different alteration zones of the Zefreh porphyry copper deposit were detected by Band Ratio, Relative Band Depth (RBD), Linear Spectral Unmixing (LSU), Spectral Feature Fitting (SFF), and Orthogonal Subspace Projection (OSP) techniques. Then, using the DP method, the exact extent of each alteration was determined. Finally, the detected alterations were used as training data to identify similar alteration zones in full scene of ASTER using SVM and Spectral Angle Mapper (SAM) methods. Several high potential zones were identified in the study area. Field surveys and laboratory analysis were used to validate the image processing results. This investigation demonstrates that the application of the SVM algorithm for mapping hydrothermal alteration zones associated with porphyry copper deposits is broadly applicable to ASTER data and can be used for prospectivity mapping in many metallogenic provinces around the world.

Keywords: porphyry copper deposits; ASTER; machine learning; DP; SVM; SAM

1. Introduction

Because of the importance of minerals in industry and other aspects of human life, appropriate methods to explore minerals are essential. The use of remote sensing data to obtain information from far objects is one of the most significant technologies in this century. Remote sensing satellite imagery is extensively used in different sectors of Earth science such as mineral mapping [1–4]. The results of remote sensing studies, by means of saving time and cost in identifying alteration zones, have greatly contributed to the exploration of minerals, especially in the reconnaissance stages [5–8].

In recent decades, remote sensing has been used successfully in the identification of lithological units, structure features, and alterations zones with the development of new algorithms and ML techniques [9–11]. Owing to the high volume of remote sensing satellite data, data mining methods to extract the desired information are necessary [12,13]. Classification algorithms undoubtedly play an essential role in analyzing multidimensional data such as multispectral and hyperspectral images. Depending on need, different classification methods have been used for mineral mapping. These methods are generally divided into three categories: supervised, unsupervised, and semi-supervised. Supervised methods such as spectral angle mapping (SAM), support vector machines (SVM), and maximum likelihood (ML) have been widely used for remote sensing data processing with the aim of geological mapping [14,15]. The SVM method in the field of mineral mapping has been considered over the past two decades [16,17]. Clustering or unsupervised methods divide the data into groups to have the most similarity in each group and the least similarity between the groups [18]. Unlike supervised methods, these methods are less commonly used for remote sensing data processing in mineral exploration. For mineral exploration, clustering methods are usually used in conjunction with supervised methods to obtain better results. Semi-supervised methods aim to improve the results by combining these two methods [19]. Different clustering methods are used in various sciences, such as data mining, pattern recognition, image clustering, etc. [20]. These methods do not require training data. These methods are divided into two main categories: model-based and non-model-based. In non-model-based methods, the only parameter that needs to be known initially is the number of clusters [21]. Determining the number of clusters is a significant challenge that can be problematic in clustering big data [22]. Model-based methods do not even need to determine the number of clusters and can cluster the data without any information.

Despite proper performance in identifying minerals and alterations using supervised methods, preparing and selecting appropriate training data from them is costly and time-consuming. In this research, an attempt has been made to determine appropriate training data based on the nature of the data, using an approach consisting of clustering and classification methods. Then, using this training data for the supervised methods, identical areas in terms of alteration were identified. In this study, the basic model of the DP method was used to cluster the alteration zones of the Zefreh porphyry copper deposit, the UDMA, central Iran, using ASTER data. Then the results of this clustering were used as training data to identify corresponding alteration zones using the SVM method. The specific objectives of this research are: (i) to detect alteration zones in the Zefreh porphyry copper deposit using RBD, LSU, OSP, SFF algorithms; (ii) to determine the exact expansion of alteration zones in the Zefreh porphyry copper deposit using the DP method and use its results as training data for supervised methods; (iii) to perform SVM and SAM methods using training data obtained from the DP method and specify analogous alteration areas in the ASTER scene; and (iv) to verify the classification results using field checking of alteration zones.

2. Geology of the Study Area

The Zefreh porphyry copper deposit is located in the UDMA belt of central Iran, northeast (65 km) of Isfahan province. The location is bounded by latitudes $33°03'9.35''$ N and $33°03'54.5''$ N and longitudes $52°13'38.48''$ and $52°14'25.88''$ E (Figure 1a,b). The copper deposit is related to the Qom-Zefreh fault and its placement in the UDMA belt. Mechanism of movement of Qom-Zefreh and Naein-Baft and tensile performance between these two tectonic lineaments has led to the creation of longitudinal sliding tensile basins in the Zefreh region [23]. Crustal stretching along these strike–slip faults has facilitated the ascent and replacement of intrusive masses and the formation of dikes and the concentration of several mineralogical hydrothermal systems such as the Zefreh porphyry copper deposits, Kahang, Zafarqand, and the Kalchoye epithermal deposits, which are related to these two tectonic lineaments [23]. Volcanic activity in the Zefreh region occurred from the Eocene to the Miocene. Pyroclastic and andesitic lavas in the eastern, southeastern, and southwestern parts of the area are the oldest rock units in the region. These units have been altered to propylitic as a result of magmatic activity. In the central part, in the Late Eocene dacites, phyllic and argillic alterations zones are observed. Granodiorite subvolcanics are presented in the northeastern, penetrating dacite, pyroclastic, and andesitic lavas. In these stocks, weak potassium alteration and abundant quartz-magnetite veins are observed [24,25].

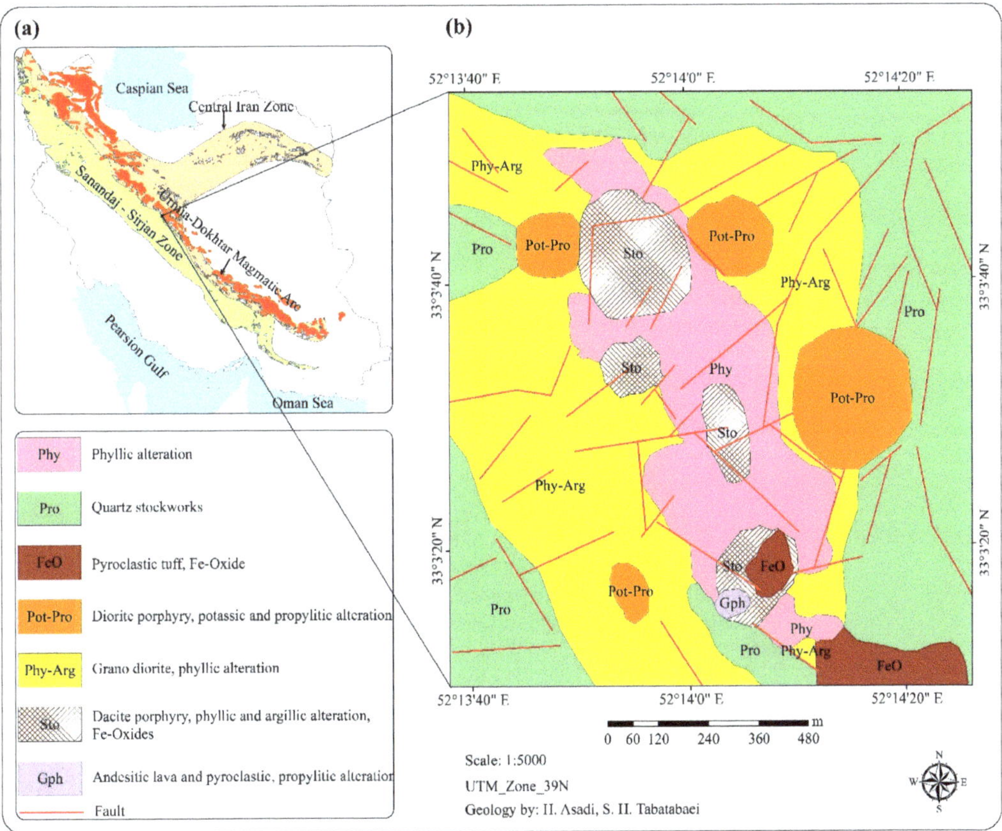

Figure 1. (**a**) The geographical location of the Zefreh area in the UDMA belt of central Iran; (**b**) geological map of the Zefreh area (scale 1:5000).

3. Materials and Methods

3.1. Data Characteristics

In this study, ASTER Level 1 T (Precision Terrain Corrected Registered At-Sensor Radiance) data were used. The ASTER sensor is a multispectral imager on NASA's Terra platform. ASTER has 14 bands in three subsystems, the visible and near infrared (VNIR) (3 bands), the shortwave infrared (SWIR) (6 bands), and the thermal infrared (TIR) (5 bands), in the range of 0.52–11.65 µm. The ASTER image has a spatial resolution of 15 m in the VNIR bands, 30 m in the SWIR bands, and 90 m in the TIR bands [26]. ASTER satellite imagery was designed based on geological needs, and it has been very efficient in this field over the last two decades [27]. ASTER data have an appropriate spectral and spatial resolution in the SWIR range, where many alteration minerals can be distinguished [8,9]. ASTER SWIR detectors are no longer functioning due to anomalously high SWIR detector temperatures. ASTER SWIR data acquired since April 2008 are not usable and show saturation of values and severe striping. However, VNIR and TIR data continue to show excellent quality, meeting all mission requirements and specifications. ASTER images can be downloaded from the "https://search.earthdata.nasa.gov/" site. To download the ASTER data, the ASTER granule ID can be found in the "https://earthexplorer.usgs.gov/" site.

The ASTER image used in this study was acquired on 11 March 2008. This ASTER scene covers the Zefreh porphyry copper deposit in the UDMA of central Iran. The image has 1% cloud coverage and is suitable for a remote sensing study. In this study, the nine bands of the VNIR and SWIR subsystems were stacked and used. The 30 m resolution SWIR of the ASTER data was re-sampled to correspond to the VNIR 15-m spatial dimensions. Nearest neighbor re-sampling method was applied to preserve the original pixel values in the re-sampled image. Radiometric and geometric corrections had been already applied on the ASTER L1T level data used in this study. ASTER data were also georeferenced and orthorectified [28]. The necessary preprocessing of this data included atmospheric correction and vegetation removal, which were subsequently done. Internal Average Relative Reflectance (IARR) correction was used to eliminate atmospheric effects. The IARR technique is recommended for mineralogical mapping as a preferred calibration technique in arid and semi-arid regions, because it does not require the prior knowledge of samples collected from the field [29]. Parts of the image that contained vegetation were identified with the NDVI index [30], and values greater than 0.3 were masked so that the results were not affected by vegetation reflectance. Figure 2 show the flowchart of the methodology used in this study.

3.2. Methods

3.2.1. Dirichlet Process (DP)

Owing to the nature of alterations, which are composed of different minerals with different values, their values can be modeled as distributions and can be separated from each other through the distribution of their compounds. In other words, different alterations can be separated into separate clusters. In this research, the DP method, which is based on the distribution over the dispersal of parameters, was used to model different alterations. In addition to the expected results, the advantage of using this method is that there was no need to determine the number of clusters.

In this study, considering that the DP clustering algorithm was implemented on the image in the Zefreh area with different lithologies, we assumed that each type of lithology was a multivariate normal distribution. Because each lithology was composed of a number of minerals with different compositions that have different spectral characteristics, we also considered their distribution to be normal. Because of the complexity of the composition of lithologies and their constituent minerals, we considered a hierarchical structure for the model parameters to fit well with the data structure.

The DP method is a non-parametric Bayesian method. DP was first introduced in 1973 by Ferguson [31]. This method was then developed and used in various sciences [32–34]. Mixed model DP uses a database distribution to model data that are mixed from several clusters. DP is generally formulated using Equation (1), but the number of model parameters is not fixed and can be changed as needed.

$$\begin{aligned} G &\sim DP(\alpha.G_0) \\ \theta_{z_i} &\sim G \\ P(z_i = k) &= \pi_k \\ z_i &\sim cat(\pi_k) \\ x_i | z_i.\theta_{z_i} &\sim F(\theta_{z_i}) \; i = 1:n \end{aligned} \tag{1}$$

where, G and G_0 are the distributions on the θ parameter. G_0 is the base distribution, and α is the concentration parameter of the Dirichlet distribution (Equation (1)). This parameter controls the degree of similarity of the G distribution to the base distribution. It is also effective in assigning a new sample to the previous cluster or being in a new cluster [35,36]. Equation (1) has a hierarchical structure so that each parameter is obtained from the posterior distribution of another parameter. θ is the parameter of data distribution. This study assumed that the values of each pixel x_i are a mixture of several clusters, and π_k is the mixing proportion of each cluster (k). The value of z_i was obtained from the categorical distribution on π_k.

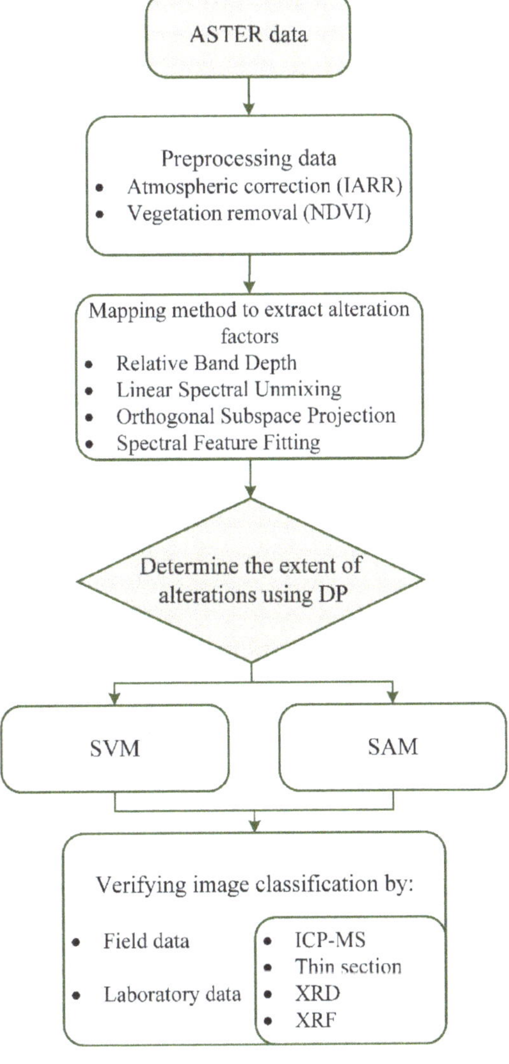

Figure 2. An overview of methodological flowchart used in this study.

Then Equation (2) was used to classify each data point (in this study, each pixel) in an existing cluster or a new cluster.

$$P(z_i = z | z_{-i}, x_{-i}, \theta) \propto \begin{cases} \frac{N_{-i,z}}{N-1+\alpha} F(x_i, \theta_c) & \text{if c exist} \\ \frac{\alpha}{N-1+\alpha} \int F(x_i, \theta) dG_0(\theta) & \text{new c} \end{cases}. \tag{2}$$

Several methods have been proposed in the literature to represent DP, including the Stick-Breaking (SB), Chinese restaurant, and, the Polya urn [35,37]. Here, the SB process was used for the probability of each cluster (Equation (3)). Each part of the SB models the probability of mixing proportions. In Equation (3), β is the beta distribution.

$$\pi_1 = \beta_1$$
$$\pi_k \sim \beta_k \prod_{j=1}^{k-1} \left(1 - \beta_j\right) \sum_{j=1}^{k} \pi_j = 1 \tag{3}$$
$$\beta_k \sim \text{Beta}(1.\alpha) \quad k = 2, 3, \ldots$$

As mentioned before, this method is non-parametric, and after constructing the model that fit the data, we were faced with several unknown parameters where the Markov chain Monte Carlo (MCMC) simulation was used to find their values. Using MCMC methods, the number of unknown quantities based on posterior probability is simulated in an acceptable way [38] (Equation (4)).

$$p(\theta, \pi | x_1, \ldots, x_n) \propto \prod_{i=1}^{n} \left\{ \sum_{j=1}^{k} \pi_j f\left(x_i | \theta_j\right) \right\} p(\theta) p(\pi) \tag{4}$$

3.2.2. Support Vector Machine (SVM)

Geo-computational methods for mapping minerals in satellite images, analysis of geochemical, geophysical data, etc., are kinds of classification because each method aims to find a prospect or non-prospect area [39]. SVM is one of the classification methods used to classify high-dimensional data and is suitable for cases where a limited number of training data are available [40].

The SVM algorithm was first used by [41] as a supervised method. Other studies have used this method as an unsupervised method [42], and a semi-supervised method [43] for clustering and classification. This method uses a hyperplane to separate the data (background value from an anomaly or desired from undesirable), which maximizes the margin between classes. SVM uses the pairwise classification strategy for multiclass classification. Suppose we have $x_i \in R^n$ $i = 1, \ldots, n$ educational data vectors (in this study, we had n as the number of pixels with dimension P) so that each pixel belongs to the class $y_i \in \{-1, 1\}$. Multiple hyperplanes can be used to separate data; a hyperplane with the maximal margin from the most external data of each class (Support vectors) is desirable. This hyperplane can be formulated as follows [39]:

$$f(X) = \text{sgn}\left(W^T X + b\right), \tag{5}$$

$$\text{sgn}(X) = \begin{cases} 1 & \text{if } x > 0 \\ 0 & \text{if } x = 0 \\ -1 & \text{if } x < 0 \end{cases} \tag{6}$$

The soft margin is used to obtain the parameters w and b by considering the variable ξ_i and the penalty function C (Equations (7) and (8)). This hyperplane permits the misclassification of some data in a controlled condition [44]:

$$\text{Minimize} \quad \tfrac{1}{2}||w||^2 + C\sum_{i=1}^{n}\xi_i$$

$$\text{Subject to} \quad \left\{ \begin{array}{l} y_i\left(W^TX_i + b\right) \geq 1 - \xi_i \ i = 1, \ldots, n \\ \xi_i \geq 0 \qquad i = 1, \ldots, n \end{array} \right. \tag{7}$$

$$\text{Minimize} \quad L(W,b,\alpha) = \frac{1}{2}||W||^2 - \sum_{i=1}^{n}\alpha_iy_i(W.X_i + b) + \sum_{i=1}^{n}\alpha_i \tag{8}$$

To minimize Equation (8) concerning W and b, we obtained the derivative of the above equation with respect to these variables (Equation (9)). Finally, we arrived at the following equations by placing the results (Equation (10)). By converting the problem to a quadratic programming problem and calculating the Lagrangian multipliers (Equation (11)), the problem is solved by finding the saddle point [39,44]:

$$\frac{\partial L}{\partial W} = 0, \quad \frac{\partial L}{\partial b} = 0 \tag{9}$$

$$W = \sum_{i=1}^{n}\alpha_iy_iX_i \quad \sum_{i=1}^{n}\alpha_iy_{i=0} \tag{10}$$

$$\text{Maximize} \ L(\alpha) = \sum_{i=1}^{n}\alpha_i - \frac{1}{2}\sum_{i,j=0}^{n}\alpha_i\alpha_jy_iy_jX_i.X_j = 0 \tag{11}$$

$$\text{Subject to} \quad \alpha_i \geq 0, \quad i = 1, \ldots, n, \quad \sum_{i=1}^{n}\alpha_iy_i = 0 \tag{12}$$

$$f(x) = \text{sgn}\left(\sum_{i,j=1}^{n}\alpha_iy_i(X_iX_j) + b\right) \tag{13}$$

In high-dimensional data, classification will be difficult. One way to overcome this problem is to use a kernel to transfer data to another feature space to make class separations easier and better. In this study, the Radial Basis Function (RBF) kernel was used (Equation (14)), which studies show has a better performance in this field. This kernel is like the K-nearest neighbor. It has all the advantages of a K-nearest neighbor. In addition, because it only needs to save support vectors instead of entire data it reduces space and complexity [45,46]. Finally, the decision function is changed as follows [44,47] (Equation (15)).

$$K(X_i, X_j) = e^{-\gamma(X_i - X_j)^2} \tag{14}$$

$$f(x) = \text{sgn}\left(\sum_{i,j=1}^{n}\alpha_iy_iK(X_i, X_j) + b\right) \tag{15}$$

3.2.3. Spectral Angle Mapper (SAM)

The SAM classification method is one of the most widely used methods in mineral mapping. The library spectrum, field spectrum, and image spectrum can be used for training or reference data in this method. Each pixel is considered a multidimensional vector with dimensions equal to the number of bands [48]. In the SAM method, the similarities between training or known data and test data in n-dimensional space are calculated with the angle between their spectra [48,49]. In this method, the direction of the spectra vectors is substantial, not their length, so the difference of light intensity in different parts of the image does not affect processing.

$$\text{SAM} \; = \; \arccos\left(\frac{\left\langle I_{\{k\}} \cdot J_{\{k\}} \right\rangle}{\|I_{\{k\}}\| \; \|J_{\{k\}}\|}\right). \tag{16}$$

In Equation (16), $I_{\{k\}}$ is the spectrum vector of the known data (in this study, Zefreh training data), and $J_{\{k\}}$ is the spectrum vector of the ASTER scene case study. $\langle \cdot \cdot \rangle$ indicates the scalar multiplication. $\| \cdot \|$ is the vector's norm [50,51].

3.2.4. Laboratory Analysis

Inductively coupled plasma–mass spectrometry (ICP-MS) analysis is one of the most accurate methods for measuring the value of elements in the selected samples. This analysis can detect and measure values less than one per billion (ppb). The input of the ICP-MS device must be a solution without suspended particles. The sample solution is sprayed into a plasma torch. The argon gas plasma ionizes the solution's molecules in the ICP. An electric field then accelerates these ions. Accelerated ions enter a magnetic field in the ICP device. The ions in the magnetic field are separated based on the charge-to-mass ratio, and the device can measure the value of each ion [52]. In this study, the collected rock samples were analyzed using a Perkin Elmer Sciex ELAN 9000 ICP-MS for some trace elements. The X-ray fluorescence (XRF) measured the value of sample compounds by bombarding the sample with X-rays or gamma rays and measuring the emission characteristic [53]. A Philips PW1480 XRF spectrometer was used in this study for measuring the percentage of major oxides in the selected rock samples. The samples were analyzed in the Zarazma Laboratory, Tehran, Iran. The results of these analysis are presented in Appendix A, Tables A1 and A2.

The thin section was a microscopic cut of rock, thickness between 25–30 μm, both sides were covered with glass slides. Thin sections were used for petrographic studies by optical microscopy. Quartz and feldspars should be gray to white in cross-polarized light in standard thin sections [54,55]. In this study, thin sections of alteration zones and lithological units were prepared. Thin sections were studied using the Kyowa ME-POL2 microscope (made in Japan) at magnification 20 in the Isfahan University of Technology, Iran. X-ray diffraction (XRD) was used to identify the crystal structure and major and minor minerals in a sample. In this method, the X-ray beam was irradiated to the sample, and the output diffraction pattern determined the type of mineral [56]. An ASENWARE/AW-XDM300 XRD diffractometer was used for measuring the important minerals in the sample collected in this study (Appendix A, Table A3). The XRD analysis was also performed in the Zarazma Laboratory, Tehran, Iran.

4. Results and Analysis

4.1. Determining the Training Data

In order to accurately determine the training data to use in the SVM and SAM algorithms, firstly, the alteration zones were identified by several mapping methods such as RBD, LSU, OSP, and SFF [57–60]. Then the exact extent of each alteration zone in the Zefreh porphyry copper deposit was determined using the DP algorithm.

4.2. Detection of the Alteration Zones

In each alteration, several indicator minerals had a specific spectral signature that made it possible to identify them in remote sensing images and determine the type of alteration. According to the kind of alteration, the location of enrichment elements and mineralization was identified. In this study, we used RBD, LSU, OSP, and SFF mapping methods to reveal phyllic, argillic, and propylitic alterations in the Zefreh porphyry copper deposit. Figure 3a shows an RGB color composite (R:3, G:2, B:1) of the ASTER full scene covering the study area.

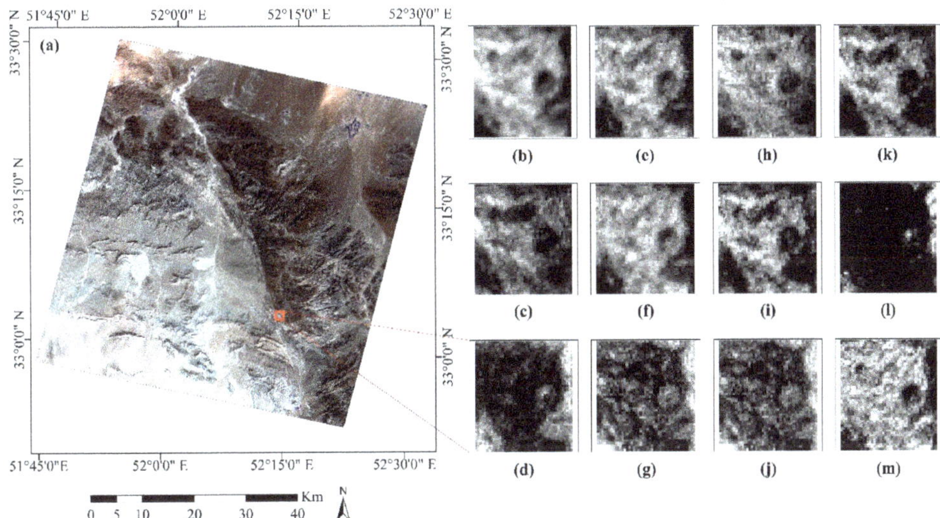

Figure 3. (**a**) Color composite of ASTER (R:3, G:2, B:1); (**b–d**) Phyllic, argillic and propylitic alteration results of the RBD method; (**e–g**) alteration results of the LSU method; (**h–j**) show mapping of phyllic, argillic and propylitic alterations using the OSP; (**k,l**) are the results of the SFF method; (**m**) shows the Fe-oxide alteration.

In the RBD method, considering the points of absorption and reflectance of mineral spectra, to determine the alterations, the band ratios (B7 + B5)/B6 for phyllic, the ratio (B7 + B4) / B5 for the argillic, and the ratio (B7 + B9)/B8 for propylitic alterations were used [61,62] (Figure 3b–d). To identify the alterations using LSU or SFF methods, the reference spectra related to the indicator minerals of each alteration zones were extracted from the USGS spectral library [63]. Figure 4 shows the USGS spectral of the indicator minerals after re-sampling to the ASTER band-passes. The phyllic alteration zone included sericite, illite, pyrite, and quartz [64]. The sericite mineral spectral signature was considered for mapping the phyllic zone. The argillic zone accumulated clay minerals, including illite, kaolinite, montmorillonite, alunite, halloysite, and quartz [64]. Argillic was identified by representative spectra of kaolinite and montmorillonite. The propylitic alteration zone consisted of epidote, calcite, and chlorite minerals, and was characterized mainly by the spectral signature of chlorite and epidote minerals [65]. Implementing the LSU method on the ASTER subset of the Zefreh porphyry copper deposit, the regions containing the indicator minerals manifested as bright pixels (Figure 3e–g). These images showed the mapping of phyllic, argillic, and propylitic alteration zones, respectively. LSU assumed that the value of each pixel was a linear combination of its endmembers in the fraction of endmembers with noise [57]. By projecting the pixel vector of the image in the subspaces, the OSP method eliminated the undesirable effects by increasing the signal-to-noise ratio, determining the spectral signature of the desired indicator mineral [66]. The results of the OSP method are shown in Figure 3h–j. The SFF method identified the desired areas by comparing the image spectrum with the spectral library spectrum, performing the least squares fitting, and selecting the best fit [48]. The SFF method showed acceptable results only for the phyllic and propylitic alteration (Figure 3k,l). The B2/B1 band ratio was used for mapping iron oxides (Figure 3m).

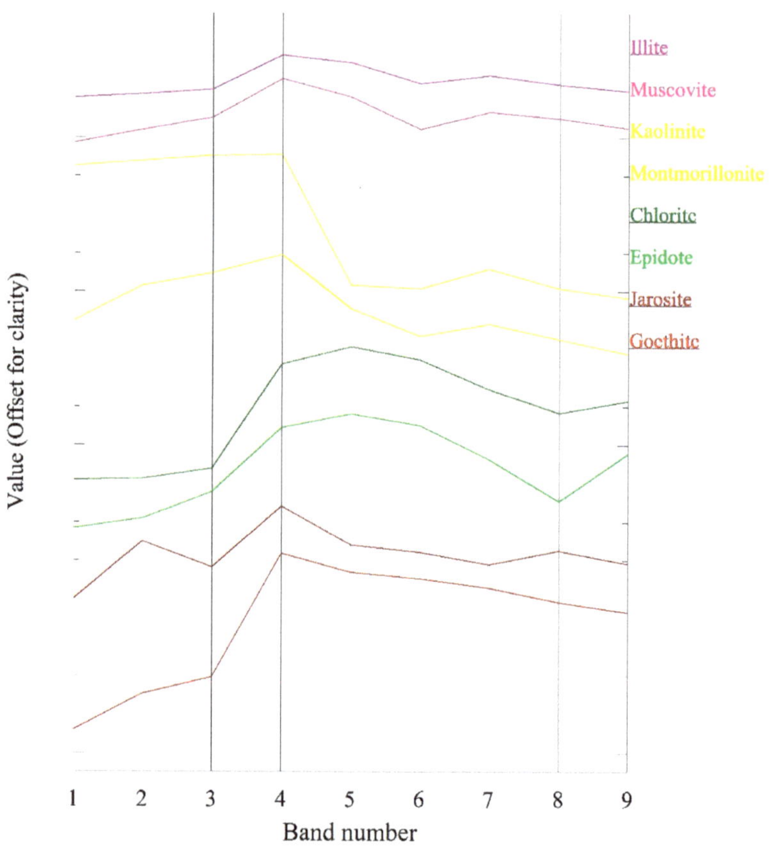

Figure 4. The spectral signatures (reflectance spectra) of indicator alteration minerals selected from the USGS spectral library that was re-sampled to the ASTER band-passes.

4.3. Implementation of the DP Method on the Zeftreh Area

The results obtained from different alteration mapping methods (Figure 3) were used as input to the implementation of the DB method. Using the DP method, the digital number (DN) value of the distribution of the pixels was assumed to be Gaussian. Considering the fact that rocks are composed of minerals and minerals are composed of elements, the DN values of each pixel were modeled as distributions over dispersals. This means that the value of each pixel (X_i) was considered a normal distribution with a distinct mean and variance. The number of different distributions was equal to the number of clusters. It was assumed that the mean value of these distributions had a normal distribution (base distribution) (Equations (17) and (18)). Gaussianness of data distribution is not required, and if the data distribution is not normal, the results will not be much different. As mentioned in Section 3.2.1, this method is a non-parametric method where, after defining the model and implementing it in the Bayesian inference Using Gibbs Sampling (BUGS) software, unknown parameters including the mean and variance of data distribution, probability of each cluster, and the number of clusters were identified by the MCMC simulation method. The BUGS was first released by Smith and Gelfand [67].

The parameter values were obtained after 3000 times MCMC simulation and removing the first 1000 unstable values (Figure 5a,b). The number of simulations varies and should continue until the value of the parameters converges.

$$X_i \sim \text{Normal}(\mu_1(C_i), \sigma_1(C_i)), \tag{17}$$

$$\mu_1(C_i) \sim \text{Normal}(\mu_2, \sigma_2), \tag{18}$$

$$C_i \sim \text{Categorical}(P(1:C)), \tag{19}$$

$$\sigma_1(C_i) \sim \text{Wishart}(R). \tag{20}$$

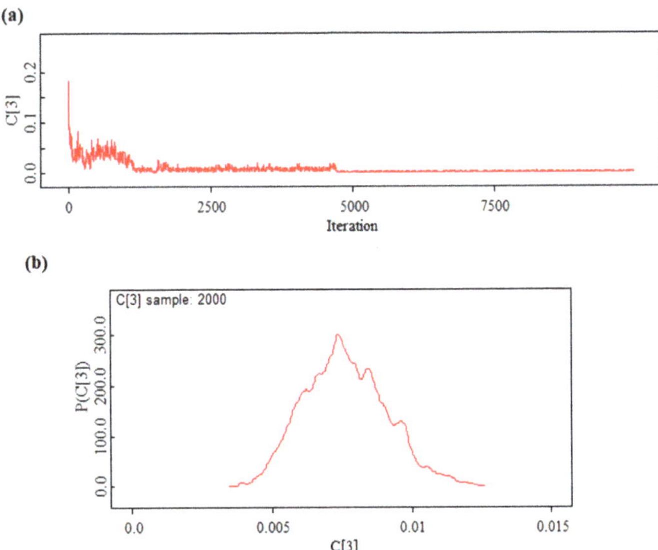

Figure 5. (**a**) MCMC simulation history of the C parameter to achieve convergence. (**b**) The posterior distribution of the parameter.

In the above equations, X_i specifies the DN value of each pixel, C_i is the categorical distribution of the probability occurrence of each cluster. μ and σ show the mean and variance of the normal distribution, respectively (in BUGS software, precision is used instead of variance). The precision value is calculated from the Wishart distribution [68,69].

The result of the implementation of the DP algorithm was the clustering of the ASTER image in Figure 6. The alteration zones in this image were more accurate than the geological map because the alteration zones were detected by the approach applied in this analysis. Therefore, the results of this clustering as training data provided more significant results compared to the existing geological map (Figure 1). Hence, the result of DP was used as the training data in the SVM and SAM supervised methods. The result of this clustering was five distinct clusters, and by adapting them to the geological map, we determined which kind of alterations defined each cluster. Figure 6 indicates that the areas showing the potassic zone were also mapped. This zone was not used as training data because the potassic spectra were not easily detectable in the SWIR and VNIR bands of ASTER. Mapping the potassic zone can be performed using the TIR bands, which was not considered in this analysis.

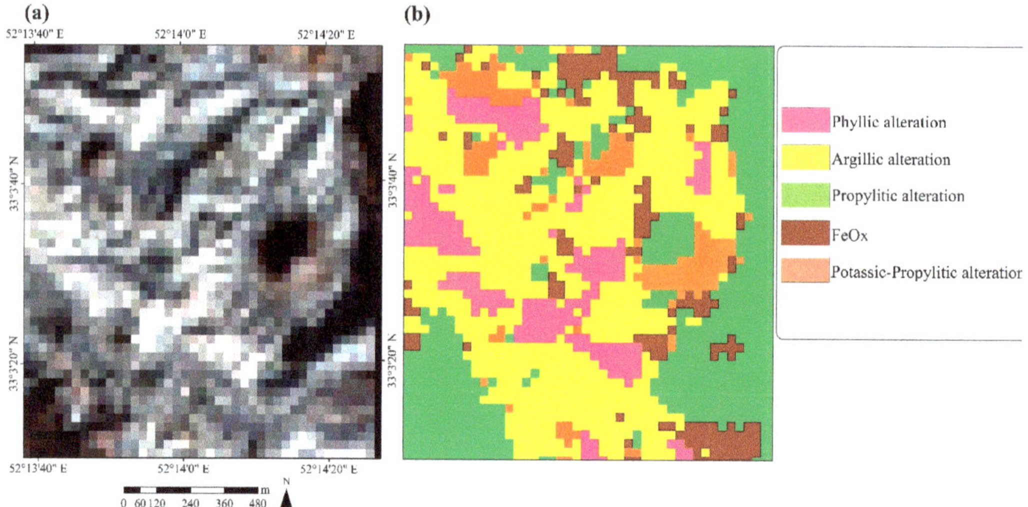

Figure 6. (**a**) The Zefreh ASTER image, (**b**) the DP clustering method results showing the spatial distribution of alteration zones in the study area.

4.4. Implementation of the SVM on ASTER Data

The alteration zones derived from the DP were used as training data to find similar alteration zones in the ASTER scene. We used the RBF kernel to transfer data to other spaces and classify the data. In the SVM method, the γ parameter and the penalty parameter must be defined. These parameters were optimized on the training data using the genetic algorithm in MATLAB software. The optimal values of the penalty and γ, 0.211 and 2, respectively, were obtained. Finally, the SVM algorithm was implemented on the ASTER image using the specified parameters and the training data using ENVI (Environment for Visualizing Images, http://www.exelisvis.com) version 5.3 software package (L3Harris Technologies, Melbourne, FL, USA). The results of this classification for phyllic, argillic and propylitic alterations are shown in Figure 7a–d.

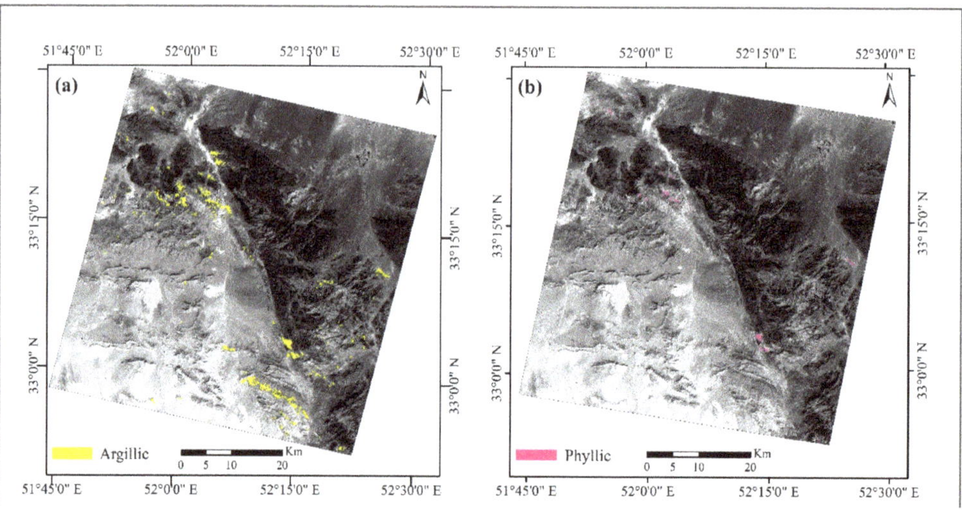

Figure 7. *Cont.*

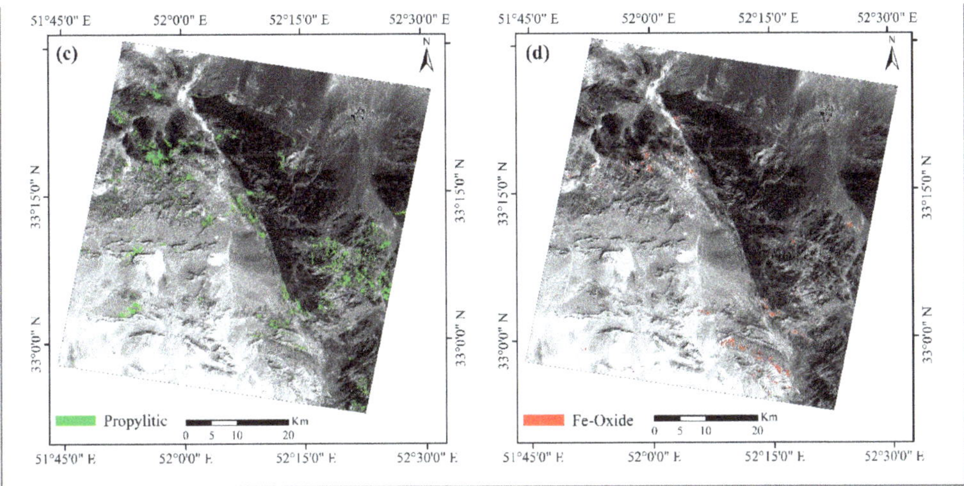

Figure 7. The results of SVM on the ASTER data. (**a**) Argillic alteration, (**b**) phyllic alteration, (**c**) propylitic alteration, and (**d**) Fe-oxide alteration.

4.5. Implementation of the SAM on ASTER Data

The SAM classification was used to compare the efficiency of the SVM results. DP clustering data was used as the training data for this algorithm. Phyllic, argillic, and propylitic alteration zones were mapped using the SAM algorithm on the ASTER image. The selected SAM spectral angles (in radians) used in this study were: $\alpha = 0.4$ for phyllic alteration, $\alpha = 0.25$ for argillic, and $\alpha = 0.3$ for propylitic. The results of the SAM classification are shown in Figure 8a–d.

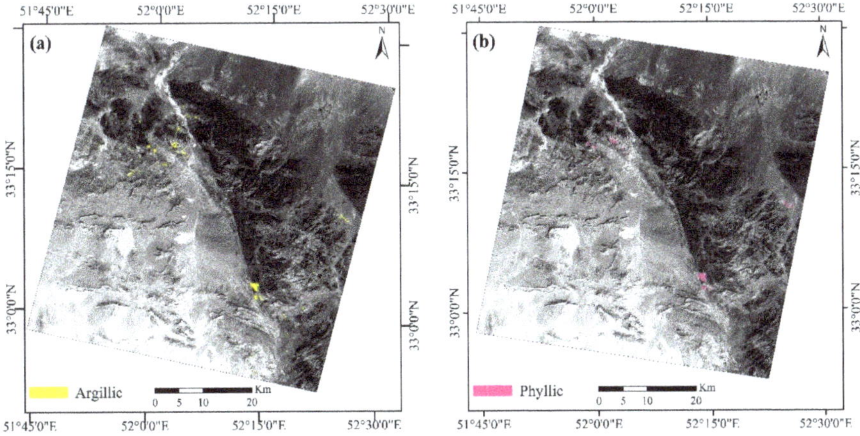

Figure 8. *Cont.*

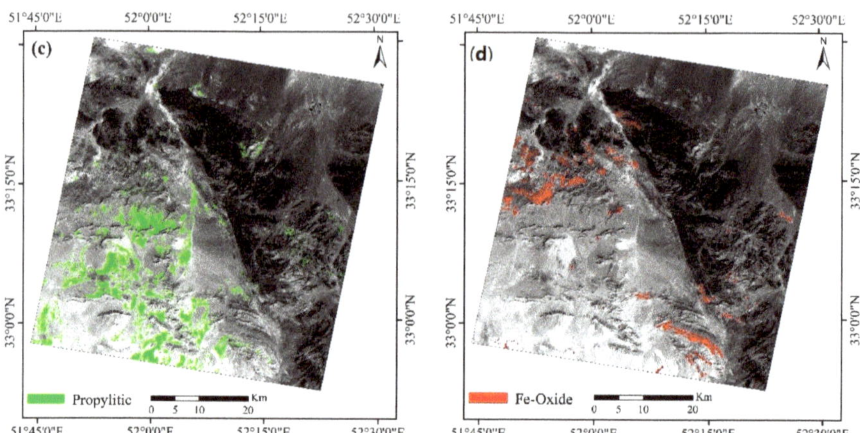

Figure 8. The results of SAM spectral mapping on the ASTER data. (**a**) Argillic alteration, (**b**) phyllic alteration, (**c**) propylitic alteration, and (**d**) Fe-oxide alteration.

5. Fieldworks

To validate the classification results, the field survey was performed by considering the following records: (i) the areas where the results of SVM showed the distribution of several alteration zones, especially phyllic and argillic; (ii) rock units of the mineralization zone; and (iii) areas where faults and ring structures were identified. In the field survey, 21 rock samples were collected by bulk sampling method for ICP-MS analysis, thin section, XRD and XRF. The location of sampling points was recorded with a handheld GPS (Garmin eTrex 30x; average accuracy of 3 m; made in Taiwan). The results of ICP-MS, XRF, and XRD are presented in Appendix A, Tables A1–A3. Figures 9 and 10 show the location of the sampling points on the SVM and SAM alteration maps, respectively.

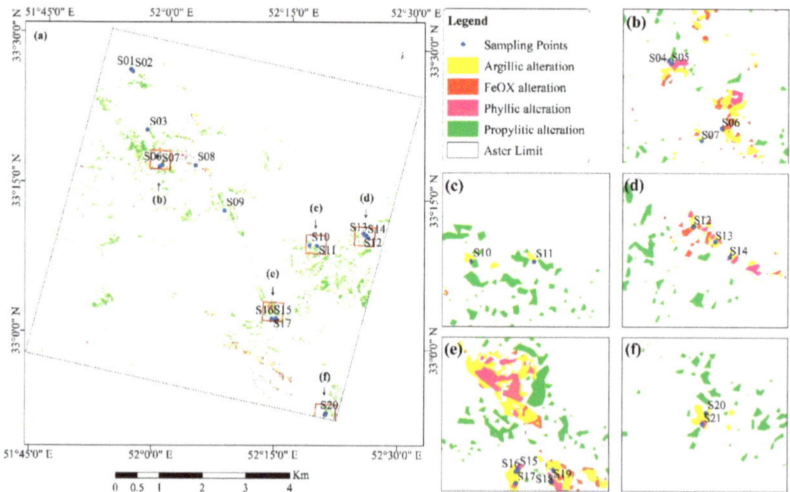

Figure 9. The results of SVM spectral mapping on the ASTER data. (**a**) ASTER full scene of the study area. (**b–f**) Selected subsets for sampling and field survey.

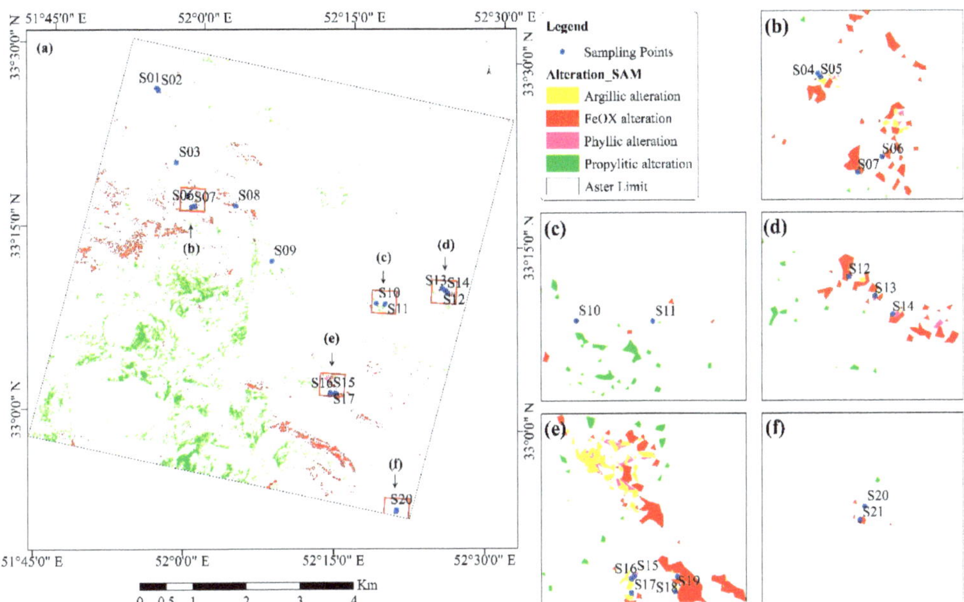

Figure 10. (**a**) The results of SAM spectral mapping on the ASTER data. (**a**) ASTER full scene of the study area. (**b–f**) Selected subsets for sampling and field survey.

The S01 and S02 samples were taken from the northwestern part of the study area. The rock of this area is diorite, which had been altered to argillic, phyllic, and iron oxides. The SVM results showed phyllic and argillic alteration, and the SAM method performed better in determining iron oxides in this area. The S03 was sampled from rhyodacite rocks. In this area, the rocks had been altered to sericite and silica. The zone of S04 and S05 sampling (Figure 11a) consisted of rhyolite and dacite rocks with calcareous interlayers altered to argillic and phyllic caused by intrusive masses. The S06 and S07 samples that were collected from marl and limestone tuffs had been severely altered by the intrusion of diorite and rhyodacite rocks. In this area, the thickness of the adjacent metamorphic zone, which consisted mainly of garnet and epidote, reached about 100 m. There were lenses made of silica and iron oxide with a thickness of 2 m among these skarns. The S08 sample was composed of rhyodacite and breccias tuff. This area incurred argillic alteration and is strongly siliceous along northwest-southeast faults. Sampling was performed from the S09 point owing to the presence of multiple faults and the detection of argillic alteration in the SVM results. During the field survey, a skarn mass was observed, and silicification and epidotization were identified in some parts of this zone. The S10 and S11 samples were taken from granodiorite and diorite, where argillic, advanced argillic, and propylitic alteration occurred. Mn dendrites were observed in this part. The S12, S13 and S14 samples were taken from the zones of argillic, propylitic, phyllic alterations and iron oxides (Figure 11b), which were identified in the SVM and SAM maps. At the field surveys, the argillic alteration was observed in a pyroclastic tuff unit, and in some parts the partial silicification alteration was recorded. The S15, S16, S17, S18 and S19 samples were taken from the southern part of the study area (Figure 11c). The field survey of these points showed that the porphyry dacites had been altered to argillic, phyllic, iron oxides, and silica. The S20 and S21 samples were collected from the southeastern part of the study area. This area is a pyroclastic complex that was influenced by dacite to diorite masses. Argillitization (Figure 11d), silicification, turmalinization, and iron oxides were seen in this zone.

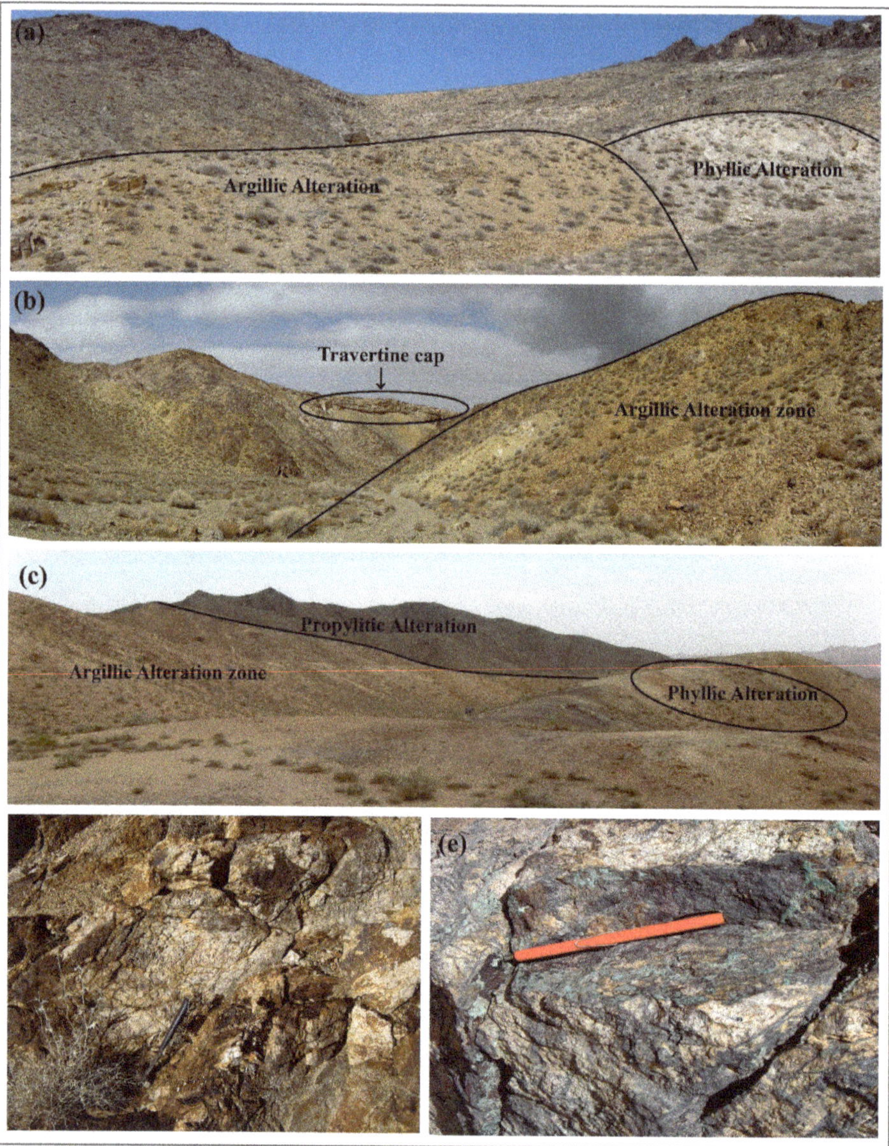

Figure 11. (**a**) Argillic and phyllic alterations in the area of the S04 and S05 sampling points; (**b**) argillic alteration in the area where the S12, S13 and S14 samples were taken; (**c**) propylitic, argillic and phyllic alteration in the S15, S16, S17, S18 and S19 sampling zones; (**d**) argillic alteration; (**e**) malachite mineralization observed in the southeastern part of the study area.

5.1. Petrography Study

The thin section in Figure 12a was prepared from the S04 sample. This thin section had a porphyry texture with a microgranular matrix. This sample had undergone pervasive alteration, and the rock had been entirely replaced by secondary minerals such as sericite, quartz, clay minerals, and iron oxides. Quartz veins were also observable. The thin section of the sample S07 (Figure 12b) was a porphyry rhyodacite with a hyalomicrogranular matrix. Rock feldspar minerals were selectively altered to sericite, illite, and muscovite.

Alteration to clay minerals, especially kaolinite, was observed throughout the thin section. Iron oxides were observed on the opaque minerals. In the thin section of the sample S13 rhyodacite porphyry with hyalomicrogranular matrix was detected (Figure 12c). Euhedral and subhedral plagioclase, alkaline feldspar, and quartz with corrosion gulf were the rock's main minerals. Tourmaline was also found among feldspar crystals. The main alterations observed in this section were argillic, sericite and phyllic. The thin section in Figure 12d was prepared from sample S19 corresponding to a lithic tuff with a volcanoclastic texture. Subhedral to amorphous plagioclase–alkali crystals, feldspar, and quartz were the main minerals, which had been altered to sericite, muscovite, iron oxides, and clay minerals.

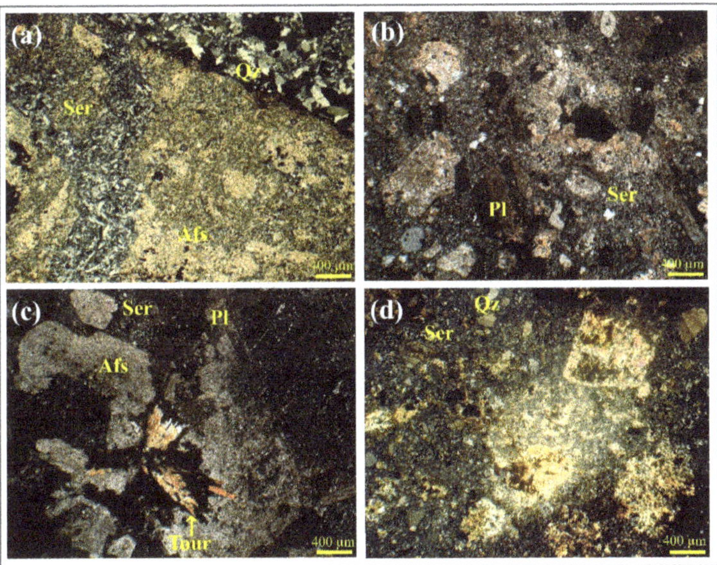

Figure 12. Thin sections of (**a**) the S04, (**b**) S07, (**c**) S13, and (**d**) S19 rock samples. The dusty surface of the images is due to the alteration and formation of clay and sericite minerals. Abbreviations: Ser = Serecite, and Qt = quartz, Afs = Alkali-feldspar, Pl = plagioclase, Tour = Tourmaline.

In the XRD results, indicator minerals for phyllic, argillic, and propylitic alteration zones such as hematite, muscovite, illite, kaolinite, montmorillonite, chlorite, epidote, and goethite were detected (Figure 13a–g and Appendix A, Table A3), which confirmed the results of the remote sensing analysis.

5.2. Geochemical Analysis

ICP-MS and XRF analyses were performed on all 21 samples taken from the study areas. The ICP-MS analysis of the S04 sample showed enrichment of Au (104 ppb), As (289 ppm), Cu (467 ppm), and Mo (21 ppm) elements (Appendix A, Table A1). In the ICP-MS results of S06 and S07 samples, the Zn enrichment (1195 and 3014 ppm) was observable (Appendix A, Table A1). In the S09 sample, Mn (3464 ppm), Cu (198 ppm) and Au (60 ppb) showed enrichment (Appendix A, Table A1). The ICP-MS results of the S11 sample analysis showed Mn (1664 ppm) enrichment (Appendix A, Table A1). Pb (280 ppm) and Cu (509 ppm) enrichment in the form of malachite were observed at the location of Figure 11e, from which the S21 sample was collected (Appendix A, Table A1). The XRF analysis was performed for all samples; the results are shown in Appendix A, Table A2. Altered samples showed high amounts of Al_2O_3 (17.00% up to 24.20%), SiO_2 (41.42% up to 56.24%), and Fe_2O_3 (2.44 % up to 9.43%) and low amounts of Na_2O (<0.1% up to 3.68%) and K_2O (1.85% up to 3.70%) owing to alteration processes (Appendix A, Table A2).

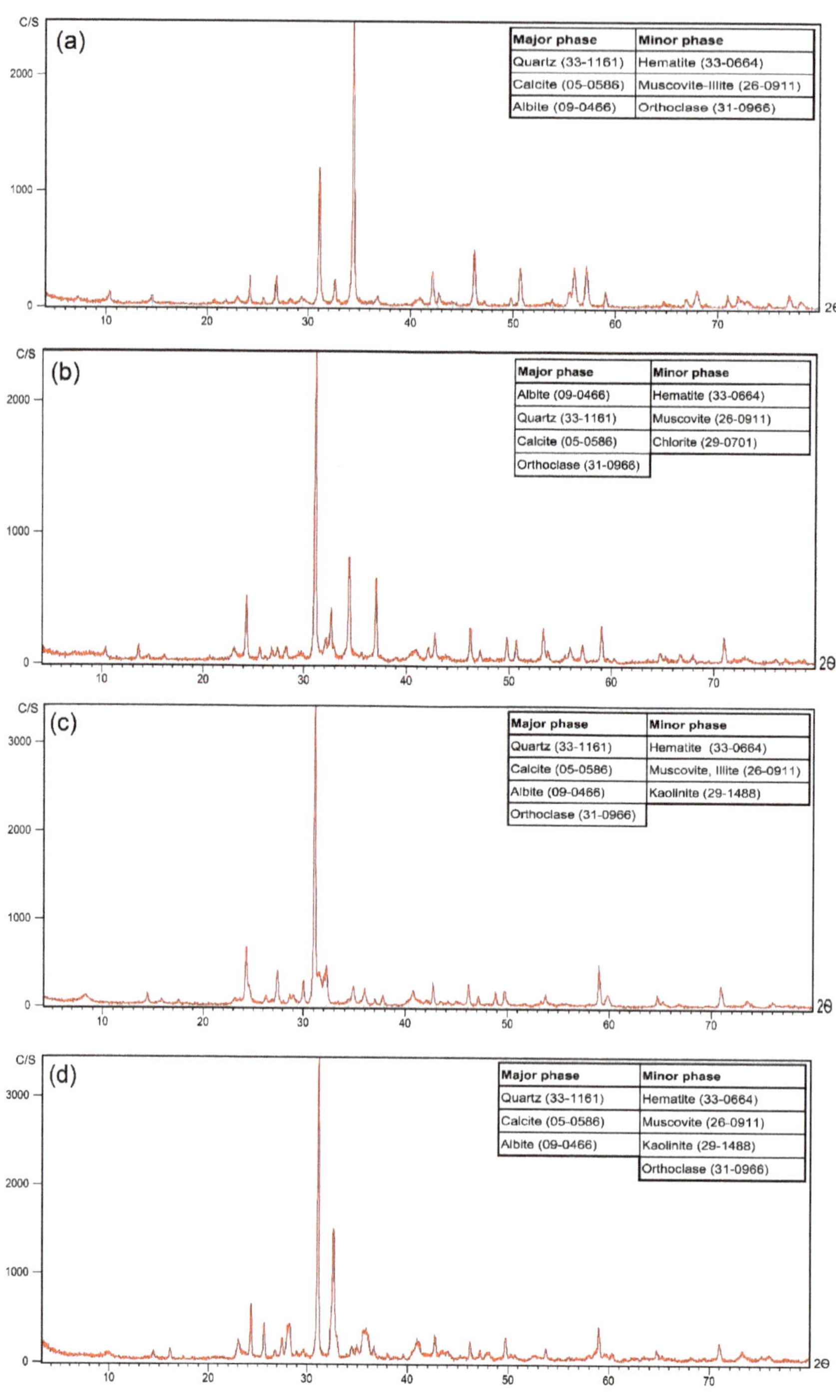

Figure 13. *Cont.*

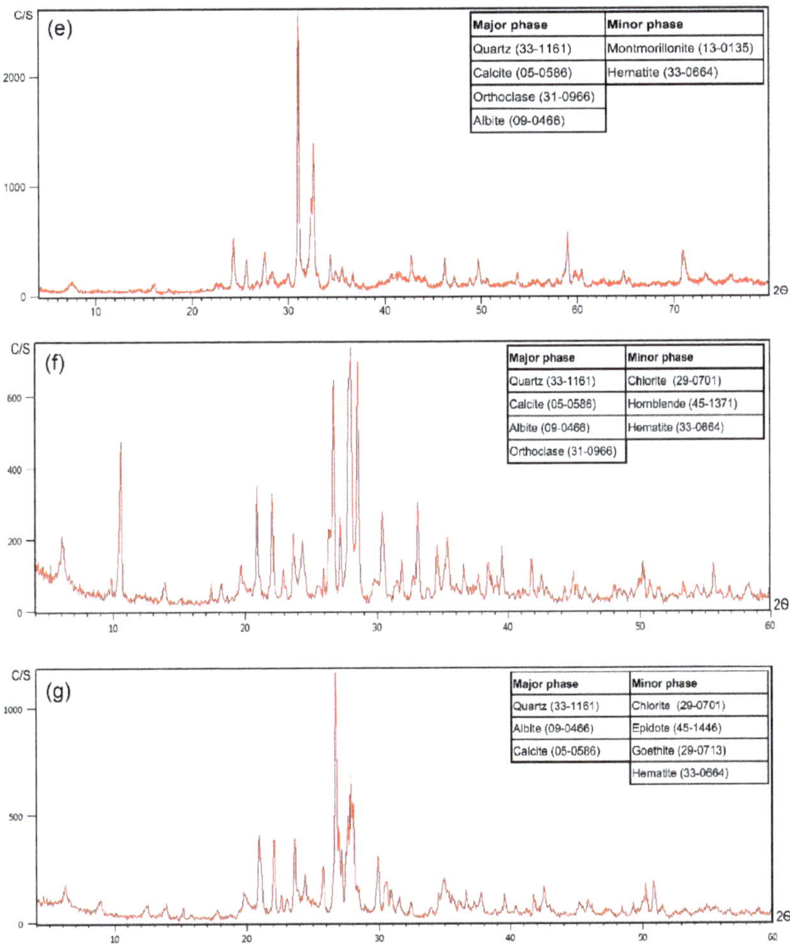

Figure 13. The XRD analysis of (**a**) the S04, (**b**) S07, (**c**) S13, (**d**) S19, (**e**) S21, (**f**) S14, and (**g**) S16 samples. The XRD results of (**a**,**b**) show peaks of phyllic alteration. (**c**,**d**) show peaks of phyllic–argillic alteration minerals. The XRD in (**e**) shows argillic alteration. The peaks in (**f**,**g**) indicate that samples S14 and S16 were collected from propylitic alteration. The C/S means Count per second is y title and 2Θ is the x title.

6. Accuracy Assessment

To evaluate the performances of the SVM and SAM methods, classification results were compared with the DP results. In this comparison, the DP results that included the alteration of phyllite, argillic, propylitic, and iron oxides were used as ground truth. User accuracy, producer accuracy, overall accuracy, and kappa coefficient [70,71] were calculated to evaluate the accuracy of the results. The results showed a total accuracy of 84.4 and 67.2% for SVM and SAM, respectively. The value of the kappa coefficient for SVM was 0.74 and for SAM it was 0.52. As can be seen from Tables 1 and 2, the classification of the phyllic, argillic, and propylitic alteration results of the SVM method were more accurate, but the Fe-oxides alteration result of the SAM classification was more consistent with ground truth. The best result was in the classification of propylitic alteration in the SVM method.

Table 1. Confusion matrix for the SVM classification.

Classes	Phyllic	Argillic	Propylitic	Fe-Oxides	Total	User's Accuracy
Unclassified	20	46	9	30	105	
Phyllic	172	23	0	0	195	88.21
Argillic	33	795	6	47	881	90.24
Propylitic	0	3	201	1	205	98.05
Fe-Oxides	0	17	0	104	121	85.95
Total	225	884	216	182	1507	
Producer's accuracy	76.44	89.93	93.06	57.14		
Overall accuracy	84.4					
Kappa coefficient	0.744					

Table 2. Confusion matrix for the SAM classification.

Classes	Phyllic	Argillic	Propylitic	Fe-Oxides	Total	User's Accuracy
Unclassified	8	102	47	23	180	
Phyllic	146	107	0	7	260	56.15
Argillic	43	586	0	15	644	90.99
Propylitic	0	1	128	0	129	99.22
Fe-Oxides	24	108	9	153	294	52.04
Total	221	904	184	198	1507	
Producer's accuracy	66.06	64.82	69.57	77.27		
Overall accuracy	67.2					
Kappa coefficient	0.52					

7. Discussion

Distinguishing hydrothermal alteration zones resulting from hydrothermal processes in the porphyry systems is a significant stage of mineral exploration [58]. Remote sensing data have a great capability for mapping hydrothermal alteration zones and are extensively and successfully used for distinguishing hydrothermal alteration minerals and zones in metallogenic provinces around the world [8,9,72–74]. Several image processing techniques are broadly applied to remote sensing imagery for classifying, identifying, and distinguishing spatial distribution of alteration minerals and zones [61,62]. Band ratios, Principal Component Analysis (PCA), Independent Component Analysis (ICA), Matched-Filtering (MF), Mixture-Tuned Matched-Filtering (MTMF), Linear Spectral Mixing (LUS), and Constrained Energy Minimization (CEM) methods have been extensively implemented on ASTER data for mapping alteration zones associated with porphyry copper deposits [75–77]. However, these techniques are conceptual (i.e., knowledge-driven) algorithms and the reconfiguration formula is used to map the desired criteria. Consequently, the zones that encounter most of the desired criteria are highlighted as prospective zones. These algorithms are provisional regarding the type of input remote sensing data and thus can be biased. By applying these algorithms, expert knowledge is used more than the proficiency of the statistical methods [78]. The application of ML algorithms to remote sensing data has high potential to produce accurate maps, especially for mapping argillic, phyllic, and propylitic zones associated with porphyry copper deposits [78–80].

In hydrothermal alteration mapping, the placement of each pixel in a cluster is essential. Hence, the image processing methods categorizing only a fraction of the pixels into a particular class are not very effective and accurate. In view of that, the use of clustering methods is highly useful in determining the ML of a pixel belonging to a cluster. This study showed that the fusion of unsupervised and supervised methods in mineral mapping leads to more accurate results. The methods and algorithms used for mineral mapping are in line with the reality of the data and provide better results. The DP method used in this study models alteration zones well because its performance is based on distribution. Consequently, in specifying training data, it is more consistent with reality than using endmembers or pure training data. More reliable results can be obtained especially when the detection methods are used to determine the extent of each alteration zone. The training data achieved from the DP method are suitable input for use in the SVM and SAM methods. The SVM method with RBF kernel and training data generated from the DP showed better results than SAM. Furthermore, the DP method can also be used to cluster all other types of data, including the results of geochemical analysis of stream sediments, lithogeochemical and geophysical data, etc., which can be applied in the future mineral exploration in metallogenic provinces.

Geological surveys were performed based on the results obtained from remote sensing imagery. The results of the fieldwork and laboratory analysis showed good accordance with the obtained remote sensing results. The presence of illite and muscovite minerals in the XRD results indicated a phyllic alteration zone in the study area. The occurrence of kaolinite and montmorionite minerals in the XRD results confirmed the occurrence of an argillic alteration zone in the study areas. The manifestation of epidote and chlorite minerals in the XRD results indicated a propylitic alteration. In the XRF results, owing to the degradation of feldspars in the alteration process, the amounts of K_2O, CaO, and Na_2O decreased and the Al_2O_3, Fe_2O_3, and SiO_2 increased. Increasing the amount of Cu, Au, Zn, and Mn obtained in the ICP-MS results was related to copper mineralization in some samples collected from different zones in the study area. Consequently, the remote sensing approach applied in this study was a valuable tool for porphyry copper exploration in the metallogenic provinces.

8. Conclusions

Mineral mapping using supervised methods requires appropriate training data to classify the data accurately and comprehensively. Considering that minerals and rocks have various compositions, the DP method was used to model phyllic, argillic, propylitic, and Fe-oxides alteration zones in the Zefreh porphyry copper deposits. The classification maps with the DP results training data were more accurate. The DP process was used to specify the training data on ASTER images of the Zefreh porphyry copper deposits, where alteration zones were detected by spectral mapping methods such as BDR, LSU, OSP, and SFF. The DP clustering results were realistic, considering the field survey and laboratory analysis. By performing the SVM and SAM methods on the ASTER data, areas including phyllic, argillic, propylitic, and iron oxide alterations in the full ASTER scene were identified. By field survey of these zones, a good coincidence was perceived between the results obtained from the SVM method and field observations. Alternation zones similar to those obtained from the SVM results were observed in the field at most of the surveyed points. With the SAM method, most of the iron oxides and propylitic alterations were identified, and in some areas, it was less compatible with the alterations observed in the field than the SVM method. This study reinforced the application of the SVM algorithm for mapping hydrothermal alteration zones associated with porphyry copper deposits, which is applicable to ASTER data for potential mapping in various metallogenic provinces around the world.

Author Contributions: Conceptualization, M.Y. and S.H.T.; methodology, M.Y. and A.B.P.; software, M.Y. and R.R.; validation, M.Y.; data curation, M.Y. and S.H.T.; writing—original draft preparation M.Y.; writing—review and editing, A.B.P.; supervision, S.H.T. and B.P.; fieldwork, M.Y. and S.H.T. All authors have read and agreed to the published version of the manuscript.

Funding: The present work has received no funding or grant.

Data Availability Statement: The data presented in this study are available on request from the corresponding author.

Acknowledgments: The authors' thanks are also extended to the Board of Directors of Parsi Kan Kav Company for providing some information on the study areas.

Conflicts of Interest: The authors declare no conflict of interest.

Appendix A

Table A1. ICP-MS analysis results of some significant elements (Au unit: part per billion, other element unit: part per million).

Row	Sample_NO	X (m)	Y (m)	Au (ppb)	Fe	Ag	As	Cu	Mn	Mo	Pb	Sb	Zn
1	S01	585,770	3,701,063	<5	5855	0	8.3	18	178	2.1	5	1.08	20
2	S02	586,064	3,700,766	<5	17,302	0	8.9	5	29	5.3	5	1.24	53
3	S03	589,488	3,690,065	70	13,866	0	150.7	15	219	6.8	63	201	39
4	S04	591,489	3,685,219	104	185,397	0	289.2	467	161	20.9	86	1.1	138
5	S05	591,544	3,685,158	<5	21,278	0	9.1	6	67	2.18	6	1.29	16
6	S06	592,703	3,683,705	7	123,344	0	8.9	112	564	4.8	32	1.09	1195
7	S07	592,237	3,683,423	<5	14,281	0	909.2	346	10,111	7.4	124	1.26	3014
8	S08	599,089	3,684,040	<5	8571	0.27	16.2	6	59	4	7	1.02	9
9	S09	605,129	3,675,998	60	81,444	0.35	28.1	198	3464	2.27	32	1.17	60
10	S10	621,599	3,670,515	<5	35,705	0.27	8.8	16	110	3.4	7	1.13	78
11	S11	623,023	3,670,502	<5	47,093	0.22	8.9	63	1664	2.31	37	1.09	234
12	S12	631,737	3,673,323	<5	15,726	0.24	11.2	28	65	2.43	7	1.12	18
13	S13	632,235	3,672,977	<5	27,364	0.28	8.4	40	49	3.2	197	1.02	119
14	S14	632,566	3,672,641	8	77,478	0.36	36.4	12	80	26.9	25	1.06	23
15	S15	615,304	3,656,553	12	37,739	0.22	8.6	8	29	2.16	5	1.01	22
16	S16	615,253	3,656,502	23	20,271	0.25	8.3	40	48	6.6	5	0.97	19
17	S17	615,249	3,656,248	25	42,181	0.27	120.8	10	39	2.27	13	1.09	22
18	S18	616,106	3,656,542	6	17,893	0.28	8.4	24	52	3.8	6	1.02	17
19	S19	616,073	3,656,271	<5	43,412	0.22	8.8	56	115	2.1	7	1.1	37
20	S20	626,512	3,639,545	<5	18,158	0	13	29	36	8.1	9	1.05	6
21	S21	626,422	3,639,297	55	39,771	0	61.7	509	63	9.6	280	1.04	16

Table A2. XRF analysis results of some main oxides (Oxides unit: percent). L.O.I: Loss Of Ignition.

Row	Sample_NO	SiO_2	Al_2O_3	CaO	MgO	TiO_2	Fe_2O_3	MnO	P_2O_5	Na_2O	K_2O	SrO	L.O.I	Total
1	S01	56.24	24.19	0.71	0.66	0.75	3.14	<0.1	0.48	<0.1	2.24	<0.1	10.70	99.11
2	S02	49.23	22.81	2.70	1.55	0.85	6.08	0.14	0.46	1.54	3.23	<0.1	10.00	98.60
3	S03	41.85	20.15	8.97	2.94	0.61	6.81	0.28	0.38	0.32	2.23	<0.1	13.50	98.04
4	S04	50.16	23.24	5.66	1.58	0.43	3.43	<0.1	0.32	<0.1	1.94	<0.1	11.30	98.06
5	S05	47.91	23.51	6.90	1.64	0.43	4.01	<0.1	0.34	<0.1	2.81	<0.1	11.49	99.03
6	S06	43.86	22.51	6.31	1.39	0.65	7.35	0.17	0.31	0.62	2.91	<0.1	12.62	98.70
7	S07	53.06	18.94	4.31	3.26	0.71	6.49	0.26	0.42	3.68	3.58	<0.1	4.14	98.85
8	S08	43.38	18.91	8.87	2.86	0.66	6.80	0.16	0.43	1.99	2.22	0.12	12.70	99.08
9	S09	52.32	22.22	5.02	2.14	0.50	2.44	<0.1	0.34	<0.1	3.85	<0.1	9.50	98.33
10	S10	41.42	17.01	8.66	3.44	0.77	9.28	0.33	0.57	2.56	1.85	0.10	12.98	98.98
11	S11	44.85	21.87	4.79	2.21	0.62	7.67	0.15	0.36	0.65	3.27	<0.1	12.34	98.77
12	S12	49.31	19.29	5.46	2.63	0.55	6.14	0.15	0.38	2.33	2.46	<0.1	9.39	98.09
13	S13	48.28	17.41	8.81	3.28	0.69	8.44	0.18	0.52	2.97	3.62	0.19	5.15	99.56
14	S14	44.26	17.00	6.57	3.07	0.75	8.95	0.17	0.37	2.91	3.70	0.10	9.84	97.69
15	S15	56.24	24.19	0.71	0.66	0.75	3.14	<0.1	0.48	<0.1	2.24	<0.1	6.70	95.11
16	S16	49.23	22.81	2.70	1.55	0.85	6.08	0.14	0.46	1.54	3.23	<0.1	10.00	98.60
17	S17	41.85	20.15	8.97	2.94	0.61	6.81	0.28	0.38	0.32	2.23	<0.1	13.50	98.04
18	S18	50.16	23.24	5.66	1.58	0.43	3.43	<0.1	0.32	<0.1	1.94	<0.1	11.30	98.06
19	S19	47.91	23.51	6.90	1.64	0.43	4.01	<0.1	0.34	<0.1	2.81	<0.1	11.49	99.03
20	S20	43.86	22.51	6.31	1.39	0.65	7.35	0.17	0.31	0.62	2.91	<0.1	12.62	98.70
21	S21	53.06	18.94	4.31	3.26	0.71	6.49	0.26	0.42	3.68	3.58	<0.1	4.14	98.85

Table A3. The results of the XRD analysis.

Samples	Major Phase	Minor Phase	Alteration
S04	Quartz, Calcite, Albite	Hematite, Muscovite, Illite, Orthoclase	Phyllic
S07	Albite, Quartz, Calcite, Orthoclase	Hematite, Muscovite, Chlorite	Phyllic
S13	Quartz, Calcite, Albite, Orthoclase	Hematite, Muscovite, Illite, Kaolinite	Phyllic–Argillic
S19	Quartz, Calcite, Albite	Hematite, Muscovite, Kaolinite, Orthoclase	Phyllic–Argillic
S21	Quartz, Calcite, Orthoclase, Albite	Montmorillonite, Hematite	Argillic
S14	Quartz, Calcite, Albite, Orthoclase	Chlorite, Hornblende, Hematite	Propylitic
S16	Quartz, Albite, Calcite	Chlorite, Epidote, Goethite, Hematite	Propylitic

References

1. El-Wahed, M.A.; Zoheir, B.; Pour, A.B.; Kamh, S. Shear-Related Gold Ores in the Wadi Hodein Shear Belt, South Eastern Desert of Egypt: Analysis of Remote Sensing, Field and Structural Data. *Minerals* **2021**, *11*, 474. [CrossRef]
2. Krupnik, D.; Khan, S.D. High-Resolution Hyperspectral Mineral Mapping: Case Studies in the Edwards Limestone, Texas, USA and Sulfide-Rich Quartz Veins from the Ladakh Batholith, Northern Pakistan. *Minerals* **2020**, *10*, 967. [CrossRef]
3. Ghezelbash, R.; Maghsoudi, A.; Carranza, E.J.M. Performance evaluation of RBF-and SVM-based machine learning algorithms for predictive mineral prospectivity modeling: Integration of SA multifractal model and mineralization controls. *Earth Sci. Inform.* **2019**, *12*, 277–293. [CrossRef]
4. Cardoso-Fernandes, J.; Teodoro, A.C.; Lima, A.; Perrotta, M.; Roda-Robles, E. Detecting Lithium (Li) mineralizations from space: Current research and future perspectives. *Appl. Sci.* **2020**, *10*, 1785. [CrossRef]
5. Peyghambari, S.; Zhang, Y. Hyperspectral remote sensing in lithological mapping, mineral exploration, and environmental geology: An updated review. *J. Appl. Remote Sens.* **2021**, *15*, 031501. [CrossRef]
6. Hewson, R.; Mshiu, E.; Hecker, C.; van der Werff, H.; van Ruitenbeek, F.; Alkema, D.; van der Meer, F. The application of day and night time ASTER satellite imagery for geothermal and mineral mapping in East Africa. *IJAEO* **2020**, *85*, 101991. [CrossRef]
7. Adiri, Z.; Lhissou, R.; El Harti, A.; Jellouli, A.; Chakouri, M. Recent advances in the use of public domain satellite imagery for mineral exploration: A review of Landsat-8 and Sentinel-2 applications. *Ore Geol. Rev.* **2020**, *117*, 103332. [CrossRef]
8. Pour, A.B.; Hashim, M.; Hong, J.K.; Park, Y. Lithological and alteration mineral mapping in poorly exposed lithologies using Landsat-8 and ASTER satellite data: North-eastern Graham Land, Antarctic Peninsula. *Ore Geol. Rev.* **2019**, *108*, 112–133. [CrossRef]
9. Pour, A.B.; Zoheir, B.; Pradhan, B.; Hashim, M. Editorial for the Special Issue: Multispectral and Hyperspectral Remote Sensing Data for Mineral Exploration and Environmental Monitoring of Mined Areas. *Remote Sens.* **2021**, *13*, 519. [CrossRef]
10. Gupta, P.; Venkatesan, M. Mineral identification using unsupervised classification from hyperspectral data. In *Emerging Research in Data Engineering Systems and Computer Communications*; Springer: Berlin/Heidelberg, Germany, 2020; pp. 259–268.
11. Pour, A.B.; Park, T.-Y.S.; Park, Y.; Hong, J.K.; Zoheir, B.; Pradhan, B.; Ayoobi, I.; Hashim, M. Application of multi-sensor satellite data for exploration of Zn–Pb sulfide mineralization in the Franklinian Basin, North Greenland. *Remote Sens.* **2018**, *10*, 1186. [CrossRef]
12. Wylie, B.K.; Pastick, N.J.; Picotte, J.J.; Deering, C.A. Geospatial data mining for digital raster mapping. *GISci. Remote Sens.* **2019**, *56*, 406–429. [CrossRef]
13. Dhingra, S.; Kumar, D. A review of remotely sensed satellite image classification. *Int. J. Electr. Comput. Eng.* **2019**, *9*, 1720–1731. [CrossRef]
14. Gemusse, U.; Lima, A.; Teodoro, A. Comparing different techniques of satellite imagery classification to mineral mapping pegmatite of Muiane and Naipa: Mozambique. In Proceedings of the Earth Resources and Environmental Remote Sensing/GIS Applications X, Strasbourg, France, 10–12 September 2019; p. 111561E.
15. Bachri, I.; Hakdaoui, M.; Raji, M.; Teodoro, A.C.; Benbouziane, A. Machine learning algorithms for automatic lithological mapping using remote sensing data: A case study from Souk Arbaa Sahel, Sidi Ifni Inlier, Western Anti-Atlas, Morocco. *ISPRS Int. J. Geo-Inf.* **2019**, *8*, 248. [CrossRef]
16. Mojeddifar, S.; Mavadati, M. Integration of support vector machines for hydrothermal alteration mapping using ASTER data–case study: The northwestern part of the Kerman Cenozoic Magmatic Arc, Iran. *Int. J. Min. Geo-Eng.* **2020**, *54*, 45–50.
17. Cardoso-Fernandes, J.; Teodoro, A.C.; Lima, A.; Roda-Robles, E. Semi-automatization of support vector machines to map lithium (Li) bearing pegmatites. *Remote Sens.* **2020**, *12*, 2319. [CrossRef]
18. Reddy, C.K. *Data Clustering: Algorithms and Applications*; Chapman and Hall/CRC: London, UK, 2018.
19. Abdi Jalebi, S.; Sharifzadeh, S.; Amiri, S. A New Method for Semi-Supervised Segmentation of Satellite Images. In Proceedings of the 2021 22nd IEEE International Conference on Industrial Technology (ICIT), Valencia, Spain, 10–12 March 2021; IEEE: New York, NY, USA, 2021.
20. Saxena, A.; Prasad, M.; Gupta, A.; Bharill, N.; Patel, O.P.; Tiwari, A.; Er, M.J.; Ding, W.; Lin, C.-T. A review of clustering techniques and developments. *Neurocomputing* **2017**, *267*, 664–681. [CrossRef]
21. Yang, M.-S.; Sinaga, K.P. A feature-reduction multi-view k-means clustering algorithm. *IEEE Access* **2019**, *7*, 114472–114486. [CrossRef]
22. Ünlü, R.; Xanthopoulos, P. Estimating the number of clusters in a dataset via consensus clustering. *Expert Syst. Appl.* **2019**, *125*, 33–39. [CrossRef]
23. Khosravi, M.; Rajabzadeh, M.A.; Qin, K.; Asadi, H.H. Tectonic setting and mineralization potential of the Zefreh porphyry Cu-Mo deposit, central Iran: Constraints from petrographic and geochemical data. *J. Geochem. Explor.* **2019**, *199*, 1–15. [CrossRef]
24. Khosravi, M.; Christiansen, E.H.; Rajabzadeh, M.A. Chemistry of rock-forming silicate and sulfide minerals in the granitoids and volcanic rocks of the Zefreh porphyry Cu–Mo deposit, central Iran: Implications for crystallization, alteration, and mineralization potential. *Ore Geol. Rev.* **2021**, *124*, 104150. [CrossRef]
25. Salehi, T.; Tangestani, M.H. Large-scale mapping of iron oxide and hydroxide minerals of Zefreh porphyry copper deposit, using Worldview-3 VNIR data in the Northeastern Isfahan, Iran. *IJAEO* **2018**, *73*, 156–169. [CrossRef]
26. Abrams, M. The Advanced Spaceborne Thermal Emission and Reflection Radiometer (ASTER): Data products for the high spatial resolution imager on NASA's Terra platform. *IJRS* **2000**, *21*, 847–859. [CrossRef]

27. Guo, Y.-D.; Shi, Z. Characteristics and Applications of ASTER. *Remote Sens. Technol. Appl.* **2003**, *5*, 346–352.
28. Erenoglu, R.C.; Arslan, N.; Erenoglu, O.; Arslan, E. Application of spectral analysis to determine geothermal anomalies in the Tuzla region, NW Turkey. *Arab. J. Geosci.* **2019**, *12*, 1–15. [CrossRef]
29. Shawkya, M. Comparative Study of Atmospheric Correction Methods of ASTER Data to Enhance the Delineation of Uranium Mineralized Zones. *Int. J. Intell. Comput. Inf. Sci.* **2019**, *19*, 48–65. [CrossRef]
30. Tucker, C.J. Red and photographic infrared linear combinations for monitoring vegetation. *Remote Sens. Environ.* **1979**, *8*, 127–150. [CrossRef]
31. Ferguson, T.S. A Bayesian analysis of some nonparametric problems. *Ann. Stat.* **1973**, *1*, 209–230. [CrossRef]
32. Escobar, M.D.; West, M. Bayesian density estimation and inference using mixtures. *J. Am. Stat. Assoc.* **1995**, *90*, 577–588. [CrossRef]
33. Duan, J.A.; Guindani, M.; Gelfand, A.E. Generalized spatial Dirichlet process models. *Biometrika* **2007**, *94*, 809–825. [CrossRef]
34. Ma, Z.; Lai, Y.; Kleijn, W.B.; Song, Y.-Z.; Wang, L.; Guo, J. Variational Bayesian learning for Dirichlet process mixture of inverted Dirichlet distributions in non-Gaussian image feature modeling. *IEEE Trans. Neural Netw. Learn. Syst.* **2018**, *30*, 449–463. [CrossRef]
35. Teh, Y.W.; Jordan, M.I.; Beal, M.J.; Blei, D.M. Sharing clusters among related groups: Hierarchical Dirichlet processes. *Adv. Neural Inf. Process. Syst.* **2005**, *17*, 1385–1392.
36. Vlachos, A.; Ghahramani, Z.; Korhonen, A. Dirichlet process mixture models for verb clustering. In Proceedings of the ICML Workshop on Prior Knowledge for Text and Language, Helsinki, Finland, 9 July 2008.
37. Lugrin, T. *Bayesian Semiparametrics for Modelling the Clustering of Extreme Values*; École polytechnique fédérale de Lausanne: Écublens, Switzerland, 2013.
38. Jain, A.; Murty, M.; Flynn, P.J. Data clustering: A review. *ACM Comput. Surv.* **2011**, *31*, 264–323. [CrossRef]
39. Zuo, R.; Carranza, E.J.M. Support vector machine: A tool for mapping mineral prospectivity. *Comput. Geosci.* **2011**, *37*, 1967–1975. [CrossRef]
40. Wang, K.; Cheng, L.; Yong, B. Spectral-similarity-based kernel of SVM for hyperspectral image classification. *Remote Sens.* **2020**, *12*, 2154. [CrossRef]
41. Vapnik, V. *The Nature of Statistical Learning Theory*; Springer: New York, NY, USA, 1995; pp. 841–842.
42. Ben-Hur, A.; Horn, D.; Siegelmann, H.T.; Vapnik, V. Support vector clustering. *J. Mach. Learn. Res.* **2001**, *2*, 125–137. [CrossRef]
43. Bennett, K.; Demiriz, A. Semi-Supervised Support Vector Machines. Available online: https://proceedings.neurips.cc/paper/1998/file/b710915795b9e9c02cf10d6d2bdb688c-Paper.pdf (accessed on 24 March 2021).
44. Widodo, A.; Yang, B.-S. Support vector machine in machine condition monitoring and fault diagnosis. *MSSP* **2007**, *21*, 2560–2574. [CrossRef]
45. Wainer, J.; Fonseca, P. How to tune the RBF SVM hyperparameters? An empirical evaluation of 18 search algorithms. *Artif. Intell. Rev.* **2021**, *54*, 4771–4797. [CrossRef]
46. Morales, R.; Wang, Y.; Zhang, Z. Unstructured point cloud surface denoising and decimation using distance RBF K-nearest neighbor kernel. In Proceedings of the Pacific-Rim Conference on Multimedia, Shanghai, China, 21–24 September 2010; pp. 214–225.
47. Saljoughi, B.S.; Hezarkhani, A. A comparative analysis of artificial neural network (ANN), wavelet neural network (WNN), and support vector machine (SVM) data-driven models to mineral potential mapping for copper mineralizations in the Shahr-e-Babak region, Kerman, Iran. *Appl. Geomat.* **2018**, *10*, 229–256. [CrossRef]
48. Khaleghi, M.; Ranjbar, H.; Shahabpour, J.; Honarmand, M. Spectral angle mapping, spectral information divergence, and principal component analysis of the ASTER SWIR data for exploration of porphyry copper mineralization in the Sarduiyeh area, Kerman province, Iran. *Appl. Geomat.* **2014**, *6*, 49–58. [CrossRef]
49. Wang, M.; Huang, Z.; Zhang, X.; Zhang, Y.; Chen, M. Altered mineral mapping based on ground-airborne hyperspectral data and wavelet spectral angle mapper tri-training model: Case studies from Dehua-Youxi-Yongtai Ore District, Central Fujian, China. *IJAEO* **2021**, *102*, 102409.
50. Choi, J.; Kim, G.; Park, N.; Park, H.; Choi, S. A hybrid pansharpening algorithm of VHR satellite images that employs injection gains based on NDVI to reduce computational costs. *Remote Sens.* **2017**, *9*, 976. [CrossRef]
51. Renza, D.; Martinez, E.; Molina, I.J.A.i.S.R. Unsupervised change detection in a particular vegetation land cover type using spectral angle mapper. *AdSpR* **2017**, *59*, 2019–2031. [CrossRef]
52. Wolf, R.E.; Adams, M. *Multi-Elemental Analysis of Aqueous Geochemical Samples by Quadrupole Inductively Coupled Plasma-Mass Spectrometry (ICP-MS)*; US Department of the Interior, US Geological Survey: Washington, DC, USA, 2015.
53. Monecke, T.; Köhler, S.; Kleeberg, R.; Herzig, P.M.; Gemmell, J.B. Quantitative phase-analysis by the Rietveld method using X-ray powder-diffraction data: Application to the study of alteration halos associated with volcanic-rock-hosted massive sulfide deposits. *Can. Mineral.* **2001**, *39*, 1617–1633. [CrossRef]
54. Raith, M.M.; Raase, P. Thin Section Microscopy: A Comprehensive Guide. Available online: http://nationalpetrographic.com/thin-section-microscopy-a-comprehensive-guide.html (accessed on 24 March 2021).
55. Pichler, H.; Schmitt-Riegraf, C. *Rock-Forming Minerals in Thin Section*; Springer Science & Business Media: Berlin, Germany, 1997.
56. Chauhan, A.; Chauhan, P. Powder XRD technique and its applications in science and technology. *J. Anal. Bioanal. Tech.* **2014**, *5*, 1–5. [CrossRef]
57. Rajendran, S.; Nasir, S. Characterization of ASTER spectral bands for mapping of alteration zones of volcanogenic massive sulphide deposits. *Ore Geol. Rev.* **2017**, *88*, 317–335. [CrossRef]

58. Jain, R.; Sharma, R.U. Mapping of Mineral Zones using the Spectral Feature Fitting Method in Jahazpur belt, Rajasthan, India. *Int. Res. J. Eng. Technol. (IRJET)* **2018**, *5*, 562–567.

59. Zoheir, B.; Emam, A.; Abdel-Wahed, M.; Soliman, N. Multispectral and radar data for the setting of gold mineralization in the South Eastern Desert, Egypt. *Remote Sens.* **2019**, *11*, 1450. [CrossRef]

60. Chang, C.-I.; Cao, H.; Chen, S.; Shang, X.; Yu, C.; Song, M. Orthogonal subspace projection-based go-decomposition approach to finding low-rank and sparsity matrices for hyperspectral anomaly detection. *ITGRS* **2020**, *59*, 2403–2429. [CrossRef]

61. Pour, A.B.; Hashim, M. Identification of hydrothermal alteration minerals for exploring of porphyry copper deposit using ASTER data, SE Iran. *JAESc* **2011**, *42*, 1309–1323. [CrossRef]

62. Moghtaderi, A.; Moore, F.; Ranjbar, H. Application of ASTER and Landsat 8 imagery data and mathematical evaluation method in detecting iron minerals contamination in the Chadormalu iron mine area, central Iran. *J. Appl. Remote Sens.* **2017**, *11*, 016027. [CrossRef]

63. Kokaly, R.; Clark, R.; Swayze, G.; Livo, K.; Hoefen, T.; Pearson, N.; Wise, R.; Benzel, W.; Lowers, H.; Driscoll, R. *Usgs Spectral Library Version 7 Data: Us Geological Survey Data Release*; United States Geological Survey (USGS): Reston, VA, USA, 2017.

64. Sabins, F.F. Remote sensing for mineral exploration. *Ore Geol. Rev.* **1999**, *14*, 157–183. [CrossRef]

65. Pour, A.B.; Sekandari, M.; Rahmani, O.; Crispini, L.; Läufer, A.; Park, Y.; Hong, J.K.; Pradhan, B.; Hashim, M.; Hossain, M.S. Identification of phyllosilicates in the antarctic environment using aster satellite data: Case study from the Mesa range, Campbell and Priestley glaciers, northern Victoria Land. *Remote Sens.* **2021**, *13*, 38. [CrossRef]

66. Harsanyi, J.C.; Chang, C.-I. Hyperspectral image classification and dimensionality reduction: An orthogonal subspace projection approach. *ITGRS* **1994**, *32*, 779–785. [CrossRef]

67. Gelfand, A.E.; Smith, A.F. Sampling-based approaches to calculating marginal densities. *J. Am. Stat. Assoc.* **1990**, *85*, 398–409. [CrossRef]

68. Spiegelhalter, D.; Thomas, A.; Best, N.; Gilks, W. *Bayesian Inference Using Gibbs Sampling Manual (Version ii) BUGS 0.5*; MRC Biostatistics Unit, Institute of Public Health: Cambridge, UK, 1996; p. 59.

69. Spiegelhalter, D.; Thomas, A.; Best, N.; Lunn, D. WinBUGS User Manual. 2003. Available online: https://www.mrc-bsu.cam.ac.uk/wp-content/uploads/manual14.pdf (accessed on 27 February 2021).

70. Foody, G.M. Explaining the unsuitability of the kappa coefficient in the assessment and comparison of the accuracy of thematic maps obtained by image classification. *Remote Sens. Environ.* **2020**, *239*, 111630. [CrossRef]

71. Congalton, R.G. A review of assessing the accuracy of classifications of remotely sensed data. *Remote Sens. Environ.* **1991**, *37*, 35–46. [CrossRef]

72. Mathieu, L. Quantifying hydrothermal alteration: A review of methods. *Geosciences* **2018**, *8*, 245. [CrossRef]

73. Frutuoso, R.; Lima, A.; Teodoro, A.C. Application of remote sensing data in gold exploration: Targeting hydrothermal alteration using Landsat 8 imagery in northern Portugal. *Arab. J. Geosci.* **2021**, *14*, 1–18. [CrossRef]

74. Chattoraj, S.L.; Prasad, G.; Sharma, R.U.; van der Meer, F.D.; Guha, A.; Pour, A.B. Integration of remote sensing, gravity and geochemical data for exploration of Cu-mineralization in Alwar basin, Rajasthan, India. *IJAEO* **2020**, *91*, 102162. [CrossRef]

75. Pour, A.B.; Park, T.-Y.S.; Park, Y.; Hong, J.K.; Muslim, A.M.; Läufer, A.; Crispini, L.; Pradhan, B.; Zoheir, B.; Rahmani, O.; et al. Landsat-8, advanced spaceborne thermal emission and reflection radiometer, and WorldView-3 multispectral satellite imagery for prospecting copper-gold mineralization in the northeastern Inglefield Mobile Belt (IMB), northwest Greenland. *Remote Sens.* **2019**, *11*, 2430. [CrossRef]

76. Shirmard, H.; Farahbakhsh, E.; Pour, A.B.; Muslim, A.M.; Müller, R.D.; Chandra, R. Integration of Selective Dimensionality Reduction Techniques for Mineral Exploration Using ASTER Satellite Data. *Remote Sens.* **2020**, *12*, 1261. [CrossRef]

77. Sekandari, M.; Masoumi, I.; Pour, A.B.; Muslim, A.M.; Rahmani, O.; Hashim, M.; Zoheir, B.; Pradhan, B.; Misra, A.; Aminpour, S.M. Application of Landsat-8, Sentinel-2, ASTER and WorldView-3 Spectral Imagery for Exploration of Carbonate-Hosted Pb-Zn Deposits in the Central Iranian Terrane (CIT). *Remote Sens.* **2020**, *12*, 1239. [CrossRef]

78. Parsa, M.; Pour, A.B. A simulation-based framework for modulating the effects of subjectivity in greenfields' Mineral Prospectivity Mapping with geochemical and geological data. *J. Geochem. Explor.* **2021**, *229*, 106838. [CrossRef]

79. Bolouki, S.M.; Ramazi, H.R.; Maghsoudi, A.; Beiranvand Pour, A.; Sohrabi, G. A remote sensing-based application of bayesian networks for epithermal gold potential mapping in Ahar-Arasbaran area, NW Iran. *Remote Sens.* **2020**, *12*, 105. [CrossRef]

80. Parsa, M. A data augmentation approach to XGboost-based mineral potential mapping: An example of carbonate-hosted Zn-Pb mineral systems of Western Iran. *J. Geochem. Explor.* **2021**, *228*, 106811. [CrossRef]

Article

ASTER-Based Remote Sensing Image Analysis for Prospection Criteria of Podiform Chromite at the Khoy Ophiolite (NW Iran)

Behnam Mehdikhani and Ali Imamalipour *

Department of Mining Engineering, Urmia University, Urmia 57561-51818, Iran; b.mehdikhani@urmia.ac.ir
* Correspondence: A.imamalipour@urmia.ac.ir

Citation: Mehdikhani, B.; Imamalipour, A. ASTER-Based Remote Sensing Image Analysis for Prospection Criteria of Podiform Chromite at the Khoy Ophiolite (NW Iran). *Minerals* **2021**, *11*, 960. https://doi.org/10.3390/min11090960

Academic Editors: Amin Beiranvand Pour, Omeid Rahmani and Mohammad Parsa

Received: 19 July 2021
Accepted: 19 August 2021
Published: 2 September 2021

Abstract: A single chromite deposit occurrence is found in the serpentinized harzburgite unit of the Khoy ophiolite complex in northwest Iran, which is surrounded by dunite envelopes. This area has mountainous features and extremely rugged topography with difficult access, so prospecting for chromite deposits by conventional geological mapping is challenging. Therefore, using remote sensing techniques is very useful and effective, in terms of saving costs and time, to determine the chromite-bearing zones. This study evaluated the discrimination of chromite-bearing mineralized zones within the Khoy ophiolite complex by analyzing the capabilities of ASTER satellite data. Spectral transformation methods such as optimum index factor (OIF), band ratio (BR), spectral angle mapper (SAM), and principal component analysis (PCA) were applied on the ASTER bands for lithological mapping. Many chromitite lenses are scattered in this ophiolite, but only a few have been explored. ASTER bands contain improved spectral characteristics and higher spatial resolution for detecting serpentinized dunite in ophiolitic complexes. In this study, after the correction of ASTER data, many conventional techniques were used. A specialized optimum index factor RGB (8, 6, 3) was developed using ASTER bands to differentiate lithological units. The color composition of band ratios such as RGB $((4 + 2)/3, (7 + 5)/6, (9 + 7)/8)$, $(4/1, 4/7, 4/5)$, and $(4/3 \times 2/3, 3/4, 4/7)$ produced the best results. The integration of information extracted from the image processing algorithms used in this study mapped most of the lithological units of the Khoy ophiolitic complex and new prospecting targets for chromite exploration were determined. Furthermore, the results were verified by comprehensive fieldwork and previous studies in the study area. The results of this study indicate that the integration of information extracted from the image processing algorithms could be a broadly applicable tool for chromite prospecting and lithological mapping in mountainous and inaccessible regions such as Iranian ophiolitic zones.

Keywords: ASTER; chromite; Khoy ophiolite; spectral angle mapper (SAM); band ratio; principal component analysis (PCA)

1. Introduction

The mapping of ophiolite sequences has become a research interest of scientists and exploration geologists in the world because they host economic minerals such as chromium, copper, manganese, gold, nickel, barium, lead, and zinc [1–3]. Ophiolitic ultramafic rocks are the hosts of podiform chromite deposits. Podiform chromite deposits are small magmatic chromite bodies formed in the lower level of an ophiolite complex. Podiform chromite mines have produced 57.4% of the world's total chromite production [4]. Ophiolite zones in Iran are widespread and are often found in different locations with varying geologic and tectonic settings. The Khoy ophiolite complex is a part of the Tethyan ophiolite belt, and it is one of the largest Iranian ophiolite complexes, covering a widespread area in northwest Iran along the Iran–Turkey border and continuing toward western Turkey [5,6]. Ultramafic rocks, which are often serpentinized, are widespread in 250 km^2 of the Khoy ophiolite [5,6]. The Khoy ophiolite is one of the most promising areas for prospecting chromite deposits because of extensive outcrops of ultramafic rocks. So far, more than

20 chromite deposits have been identified in this area. These chromite occurrences have lenticular, tubular, and vein-like shapes hosted by serpentinized harzburgite. The chromite deposits in the Khoy ophiolite can be clearly classified into two groups: high-Al chromites (Cr# = 0.38–0.44) from the eastern ophiolite, and high-Cr chromites (#Cr = 0.54–0.72) from the western ophiolite [5,6]. Most Iranian ophiolitic zones are located in mountainous and inaccessible regions. Thus, prospecting for chromite deposits with geological mapping is challenging and time-consuming.

Remote sensing analysis plays an important role in the exploration of mineral deposits, as well as in lithological mapping and detection of associated hydrothermal mineralization, in Iran. The Advanced Spaceborne Thermal Emission and Reflection Radiometer (ASTER) is an advanced multispectral satellite imaging system that has created new tools for the mapping of geological structures and detecting certain alteration minerals or assemblages [2,3,7].

The ASTER sensor launched the TERRA platform in December 1999. The ASTER platform travels in a near-circular, sun-synchronous orbit with an inclination of approximately 98.2°, an altitude of 705 km, and a repeat cycle of 16 days, offering relatively improved spatial, spectral, and temporal resolutions. It is made from three visible and near-infrared spectral bands (VNIR, between 0.52 and 0.86 μm, with 15-m spatial resolution) and infrared, reflecting radiation in six short-wavelength infrared spectral bands (SWIR, between 1.6 and 2.43 μm, with 30 m spatial resolution). Sensor characteristics of the ASTER instruments are shown in Table 1 [8,9].

Table 1. Sensor characteristics of ASTER instruments [8].

Sensor Characteristics	ASTER					
	VNIR		SWIR		TIR	
Spectral bandswith range (μm)	Band01	0.52–0.60	Band04	1.6–1.7	Band10	8.125–8.475
	Band02	0.63–0.69	Band05	2.45–2.185	Band11	8.475–8.825
	Band03N	0.76–0.86	Band06	2.185–2.225	Band12	8.925–9.275
	Band03B	0.76–0.86	Band07	2.235–2.285	Band13	10.95–10.95
	Backward-looking		Band08	2.2295–2.365	Band14	10.95–11.65
			Band09	2.36–2.43		
Spatial resolution (m)	15		30		90	
Swathwidth (km)	60		60		60	

Due to the great extent of ultramafic rocks, which are the host of chromite deposits in the Khoy ophiolite, the possibility of discovering new chromite deposits is high and more exploration and investigation is needed. Given the extremely rugged topography with difficult access, new exploration methods such as the remote sensing method can be useful for this purpose. The present study evaluates the discrimination of chromite-bearing mineralized zones within the Khoy ophiolite complex by analyzing the capabilities of ASTER satellite data. Advanced Spaceborne Thermal Emission and Reflection Radiometer (ASTER) data can easily separate various rock units, the extent of the ultramafic rocks, and it can provide detailed geological maps of the area [8,9].

The extraction of spectral information related to ophiolite mapping can be achieved through image processing techniques such as band ratio (BR) and principal component analysis (PCA) on ASTER bands [8,10]. The color composition of the band ratio (4/1, 4/5, 4/7) is an effective means of determining the lithological ophiolite complexes [11].

Principle component analysis and band ratio methods are very useful for determining the serpentinized dunite that is the host of the chromite veins [7,11–14]. Abdeen used ASTER spectral band ratios RGB color composite of 4/7, 4/1, 2/3, 4/3, and RGB (4/7, 3/4, 2/1) for mapping ophiolitic units, metasediments, volcanoclastic, and granitoids in the southeastern desert of Egypt [15]. Amer used principal component analysis of ASTER data to determine the lithologic units of the ophiolite complexes in Pakistan. In the eastern ophiolites of Egypt, Amer used band ratios of $(7 + 9)/8$, $(5 + 7)/6$, $(2 + 4)/3$, and PCA (4,5,2) for the lithological mapping of several units [7]. Hashem and Pournamdari conducted research using ASTER data on the Abdasht ophiolites in northeastern Iran [12]. Thermal infrared (TIR) bands in the thermal range of spectral absorption can be used for the detection of silicate formations [16].

2. Description of the Study Area

The Khoy ophiolite covers an area of about 3900 km^2 in northwest Iran along the Iran–Turkey boundary. This ophiolitic complex is limited on the west and north by the Iran–Turkey border and on the east and south by a southeastern-northeastern fault (Figure 1). This zone reaches the Urmia Lake platform on the south. Precambrian metamorphic rocks including meta-volcanic, amphibolite, gneiss, and the Precambrian Kahar formation with the Rb-Sr age of 663 Ma [17] are the oldest rocks in this area and are located in the eastern portion of the ophiolite zone. It seems that this ophiolite is the remnant of a branch of the Neotethyan oceanic basin. It is joined to the northeast ophiolite of Turkey in the Western Pontides. The only reported age for this ophiolite is 81.2 ± 2.1 to 69.4 ± 1.6 Ma [18].

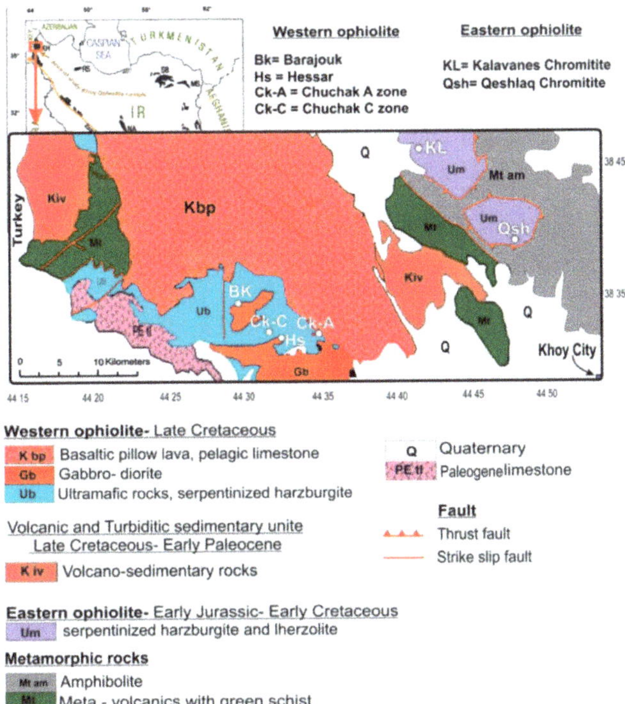

Figure 1. Geological map of the study area (modified after Khoy 1:250,000 geological map) and sample locations associated with a sketched map of Iran showing locations of some of the most important ophiolites in Iran [6,19]. KH, Khoy; MS, Mashhad; RS, Rasht; SB, Sabzevar.

New geochemical and field studies on the ophiolite of Khoy indicate that there are two ophiolite complexes in this area with different geological ages: (i) the early Jurassic to early Cretaceous eastern Khoy ophiolite and (ii) the late Cretaceous western Khoy ophiolite. The second one is a remnant of the Neotethyan oceanic crust [18,19]. The Khoy ophiolite has all the parts of an ophiolite sequence. It is composed of serpentinized peridotite, layered and isotropic gabbro, isolated diabasic dike, pillow basalt, massive sheet flow, and interbedded hyaloclastic breccia and tuffs. Ultramafic rocks have been cut by rodengitic dikes. The Khoy ophiolite was unconformably covered by Eocene rocks, including limestone, marl, and conglomerate. Associated with ophiolitic rocks are found flysch-type sediments with Paleocene-lower Eocene age that have syn-orogenic characteristics. After emplacement of the ophiolitic complex at the end of post-lower Eocene age, acidic to intermediate magmatic activity, as small granitoid intrusive rocks and andesitic-dacitic volcanic and their sub-volcanic equivalents, occurred [5,6].

The serpentinized harzburgites and related rocks in the western Khoy ophiolite are intruded on by gabbro–diorite intrusions, which appear as a spot inside and/or around serpentinized harzburgites and cannot be a member of the ophiolite sequence [6]. Ultramafic rocks of the western Khoy ophiolite host several podiform chromitite bodies. The chromite deposits have lenticular, tabular, and irregular vein shapes and are emplaced in depleted mantle harzburgite [5,6]. The recognized outcrops altogether are discordant with their harzburgite host rocks. Chromite bodies are surrounded by dunitic envelopes with variable thicknesses. The existence of a dunitic envelope with various thicknesses is a common characteristic of all chromite ore bodies in this area. Most of them are small and contain little reserves, and only the Aland, Qeshlag, and Kochek deposits, with several tens of thousands of reserves, are minable [5,6].

3. ASTER Satellite Data

This paper aims to evaluate the accuracy of ASTER images for targeting the discrimination of chromite-bearing mineralized zones within the serpentinized harzburgite rocks in an extensive area of the Khoy ophiolite complex.

The ASTER data in this study were obtained from the Earth and Remote Sensing Data Analysis Center (ERSDAC) in Japan and consist of a level 1B scene acquired in 2002. The images have been georeferenced to UTM zone 38 North projections with the WGS-84 datum. Atmospheric correction on the VNIR and SWIR bands was applied by the log residual method. Finally, correlation coefficient, optimum index factor, principal component analysis, and band ratio were evaluated for lithological mapping in this study. Figure 2 shows the serpentinite, chromite, and pillow lava in the study area.

Figure 2. Field photographs show: (**A**) serpentinized harzburgite; (**B**) lens-shaped chromite within serpentinized harzburgite; (**C**) gabbroic intrusion within the ultramafic rocks; (**D**) chromite ore body; (**E**) chromite ore body within serpentinized harzburgite covered by overburden; (**F**) basaltic pillow lava.

4. Ophiolite Spectral Properties

The spectral reflectance of a rock depends on the type of mineralogical composition of the whole rock. The absorption of minerals also depends on the number of electronic processes occurring in these rocks [20]. Recent studies on the number of reflections from the surface of rocks have provided very important aspects of the study of remote sensing data. Many researchers utilized remote sensing and GIS techniques for lithological mapping as well as identifying mineral deposits [9,10,21–24].

Sabins concluded that remote sensing techniques can be used for mineral explorations in four ways: (1) Mapping of faults and structures that deposits can form in that trend; (2) mapping local fractures that may control ore deposits individually; (3) alteration of mapping in altered rocks associated with mineralization; and (4) providing geological base maps to start explorations [10]. In laboratory studies, the reflection spectrum of some of the rocks of various ophiolite units was studied by Abrams [21]. Figure 3 shows the spectral measurements of minerals found in harzburgites and gabbros [25].

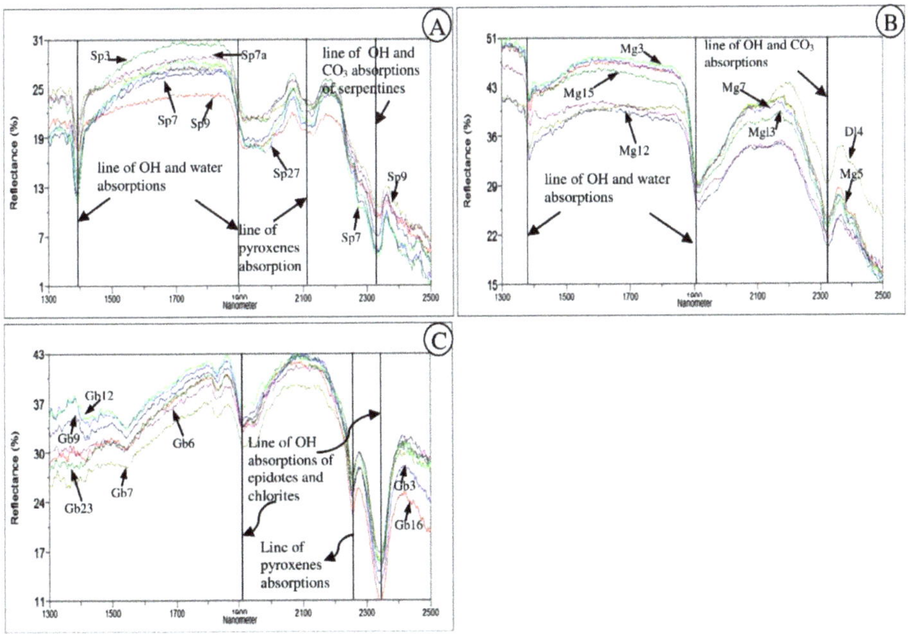

Figure 3. Spectral plots of (**A**) the harzburgites; (**B**) the harzburgites with carbonates; (**C**) the gabbros [25].

5. Methodology

To delineate the area of chromite mineralized zones within the serpentinized harzburgite, a host of chromitite in the Khoy ophiolite area, the ASTER satellite's images were processed. The first stage in the exploration program was finding the harzburgite and dunite lithologies. Methods such as the different band rationing method and principal component analysis techniques, which were tested in scientific publications on the mapping of ophiolites, were used. In this study, a regional geology map was used to support the remote sensing studies. Before any processing on ASTER data, some preprocessing—including topography, atmospheric, radiometric, and geometric corrections—had to be carried out on their bands.

The ASTER, with two data series of VNIR and SWIR for the separation of the lithology units, yielded very good results [15,24]. The selection of different band ratios was based on the spectral reflectance of rocks and their minerals, and such band ratio images, designed to display the spectral contrast of specific absorption features, can be used extensively in geological remote sensing. The band ratio method is frequently used in lithological mapping and mineral exploration using remote sensing data [7,23,26–31]. Additionally, the ASTER band ratio is suitable for the exploration and detection of serpentinite dunite and harzburgite of ophiolite [9,12,13,15]. Iron oxides, clay minerals, sulfates, and carbonates are some rocks and minerals that can be identified and separated by ASTER data [13]. Abdeen used ASTER band ratios of 4/7, 4/1, 2/3 × 4/3 and 4/7, 3/4, 2/1 in RGB for mapping ophiolites, metasediments, volcanoclastic, and granitoids, which are lithologic units of the Neoproterozoic-Allaqi suture in the southeastern desert of Egypt [15].

Amer used band ratios of $(2 + 4)/3$, $(5 + 7)/6$, and $(7 + 9)/8$ to distinguish between ophiolite and granite rocks, and was able to map ophiolite rocks, metabasalt, and metagabbro units [7]. They concluded that these new ratios are much better to separate the lithological units of the ophiolites, so in the present study, these new band ratios were used. PCA is a well-known method for lithological and alteration mapping in metallogenic provinces [7,12–14,24,31]. In this technique, the relationship between the spectral responses of target minerals or rocks and numeric values extracted from the eigenvector matrix was used to calculate the principal component images. Using this relationship, one can determine which PCs contain spectral information due to minerals and whether the digital numbers (DNs) of the pixels containing the target minerals had high (bright) or low (dark) values. Crosta and Amer noted that combining the analysis of the principal components that contain the most information and the principal components that contain the least information can provide much more useful data on the separation of lithology and mineralized zones [7,13].

5.1. Optimum Index Factor (OIF)

The total VNIR and SWIR bands of the ASTER data included 63 different band combinations, with bands 3, 6, and 8 having the highest OIF. Using the combination of different bands caused an increase in the spectral accuracy of the low-correlation bands, especially the thermal bands. Calculations of OIF are required to obtain the best false-color composites (higher OIF color combinations contain more information):

$$OIF = \left(\sum_{i=1}^{3} S_i \right) / \left(\sum_{i=1}^{3} r_i \right) \tag{1}$$

where S_i is the standard deviation in each band, and r_i is a correlation of bands of two to two. Often, the false-color combinations containing the most important information are determined from the variety of colors.

5.2. Spectral Angular Mapper Algorithm

The spectral angle mapping algorithm assumes that a pixel of remote sensing images represents certain ground cover material, which can be uniquely assigned to only one ground cover class. The SAM algorithm is measured based on the degree of similarity between the two spectra. A spectral similarity can include any number of measured spectra (Figure 4). The spectral similarity between two spectra is measured by calculating the angle between the two spectra, treating them as vectors in a space with dimensionality equal to the number of bands [32].

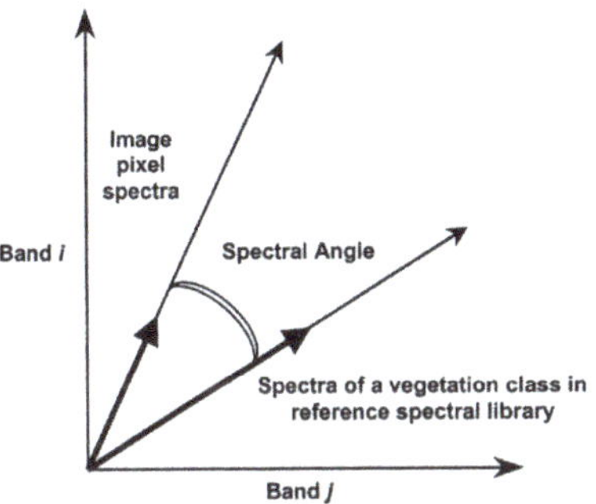

Figure 4. Representation of reference angle [33].

6. Remote Sensing in the Study Area

Optimum index factor (OIF), principal component analysis (PCA), and band ratio (BR) techniques are the spectral angle mapper techniques that were evaluated for lithological mapping in this study [34,35]. The color composition of RGB (8, 6, 3) showed that the spectral accuracy of all bands increased due to the 15 m spectral accuracy of the VNIR band. Figure 5 shows the color composite that distinguishes the serpentine dunites (light green), colored mélange (pink), vegetation (red), and carbonate rocks (yellow).

Figure 5. The color composition of RGB (8, 6, 3) from ASTER data after necessary corrections.

The satellite images were projected in the UTM Zone N38 and WGS 1984 ellipsoid (oblate spheroid) datum. For mapping the geology units, we can classify similar pixels using the optimum index factor (OIF), band ratio (BR), etc., and obtain the initial map of the lithology units. By using all available data in the study area, a map of lithological units was obtained (Figure 6).

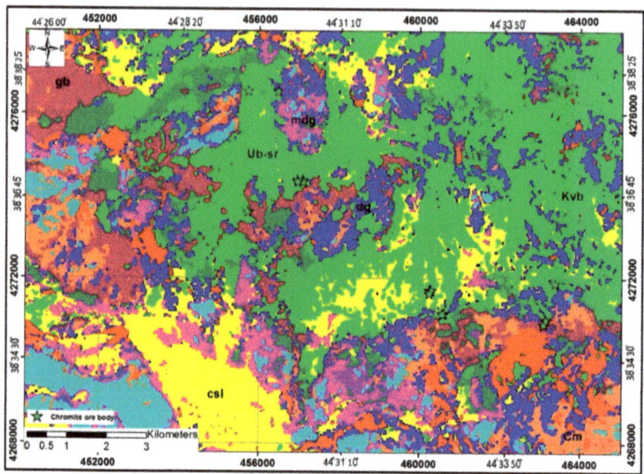

Figure 6. Classified map by using unsupervised classification method from pure pixels.

6.1. Band Ratio

The band ratio method is a suitable technique for lithological mapping, especially to discriminate rock units in ophiolite complexes [7,12,28,30]. For the discrimination of harzburgite rocks (contains more serpentine) and chromite bearing mineralized zones in the study area, all band ratios and their color composites were used, as in Sultan et al. (1986) (5/7, 5/1, 5/4 × 4/3), Sabins (1999) (3/5, 3/1, 5/7), and Gad and Kusky (2007) ((5/3, 5/1, 7/5) and (7/5, 5/4, 3/1)) [10,22,27]. This technique has been used successfully in lithological mappings for other ophiolite areas [7,9,23,26,27,29,30,36]. In the study area, based on the spectral information obtained from the ASTER bands, the color composition of the band ratios (4 + 2)/3, (7 + 5)/6, (7 + 9)/8 in Figure 7 provides the best results in the separation of ophiolite complex lithology units. In this color combination, ultrabasic rocks are pink, and the more serpentinized rocks are reddish. In Khoy ophiolites, there is no specific band ratio for the separation of all units, and several band ratios should be used to distinguish between different lithological units.

An interesting point shown in this figure is the separation of ultrabasic rocks based on the severity of serpentinization. Near the serpentine sections, the serpentinization rate increased sharply and is more reddish. The blue sections are ultrabasic with low serpentinization that is seen far from chromite lenses. The difference between the two types of serpentine spectra is shown in Figure 8. Thus, the severity of serpentinization of ultrabasic rocks can also be considered for the exploration of chromite lenses.

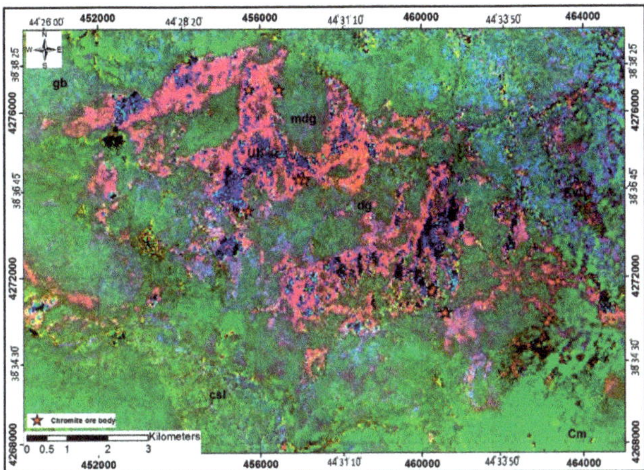

Figure 7. Harzburgite (with slight serpentinization) and highly serpentinized harzburgites (serpentinite) separation.

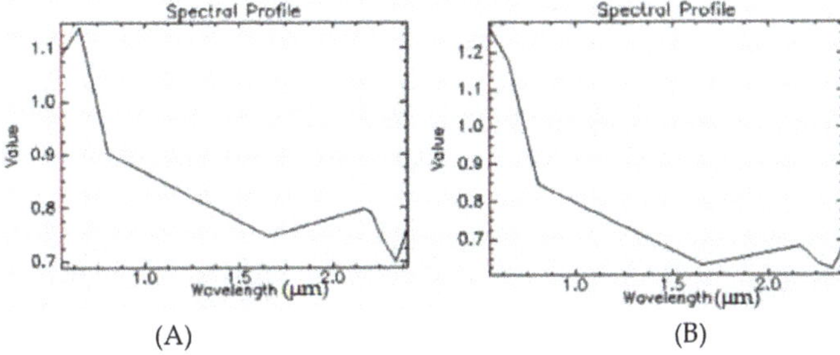

(A)   (B)

Figure 8. Spectral plots of two types of serpentinization intensity: (**A**) high serpentinization; (**B**) low serpentinization.

Therefore, the spectral reflectance in bands two and one is different. As a result, a band ratio of 2/1 can be used to differentiate ultrabasics with different serpentinization intensities. The other band ratio that was considered in this study and can be used to distinguish potential chromite areas is RGB (4/5, 4/7, 4/1), an ultrabasic area characterized by an olive color. Dioritic gabbro is mostly indigo blue, which in the vicinity of ultrabasics is yellowish. Pixels seen at the intersection of these two colors are the best place for points of chromite lenses. All the chromite outcrops in the study area comply with this rule and can be optimized by creating information layers in the GIS software and prioritizing these areas. In this ratio, the conglomerate is mainly purple and is exposed in the southwestern part of the region. In the northwest part of the study area, a gabbro unit is yellow, which is distinguished according to the spectrum obtained from micro gabbro diorite and gabbro-diorite units (Figure 9).

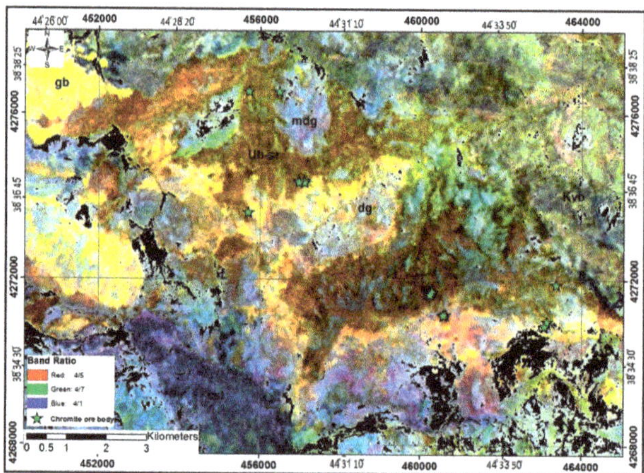

Figure 9. ASTER RGB images of band ratios (4/5, 4/7/4/1) show gabbro-diorite in yellow, serpentinized ultrabasic in purple.

6.2. PCA Analysis

Principal component analysis (PCA) was used to summarize the information in a data set described by multiple variables. Using this technique makes it possible to separate pixels that have good spectral information [13]. In this method, components that have less than 1% of information are deleted due to the high noise in the data. In this study, PCA analysis was applied to all nine bands, and the PCs (1, 2, 3) which included the most information were selected for separation. After PCA calculation, it was found that PC1, PC2, and PC3 had the greatest variances in the data. The eigenvalues for the main components of all ASTER bands are provided in Figure 10.

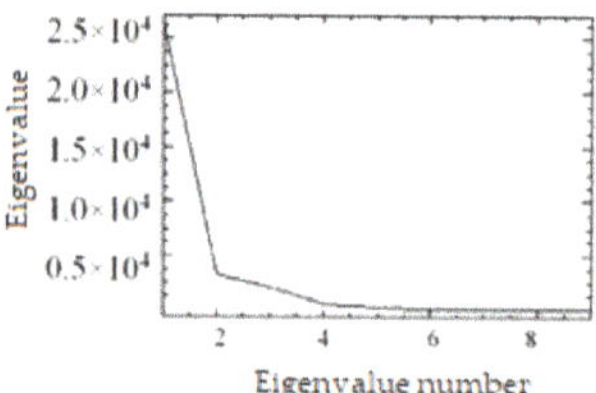

Figure 10. The PCA eigenvalue plot for the VNIR+ SWIR bands of ASTER data.

As a result, the PCA analysis of the VNIR and SWIR bands was used to determine the lithology units of the Khoy ophiolite complexes due to more spectral information. The results showed that PC1 had the highest positive variance. Thus, the PC1 component can provide more information about the lithology and mineralogy of rock units. The PC2 component had the most information from bands three and one, and its bright pixels indicated quartzites. Principal component analysis was also performed on SWIR bands that had information that was not VNIR + SWIR. In the PC4 component, iron and magnesium silicates were distinguished as light pixels. Iron and magnesium silicates such as olivine, iron, and magnesium hydrated phyllosilicates, such as serpentine, have low reflectivity in the visible region and high reflectivity in the NIR [30]. The electron processes cause high absorption in the VNIR, since cations such as Fe^{2+} and Fe^{3+}, which are often replaced by Mn, Cr, and Ni, are more frequent in the crystalline structure of minerals [20]. PC5 is also very suitable for vegetation mapping because vegetation has a low reflection in band two and a high reflection in band three. In addition, the results showed that PC6 to PC9 was very noisy and lacked proper information. Finally, the color composites of the analysis of PC1, PC2, and PC3 yielded excellent results for the separation of rock units (Figures 11–15).

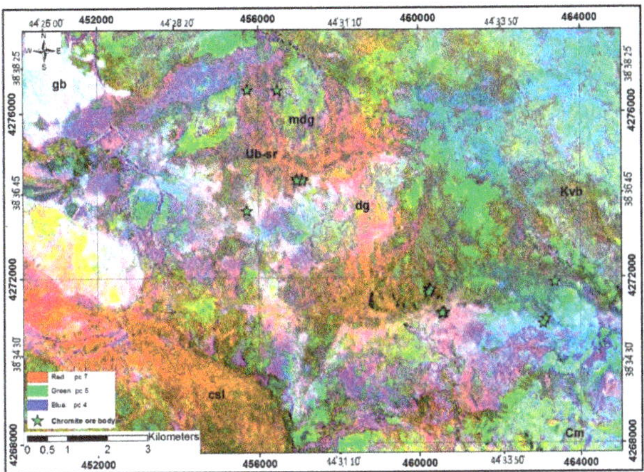

Figure 11. The RGB image of PC7, PC5, and PC4 of PCA bands in the study area.

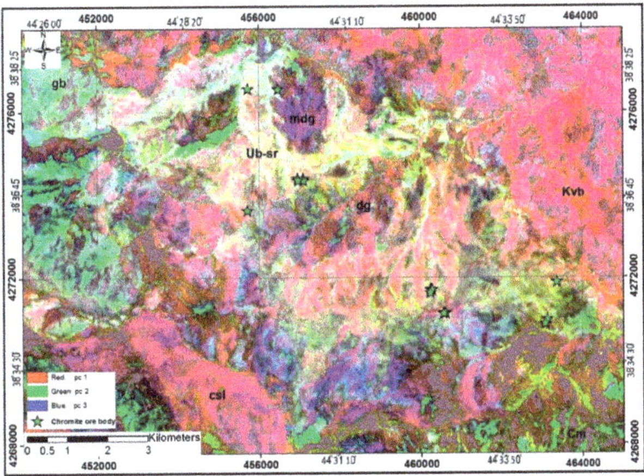

Figure 12. The RGB image of PC1, PC2, and PC3 of PCA bands in the study area: Ub-sr—serpentinized ultrabasic; mdg—microdiorite gabbro; dg—diorite gabbro; Cm—conglomerate; Kvb—basalt pillow lava; csl—shale and conglomerate.

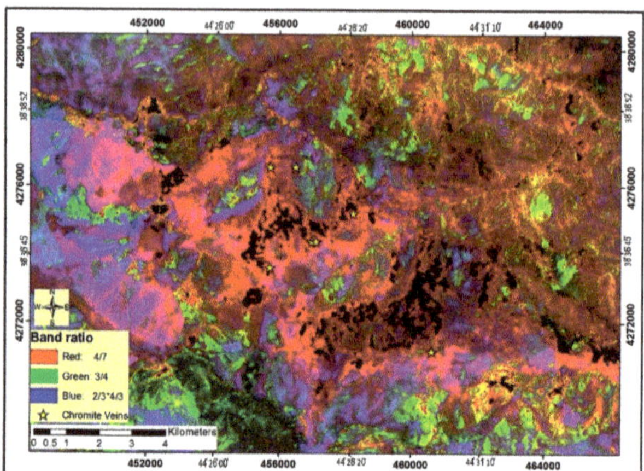

Figure 13. The RGB image of band ratios (4/7), (3/4), (2/3× 4/3) in the study area. Chromite outcrops and veins are shown on the map.

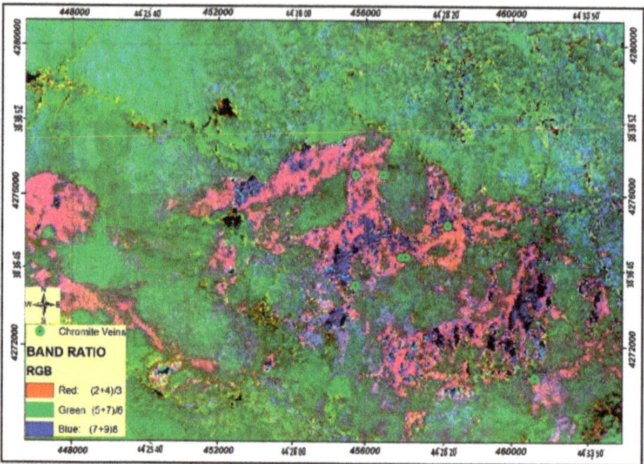

Figure 14. The RGB image of band ratios (2 + 4)/3, (5 + 7)/6, (7 + 9)/8 in the study area. Chromite outcrops and veins are shown on the map.

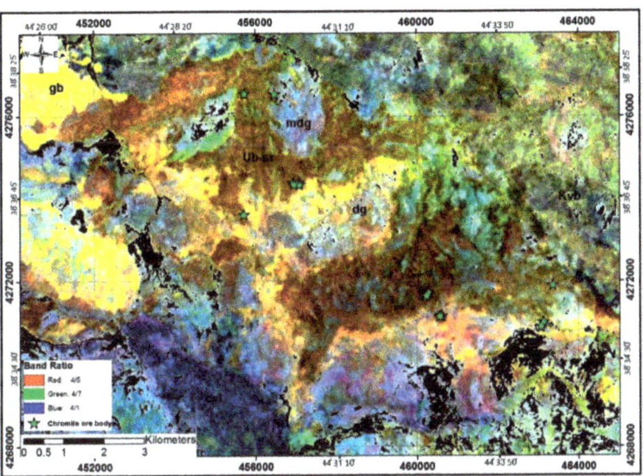

Figure 15. SATER RGB image of band ratios (4/1), (4/5), (4/7) in the study area: Ub-sr—serpentinized ultrabasic; mdg—microdiorite gabbro; dg—diorite gabbro; Cm—conglomerate; Kvb—basalt pillow lava; csl—shale and conglomerate.

6.3. Spectral Angle Mapper

The spectral angle mapper was one of the most useful tools used in this research study. The spectral library, or the spectrum of one of the sufficiently widespread outcrops in the region, was used for prospecting similar spectral pixels. In this method, all pixels were processed and the spectrum of pixels similar to chromite spectra or any other mineral in the region was considered as the objective function.

In the remote sensing studies of Khoy ophiolites there are two major problems that may affect the conclusions of these studies. All the outcrops in this region are very limited, and the chromite masses are in the halo of the dunite, which themselves are enclosed within the harzburgite. Due to the tectonic conditions of this region and the great fractures within it, the dunites and harzburgite are both serpentinized and their separation is practically impossible. In further remote sensing studies, pure spectral pixels were obtained first, and then from five existing anomalies, which were already being mined, the chromite spectra was selected. The result of these procedures is presented in Figures 16–19.

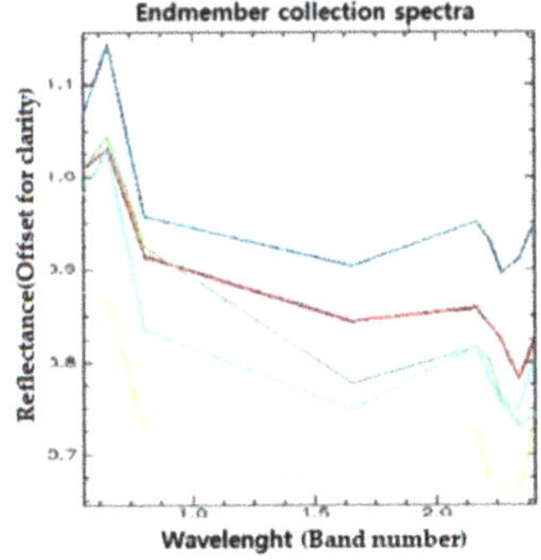

Figure 16. Reflectance spectra of harzburgite exposed at the Khoy ophiolite zone: spectra resampled to ASTER VNIR–SWIR band passes.

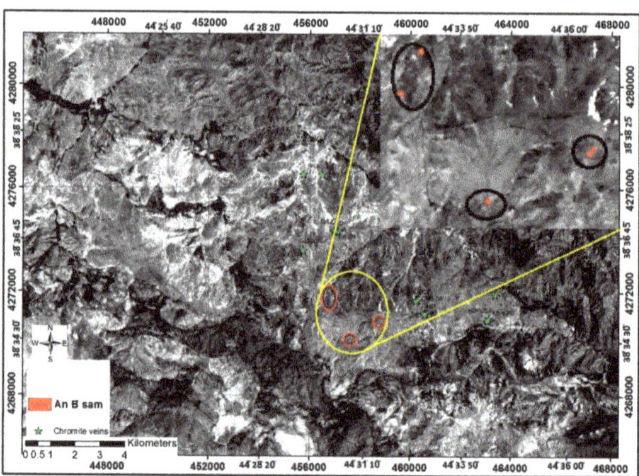

Figure 17. Chromitite-bearing pixels obtained from Anomaly B spectra and SAM in ASTER data.

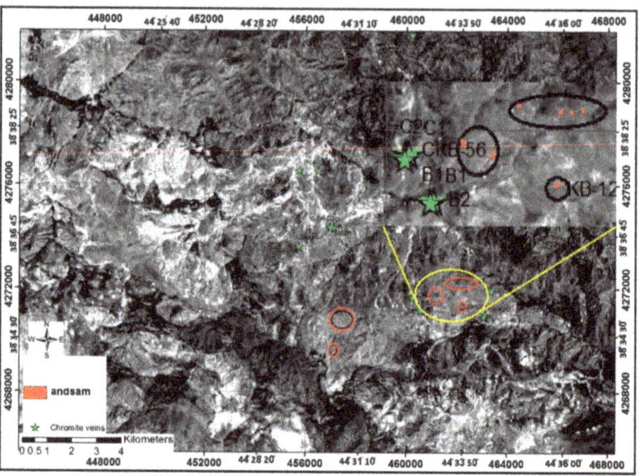

Figure 18. Chromitite-bearing pixels obtained from Anomaly D spectra and SAM in ASTER data.

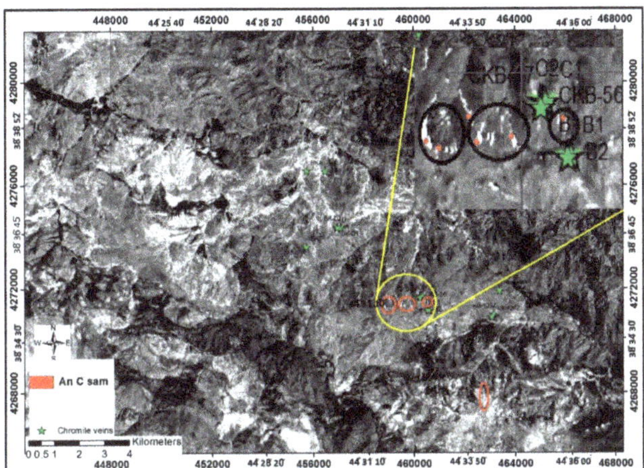

Figure 19. Chromitite-bearing pixels obtained from Anomaly C spectra and SAM in ASTER data.

Finally, with the integration of the obtained data such as the fault map, the separated lithologies, and suitable points from the remote sensing studies and chromite outcrop maps, the most suitable geological traverse lines to continue prospecting in the Khoy ophiolite complex were obtained (Figure 20).

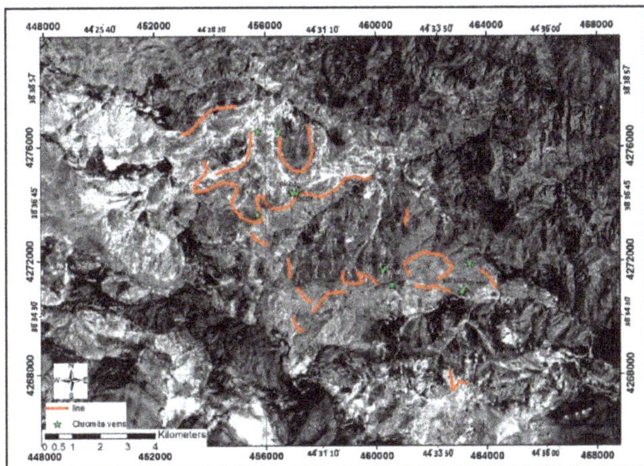

Figure 20. The most suitable geological traverse lines to continue prospecting in the Khoy ophiolite complex.

This leads to the suggestion that geophysical and geochemical studies be conducted in these paths, which pass through some outcrops, for exploration of the greatest number of chromite bodies. As a result, by using remote sensing studies, chromite exploration in ophiolites can be done economically. The prospecting paths, due to the topography of the Khoy ophiolite, are also designed to explore the greatest possible number of chromite lenses. In this way, with the integration of geophysical methods such as gravity and magnetic measurements in the designed paths, desired results can be economically achieved.

7. Conclusions

In this research, VNIR and SWIR bands of ASTER data were used to distinguish lithological units and delineate high-potential chromite mineralized zones in the Khoy ophiolites complex. Harzburgite and dunite are the main units of chromite lens hosts. During this study, using image processing techniques such as the band ratio method, principal component analysis, and the spectral angle mapper algorithm, a large area of these ophiolites was investigated. Consequently, integration of the results derived from the image processing algorithms and other data sets, such as geological maps, can produce accurate information for the reconnaissance stages of chromite exploration at both regional and district scales. This research demonstrates the remote sensing capabilities for the identification of dunite/serpentine or peridotite as host rocks for chromite mineralization in the transition zone of Iranian ophiolitic sequences and lithological mapping in mountainous and inaccessible regions.

Author Contributions: Conceptualization, A.I. and B.M.; methodology, A.I. and B.M.; software, B.M.; validation, B.M. and A.I.; formal analysis, B.M.; investigation, B.M.; resources, B.M.; data curation B.M.; writing—original draft preparation, A.I. and B.M.; writing—review and editing, A.I. and B.M.; visualization, B.M.; supervision, A.I.; project administration, A.I. All authors have read and agreed to the published version of the manuscript.

Funding: This research received no external funding.

Data Availability Statement: Not applicable.

Acknowledgments: This research was made possible with the help of the office of vice-chancellor for Research and Technology, Urmia University. We acknowledge their support.

Conflicts of Interest: The authors declare no conflict of interest.

References

1. Khan, S.D.; Mahmood, K.; Casey, J.F. Mapping of Muslim Bagh ophiolite complex (Pakistan) using new remote sensing, and field data. *J. Asian Earth Sci.* **2006**, *30*, 333–343. [CrossRef]
2. El-Desoky, H.M.; Soliman, N.; Heikal, M.A.; Abdel-Rahman, A.M. Mapping hydrothermal alteration zones using ASTER images in the Arabian–Nubian Shield: A case study of the northwestern Allaqi District, South Eastern Desert, Egypt. *J. Asian Earth Sci. X* **2021**, *5*, 100060.
3. Hewson, R.D.; Cudahy, T.J.; Mizuhiko, S.; Ueda, K.; Mauger, A.J. Seamless geological map generation using ASTER in the Broken Hill-Curnamona province of Australia. *Remote Sens. Environ.* **2005**, *99*, 159–172. [CrossRef]
4. Mosier, D.L.; Singer, D.A.; Moring, B.C.; Galloway, J.P. Podiform chromite deposits—Database and grade and tonnage models. *US Geol. Surv. Sci. Investig. Rep.* **2012**, *5157*, 45.
5. Imamalipour, A. Geochemistry and geological setting of chromitites of Aland area from the Khoy ophiolite complex, NW Iran. *J. Geosci.* **2011**, *20*, 47–56.
6. Zaeimnia, F.; Kananian, A.; Arai, S.; Mirmohammadi, M.; Imamalipour, A.; Khedr, M.Z.; Miura, M.; Abbou-Kebir, K. Mineral chemistry and petrogenesis of chromitites from the K hoy ophiolite complex, Northwestern Iran: Implications for aggregation of two ophiolites. *Isl. Arc* **2017**, *26*, e12211. [CrossRef]
7. Amer, R.; Kusky, T.; Ghulam, A. Lithological mapping in the Central Eastern Desert of Egypt using ASTER data. *J. Afr. Earth Sci.* **2010**, *56*, 75–82. [CrossRef]
8. Rajendran, S.; Al-Khirbash, S.; Pracejus, B.; Nasir, S.; Al-Abri, A.H.; Kusky, T.M.; Ghulam, A. ASTER detection of chromite bearing mineralized zones in Semail Ophiolite Massifs of the northern Oman Mountains: Exploration strategy. *Ore Geol. Rev.* **2012**, *44*, 121–135. [CrossRef]

9.   Rowan, L.C.; Hook, S.J.; Abrams, M.J.; Mars, J.C. Mapping hydrothermally altered rocks at Cuprite, Nevada, using the Advanced Spaceborne Thermal Emission and Reflection Radiometer (ASTER), a new satellite-imaging system. *Econ. Geol.* **2003**, *98*, 1019–1027. [CrossRef]

10.   Sabins, F.F. Remote sensing for mineral exploration. *Ore Geol. Rev.* **1999**, *14*, 157–183. [CrossRef]

11.   Pournamdari, M.; Hashim, M.; Pour, A.B. Spectral transformation of ASTER and Landsat TM bands for lithological mapping of Soghan ophiolite complex, south Iran. *Adv. Space Res.* **2014**, *54*, 694–709. [CrossRef]

12.   Pournamdari, M.; Hashim, M. Detection of chromite bearing mineralized zones in Abdasht ophiolite complex using ASTER and ETM+ remote sensing data. *Arab. J. Geosci.* **2014**, *7*, 1973–1983. [CrossRef]

13.   Crosta, A.P.; De Souza Filho, C.R.; Azevedo, F.; Brodie, C. Targeting key alteration minerals in epithermal deposits in Patagonia, Argentina, using ASTER imagery and principal component analysis. *Int. J. Remote Sens.* **2003**, *24*, 4233–4240. [CrossRef]

14.   Massironi, M.; Bertoldi, L.; Calafa, P.; Visonà, D.; Bistacchi, A.; Giardino, C.; Schiavo, A. Interpretation and processing of ASTER data for geological mapping and granitoids detection in the Saghro massif (eastern Anti-Atlas, Morocco). *Geosphere* **2008**, *4*, 736–759. [CrossRef]

15.   Abdeen, M.M.; Allison, T.K.; Abdelsalam, M.G.; Stern, R.J. Application of ASTER band-ratio images for geological mapping in arid regions; the Neoproterozoic Allaqi Suture, Egypt. *Abstr. Program Geol. Soc. Am.* **2001**, *3*, 289.

16.   Yamaguchi, Y.; Naito, C. Spectral indices for lithologic discrimination and mapping by using the ASTER SWIR bands. *Int. J. Remote Sens.* **2003**, *24*, 4311–4323. [CrossRef]

17.   Crawford, A.R. Possible impact structure in India. *Nature* **1972**, *237*, 96. [CrossRef]

18.   Khalatbari-Jafari, M.; Juteau, T.; Bellon, H.; Emami, H. Discovery of two ophiolite complexes of different ages in the Khoy area (NW Iran). *Comptes Rendus Geosci.* **2003**, *335*, 917–929. [CrossRef]

19.   Pessagno, E.A.; Ghazi, A.M.; Kariminia, M.; Duncan, R.A.; Hassanipak, A.A. Tectonostratigraphy of the Khoy complex, northwestern Iran. *Stratigraphy* **2005**, *2*, 49–63.

20.   Hunt, G.R.; Ashley, R.P. Spectra of altered rocks in the visible and near-infrared. *Econ. Geol.* **1979**, *74*, 1613–1629. [CrossRef]

21.   Abrams, M.; Hook, S.J. Simulated ASTER data for geologic studies. *IEEE Trans. Geosci. Remote Sens.* **1995**, *33*, 692–699. [CrossRef]

22.   Gad, S.; Kusky, T. ASTER spectral for lithological mapping in the Arabian–Nubian Shield, the Neoproterozoic Wadi Kid area, Sinai, Egypt. *Gondwana Res.* **2007**, *11*, 326–335. [CrossRef]

23.   Gabr, S.; Ghulam, A.; Kusky, T. Detecting areas of high-potential gold mineralization using ASTER data. *Ore Geol. Rev.* **2010**, *38*, 59–69. [CrossRef]

24.   Moore, F.; Rastmanesh, F.; Asadi, H.; Modabberi, S. Mapping mineralogical alteration using principal-component analysis and matched filter processing in the Takab area, north-west Iran, from ASTER data. *Int. J. Remote Sens.* **2008**, *29*, 2851–2867. [CrossRef]

25.   Rajendran, S.; Nasir, S. Mapping of Moho and Moho Transition Zone (MTZ) in Samail ophiolites of Sultanate of Oman using remote sensing technique. *Tectonophysics* **2015**, *657*, 63–80. [CrossRef]

26.   Goetz, A.F.; Rock, B.N.; Rowan, L.C. Remote sensing for exploration; an overview. *Econ. Geol.* **1983**, *78*, 573–590. [CrossRef]

27.   Sultan, M.; Arvidson, R.E.; Sturchio, N.C. Mapping of serpentinites in the Eastern Desert of Egypt by using Landsat thematic mapper data. *Geology* **1986**, *14*, 995–999. [CrossRef]

28.   Nidamanuri, R.R.; Ramiya, A.M. Spectral identification of materials by reflectance spectral library search. *Geocarto Int.* **2014**, *29*, 609–624. [CrossRef]

29.   Galvao, L.S.; Formaggio, A.R.; Tisot, D.A. Discrimination of sugarcane varieties in Southeastern Brazil with EO-1 Hyperion data. *Remote Sens. Environ.* **2005**, *94*, 523–534. [CrossRef]

30.   Ninomiya, Y.; Fu, B.; Cudahy, T.J. Detecting lithology with Advanced Spaceborne Thermal Emission and Reflection Radiometer (ASTER) multispectral thermal infrared "radiance-at-sensor" data. *Remote Sens. Environ.* **2005**, *99*, 127–139. [CrossRef]

31.   Pour, A.B.; Hashim, M. ASTER, ALI, and Hyperion sensors data for lithological mapping and ore minerals exploration. *SpringerPlus* **2014**, *3*, 1–9.

32.   Rashmi, S.; Addamani, S.; Ravikiran, A. Spectral Angle Mapper Algorithm for Remote Sensing Image Classification. *Ijiset Int. J. Innov. Sci. Eng. Technol.* **2014**, *1*, 5481.

33.   Kamal, M.; Phinn, S. Hyperspectral data for mangrove species mapping: A Comparison of pixel-based object-based approach. *Remote Sens.* **2011**, *3*, 2222–2242.

34.   Benomar, T.B.; Fuling, B. Improved geological mapping using Landsat-5 TM data in Weixi area, Yunnan province China. *GeoSpat. Inf. Sci.* **2005**, *8*, 110–114. [CrossRef]

35.   Harrison, S. Standardized principal component analysis. *Int. J. Remote Sens.* **1985**, *6*, 883–890.

36.   Kruse, F.A.; Lefkoff, A.B.; Boardman, J.W.; Heidebrecht, K.B.; Shapiro, A.T.; Barloon, P.J.; Goetz, A.F. The spectral image processing system (SIPS)—interactive visualization and analysis of imaging spectrometer data. *Remote Sens. Environ.* **1993**, *44*, 145–163. [CrossRef]

Article

Spatial Component Analysis to Improve Mineral Estimation Using Sentinel-2 Band Ratio: Application to a Greek Bauxite Residue

Roberto Bruno [1,*], **Sara Kasmaeeyazdi** [1,*], **Francesco Tinti** [1], **Emanuele Mandanici** [1] **and Efthymios Balomenos** [2]

[1] Department of Civil, Chemical, Environmental and Materials Engineering, University of Bologna, 40136 Bologna, Italy; francesco.tinti@unibo.it (F.T.); emanuele.mandanici@unibo.it (E.M.)

[2] Metallurgy Business Unit, MYTILINEOS S.A., Ag. Nikolaos, 320 03 Viotia, Greece; efthymios.balomenos-external@alhellas.gr

* Correspondence: roberto.bruno@unibo.it (R.B.); sara.kasmaeeyazdi2@unibo.it (S.K.); Tel.: +39-05-1209-0241 (S.K.)

Abstract: Remote sensing can be fruitfully used in the characterization of metals within stockpiles and tailings, produced from mining activities. Satellite information, in the form of band ratio, can act as an auxiliary variable, with a certain correlation with the ground primary data. In the presence of this auxiliary variable, modeled with nested structures, the spatial components without correlation can be filtered out, so that the useful correlation with ground data grows. This paper investigates the possibility to substitute in a co-kriging system, the whole band ratio information, with only the correlated components. The method has been applied over a bauxite residues case study and presents three estimation alternatives: ordinary kriging, co-kriging, component co-kriging. Results have shown how using the most correlated component reduces the estimation variance and improves the estimation results. In general terms, when a good correlation with ground samples exists, co-kriging of the satellite band-ratio Component improves the reconstruction of mineral grade distribution, thus affecting the selectivity. On the other hand, the use of the components approach exalts the distance variability.

Keywords: resources characterization; bauxite residues; band ratio; kriging of component; mineral grade

Citation: Bruno, R.; Kasmaeeyazdi, S.; Tinti, F.; Mandanici, E.; Balomenos, E. Spatial Component Analysis to Improve Mineral Estimation Using Sentinel-2 Band Ratio: Application to a Greek Bauxite Residue. *Minerals* **2021**, *11*, 549. https://doi.org/10.3390/min11060549

Academic Editors: Amin Beiranvand Pour, Omeid Rahmani and Mohammad Parsa

Received: 12 March 2021
Accepted: 17 May 2021
Published: 21 May 2021

Publisher's Note: MDPI stays neutral with regard to jurisdictional claims in published maps and institutional affiliations.

1. Introduction

1.1. Recovery of Minerals from Stockpiles and Tailings

Raw material and metal extractions have been conducted since pre-historic times. Mining has been present everywhere in Europe, although nowadays the majority of sites are closed. This does not mean that resources have been completely depleted. Ancient mining could not benefit from the most modern extraction and processing techniques and has left significant amounts of mining residue (including tailings and stockpiles) currently present the territory, in the forms of semi-artificial hills, lakes, and ponds. Some of them were completely stable and never reacted with the environment, while others (especially those coming from metal mining) significantly modified the environment where they were stocked. According to "Mining and Metal in a Sustainable World 2050" [1], a major gap exists in effective retreatment technology (reuse, resize, or remove) of mining residues to meet the sustainability objectives of United Nations Development Program ($UNDP$) in 2030 [2].

Moreover, the depletion of the in-situ reserves, the increasing need of using lower grade materials, and advances in recovery and processing technologies are the main reasons why mining wastes are considered as recoverable resources. Moreover, the environmental aspects have caused a strong push for more effective management of mining residuals in many mining sites [3,4]. As a first step, the raw material concentrations in tailings must be quantified and classified and a reliable expected revenue model should be developed to assess the feasibility of production, by giving particular attention to the presence of critical raw materials (*CRMs*) for the European Union (*EU*), with high impact on the economy and, at the same time, high risk of supply shortage [5,6].

A quantitative evaluation of the available resources in mining residues requires an exhaustive sampling that must be justified, so that a preliminary characterization phase is necessary for deciding if it is justified, proceeding with a full resource evaluation. Earth observation (*EO*) data can be useful for the abovementioned preliminary mapping and quantification of these mining residues, usually abandoned, in harsh environment, and with limited possibilities for full and fast sampling campaigns [7]. The main potentials of EO include the large number of easy to access data over large areas and their continuous acquisition over time, which may allow continuous land monitoring.

1.2. Exploitation of Remote Sensing Information

The most popular applications of satellite imagery refer to mapping problems, where the spectral content of images is used to recognize and characterize the observed surface, for instance, in terms of land cover or physical and chemical properties, among others. For these purposes, satellite images often require some kinds of geometrical and spectral calibration [8]. Before calibration, indeed, images are affected by artefacts, which depend on the sensor characteristics and the conditions at the time of acquisition, and that is to be removed to enhance the information considered useful. The general problem for mineral exploration and reserve characterization is the spatial distribution of the target variable, because, in most practical cases, in situ information is limited and sparse. Satellite images, instead, can provide dense additional information over the full area of interest, which can be used, especially when correlation is found with the in situ data.

Because of the fast and accessible information, many *EO* analyses have been used in mining areas and abandoned tailings, mainly for mapping pollution and environmental variables, beginning many years ago [9–13]. In the presented case studies, authors used remote sensing data and tools (such as imaging spectrometer data, and/or hyperspectral imagery combined with in situ data) to improve map accuracy of environmental pollutants affected by mining activities and their abandoned residue. In many applications, to improve map accuracy and to validate results, geostatistical approaches were used with integration of remote sensing [14–16]. The base of the geostatistical theory is the spatial correlation among georeferenced data, correlations exploitable for a correct and optimal numerical modeling of the regionalized variable (*RV*). Besides mapping the *RV* with high accuracy, as an important advantage, using geostatistical approaches provides an extra tool to improve the quality of the estimation, measured by the estimation variance maps [17]. In addition, multivariate geostatistical approaches provide an extra possibility to use more than one RV and model the correlations between RVs within so called co-regionalization modeling [18]. This analysis is able to not only verify some global results as the statistical correlations, but also to identify possible correlations between spatial components of the main variable with extra variables (the so-called secondary or auxiliary variables), including spatial anisotropies. Therefore, independent of the variables at hand (temperature, concentration, discovery probability, etc.) and of the spatial distribution model (estimation, simulation), geostatistics allow tackling the central problem: finding meaningful correlations and spatial modeling the unknown surface distribution of the interest variable, by extra information (which can be satellite data for example) [19,20].

As mentioned above, in the mining sector, while direct information is expensive, a good opportunity can be to exploit indirect information, much denser, but with good correlation with the direct variable. This is the case of the low-cost data provided by satellite images. There are many examples of mineral characterizations, where, by knowing the spectral properties of a surface feature, simple mathematical operations among spectral bands (called band ratio) have contributed to localize outcrops and surface deposits [21–26].

To detect a specific feature or mineral, usually at least two spectral bands are necessary, one band with higher reflectance features of the given material and another one with strong absorption features for the same material [23]. There are many studies on spectral analysis of minerals based on different satellite images and specific band ratios for different minerals in a target area mainly in geological mapping [27,28]. Iron oxides are among the most studied materials using band ratio techniques [25], because of their selective absorption of light in the visible and near-infrared range caused by transitions in the electron shell [29]. For example, for detection and mapping of an iron ore formation, ferric (Fe^{3+}) and ferrous (Fe^{2+}) iron oxide specific band ratios are suggested, and thanks to their correlation with iron samples, some iron ore grades maps can be obtained [26]. Hence, the band ratio, as an appropriate index, can be considered as additional information while mapping a specific mineral of a geological feature on the surface.

This paper proposes to map the iron concentration as the strategic metal within a bauxite residue in Greece. In the first step, only direct samples from the site were used (performing ordinary kriging—OK estimation). Then, the band ratio identified for iron detection was used as additional information to map the iron variability within the bauxite residues (performing co-kriging—CK estimation). To improve the map accuracies, a new method (component co-kriging—CCK estimation) was proposed to re-construct the co-regionalization model between the sample data and the band ratio information, by exploiting the possibility of extracting a specific component from the satellite data and using it in the co-regionalization models. Finally, all three models and their results were compared to check the improvement given by the proposed model in iron estimation maps.

2. Materials and Methods

2.1. Co-Regionalization Model and Application: Current Practice

The most classical estimation method to map the spatial variability of a *RV* variable is *OK*, which uses available samples to predict in "not-available" points [17]. To add an additional variable in the kriging system, and moving into multivariate geostatistics, which in many cases improves the target variable prediction, there is a need of, at least, a second variable, with the spatial correlation among them [18]. This property can be calculated within the cross-covariance. Given two stationary random functions, $Z_1(x)$, $Z_2(x)$, with the means of m_1 and m_2, the spatial cross-covariance $C_{12}(h) \neq C_{21}(h)$, are defined in Equation (1):

$$\begin{cases} C_{12}(h) = E[Z_1(x) \times Z_2(x+h)] - m_1 \times m_2 \\ C_{21}(h) = E[Z_2(x) \times Z_1(x+h)] - m_1 \times m_2 \end{cases} \tag{1}$$

The secondary variable can be known in all points of the domain and displaced in a regular grid. If remote sensing data are used as a secondary variable, the space-time concept should be considered because ground samples are taken in a certain time, while satellite information is repeated over time; this produces photographs of "different" stockpiles.

In this study, an iso-time framework was adopted. An image of the stockpile at the time zero (t_0) was considered, as well as all the ground samples referred to the same surface remote-sensed. Therefore, the 3D reconstruction of the distribution of concentrations is relevant to a constant time.

The objective was to map the distribution of the target variable (iron concentration as the strategic metal), using Sentinel-2 satellite data as the secondary variable. Sentinel-2 is a European satellite mission for Earth observation, which is part of the Copernicus program. It provides global coverage of multispectral imagery, composed by 13 bands in the spectral range between the visible and the short-wave infrared, with a revisiting time of five days at the equator [30]. It was selected for the present study because of the availability of the images at the date of sampling, and the good spatial resolution (pixel size from 10 to 60 m, depending on the band). Moreover, the spectral bands available in Sentinel-2 data allow the computation of different band ratios useful for mineral exploration. For iron deposits, in particular, different band ratios were proposed in the literature for other multispectral sensors [21–26], which can also apply to Sentinel-2 images [31].

To map the iron variability, the classical steps are:

- Using one variable: iron samples, spatial variability analysis of target variable (sample variogram and its model), and finally using the *OK* estimation method;
- Adding extra information (as an example band ratio of iron as the secondary variable): spatial variability analysis of target variable (sample variogram and its model) and the secondary variable, the cross-correlation analysis between the target variable and the secondary variable, and finally using the *CK* estimation method.
- At the end, to compare the map accuracies results, cross-validation should be performed to check if adding information can improve the results.

The first approach for mapping the mineral concentration is using just the in situ samples, which in this case are represented by the mineral concentrations from the mining residues sampling. The ordinary kriging (*OK*) method can be used and the estimated values in all nodes of the grid can be found by using Equation (2):

$$A^{OK}(x_i) = \sum_{\alpha} \lambda_{\alpha}^{OK} \times A(x_{\alpha}) \tag{2}$$

where: $A(x_\alpha)$ is the variable known in the points x_α (mineral grade from samples); $\lambda_\alpha{}^{OK}$ are the weights calculated with ordinary kriging method; $A^{OK}(x_i)$ is the estimate of the main variable in the points x_i (grid nodes).

Regarding the *OK* model definition, its spatial variability is object of the variogram analysis. The standard model of a stationary random function with nested structures is presented in Equation (3):

$$\gamma_A(h) = a_{nug} + \sum_u a_u \times \gamma_u(h) \tag{3}$$

where: $\gamma_A(h)$ is the variogram model of the main variable (mineral concentration from sampling); a_{nug} is the nugget effect; $\gamma_u(h)$ are the models of different nested structures (spatial components); a_u are the sills of each model component.

The second approach attempts to improve the estimation of the main variable, using the secondary (auxiliary) variable. The prerequisite is to verify if a correlation exists between two variables. The value of the correlation coefficient is defined by Equation (4):

$$\rho_{AB} = \frac{\sigma_{AB}}{\sqrt{\sigma_A^2 \times \sigma_B^2}} \tag{4}$$

where: ρ_{AB} is the correlation coefficient between the primary (mineral's grade from sampling) and secondary variable; σ_{AB} is the covariance between the primary and secondary variable; $\sigma^2{}_A$ is the variance of the primary variable; $\sigma^2{}_B$ is the variance of the secondary variable.

The *CK* variance allows theoretically verifying the effect of the secondary variable on reducing the estimation smoothing. The estimates by *CK* can be found by applying Equation (5):

$$A^{CK}(x_{i0}) = \sum_{\alpha} \lambda_{\alpha}^{CK} \times A(x_{\alpha}) + \sum_i v_i^{CK} \times B(x_i) \tag{5}$$

where: $B(x_i)$ is the auxiliary variable known in the points x_i (satellite grid nodes); $\lambda_\alpha{}^{CK}$ and v_i^{CK} are the weights for the primary and secondary variables calculated by *CK*; $A^{CK}(x_{i0})$ is the estimate of the main (primary) variable in the points x_{io} one of the grid nodes.

Moreover, in the case of *CK*, often a linear co-regionalization model is expected (Equation (6)):

$$\begin{cases} \gamma_A(h) = a_{nug} + \sum_u a_u \times \gamma_u(h) \\ \gamma_B(h) = b_{nug} + \sum_u b_u \times \gamma_u(h) \\ \gamma_{AB}(h) = c_{nug} + \sum_u c_u \times \gamma_u(h) \end{cases} \tag{6}$$

where: $\gamma_A(h)$, $\gamma_B(h)$ are the variogram models of the first and secondary variables; $\gamma_{AB}(h)$ is cross-variogram model between the primary and the secondary variable; γ_u (h) is the structural components of variogram models; b_{nug} is the nugget effect for the secondary variable; b_u are the sills of each component of the variogram model for the secondary variable; c_{nug}, c_u are the sills of each component for the cross-variogram model.

The comparison of estimations obtained by the *OK* and *CK* methods is performed by the following analyses [19]:

- The cross-validation;
- The estimation maps of minerals;
- The maps of estimation variance.

Usually, when the secondary variables are dense, namely available at more points than the main variable, and sufficiently correlated with the main variable, *CK* is typically of advantage [18].

2.2. New Perspective: Use of Spatial Components

In the case of multivariate geostatistics and in presence of a linear co-regionalization model, the selected variables (*A* as the target variable and B as the secondary variable) can be considered as a linear combination of independent random variables (factors) Y_{nug}, Y_i monostructure, called scale components, in addition to the mean (Equation (7)):

$$\begin{aligned} A &= m_A + \sqrt{a_{nug}}Y^A_{nug} + \sum_u \sqrt{a_u} \times Y^A_u \\ B &= m_B + \sqrt{b_{nug}}Y^B_{nug} + \sum_u \sqrt{b_u} \times Y^B_u \end{aligned} \tag{7}$$

where m_A, m_B are the means of the variables *A* and *B*; Y^A_{nug}, Y^A_u, Y^B_{nug}, Y^B_u are the structural components of the main and auxiliary variables, each of them being a {0,1} standard variable with a specific variogram structure.

The correlation coefficient between the iso-structure components of the main and auxiliary variable are presented in Equation (8):

$$\begin{aligned} \rho^{AB}_{nug} &= E\left[Y^A_{nug} Y^B_{nug}\right] \\ \rho^{AB}_u &= E\left[Y^A_u Y^B_u\right] \end{aligned} \tag{8}$$

Given the independence of factors, the total variance of the variables is just the sum of the sills of each component. Variances and covariances are presented in Equation (9):

$$\begin{cases} \sigma^2_A = a_{nug} + \sum_u a_u \\ \sigma^2_B = b_{nug} + \sum_u b_u \\ \sigma_{AB} = \sqrt{a_{nug} \times b_{nug}}\rho^{AB}_{nug} + \sum_u \sqrt{a_u \cdot b_u}\rho^{AB}_u = c_{nug} + \sum_u c_u \end{cases} \tag{9}$$

The correlation coefficient between the main (primary) and the secondary variable is presented in Equation (10):

$$\rho_{AB} = \frac{\sigma_{AB}}{\sqrt{\sigma^2_A \times \sigma^2_B}} = \frac{c_{nug} + \sum_u c_u}{\sqrt{\left(a_{nug} + \sum_u a_u\right) \times \left(b_{nug} + \sum_u b_u\right)}} \tag{10}$$

In the case of having several scale components, there could be an advantage of using as auxiliary variable just one component instead of the whole initial variable. The justification for that derives by the observation that the correlation between the main and the auxiliary variables exists only at one scale (Equation (11)):

$$\rho^{AB}{}_{u_c} = \frac{C_{u_c}}{\sqrt{a_{u_c} \times b_{u_c}}} \quad \rho^{AB}{}_{u,u' \neq u} = 0 \tag{11}$$

Such observation derives by the co-regionalization model, showing only one structure in the cross variogram, or the independence of any other structure common to the main and auxiliary variables (more generally, this approach allows also the filtering the effect of a noise [18]).

In this paper, the methodology is performed to check the improvements of the iron concentration maps, due to the increase of correlation when the auxiliary variable is just the correlated component, as shown in the following relationships (Equation (12)):

$$\begin{aligned}
\sigma_{AY_{u_c}} &= E[AY_{u_c}] = \mathrm{cov}\{AB\} = \sigma_{AB} \\
b_{u_c} &< b_{nug} + \sum_u b_u \to \sigma_{Y_{u_c}}^2 < \sigma_B^2 \\
\rho_{AY_{u_c}} &= \frac{\sigma_{AY_{u_c}}}{\sqrt{\sigma_A^2 \times \sigma_{Y_{u_c}}^2}} > \frac{\sigma_{AB}}{\sqrt{\sigma_A^2 \times \sigma_B^2}} = \rho_{AB}
\end{aligned} \tag{12}$$

In terms of variograms (Equation (13)):

$$\begin{cases} \gamma_B > \gamma_{Y_{u_c}} \\ \gamma_{AB} = \gamma_{AY_{u_c}} \end{cases} \tag{13}$$

In terms of *CK*, using one of the components, Y_{u_c}, the estimation results are different with respect to using the original auxiliary variable *B*, since the new secondary variable and the weights differ from the original one. In fact, the component co-kriging system (*CCK*) is similar to the original *CK*, but with different coefficient matrix, since the submatrix of variogram of secondary variable changes (Equation (14)):

$$\begin{bmatrix} \gamma_{AA} & \gamma_{AB} & 1 & 0 \\ \gamma_{AB} & \gamma_{BB} & 0 & 1 \\ 1 & 0 & 0 & 0 \\ 0 & 1 & 0 & 0 \end{bmatrix} \neq \begin{bmatrix} \gamma_{AA} & \gamma_{AY_{u_c}} & 1 & 0 \\ \gamma_{AY_{u_c}} & \gamma_{Y_{u_c}Y_{u_c}} & 0 & 1 \\ 1 & 0 & 0 & 0 \\ 0 & 1 & 0 & 0 \end{bmatrix}$$
$$A^{CCK}(x_i) = \sum_a \lambda_a^{CCK} \times A(x_a) + \sum_i v_i^{CCK} \times Y_{u_c}(x_i) \tag{14}$$

where: $Y_{u_c}(x_i)$ is the structural component of the auxiliary variable known in the points x_i (satellite grid nodes); λ_α^{CKY} and v_i^{CCK} are the weights for the primary and auxiliary variables calculated by *CCK*; $A^{CCK}(x_i)$ is the estimation of the main (primary) variable in the points x_i (grid nodes).

Note that $Y^B(x_i)$ is the true component that we do not know, so that we can implement the *CCK* if we estimate it by factorial kriging (*FK*) which respects the actual data [18] in Equation (15):

$$\begin{aligned}
Y_u^{FK}(x_i) &= \sum_j \lambda_j^{FK} B(x_j) \\
B &= m^{FK} B + \sqrt{b_{nug}} Y_{nug}^{FK}(x_i) + \sum_u \sqrt{b_u} Y_u^{FK}(x_j)
\end{aligned} \tag{15}$$

The second part of Equation (15) can be checked to control the estimated value of components $Y^{FK}(x_i)$. We can consider that, in case of an image band, the information is dense and the estimation quality is satisfying, so that it looks justified to use the estimated value of components in Equation (15).

The estimation variances σ^{2CK} and σ^{2CCK} allow comparing the precision of the estimations. The two estimation variances are presented in Equation (16):

$$\begin{cases} \sigma_e{}^2(A^{CK}(x_{i0}) \rightarrow A(x_{i0})) = \sum_{\alpha} \lambda_{\alpha}^{CK} \times \gamma_A(h_{\alpha i_0}) + \mu_A^{CK} + \sum_{i} v_i^{CK} \times \gamma_{AY}(h_{ij_0}) + \mu_B^{CK} \\ \sigma_e{}^2(A^{CCK}(x_{i0}) \rightarrow A(x_{i0})) = \sum_{\alpha} \lambda_{\alpha}^{CCK} \times \gamma_A(h_{\alpha i_0}) + \mu_A^{CCK} + \sum_{i} v_i^{CCK} \times \gamma_{AY}(h_{ij_0}) + \mu_B^{CCK} \end{cases} \tag{16}$$

Moreover, to check the adopted variogram models and to check if *CCK* could improve the estimation results, cross-validation can be performed on data. The principle of cross-validation is to remove the target variable at each sample point x_α and then predict by kriging with the proposed model. Therefore, since the true values are available, it is possible to compute the kriging error [19].

2.3. Case Study: The Bauxite Residuals of Greece

Bauxite residues (*BR*) remained from Bayer processing of bauxite (also commonly known as "red mud") represents important strategic wastes from mining and processing activities and they were inserted in 2020 in the list of critical raw materials for the European Union [5]. The significant amount of raw materials within these types of residues can be used as a new source of materials, specifically critical metals and rare earth elements [32]. Due to the analysis done on the BR, it has potential as a secondary resource for REE extraction [33] and TiO_2, V_2O_5, Al_2O_3, Fe_2O_3, CaO, Na_2O, SiO_2 resources.

The case study used in this research is the bauxite residue from the alumina refinery of Mytilineos S.A. in Greece, located on the Gulf of Corinth, 136 km from Athens (Figure 1). The exact location is at latitude 38.354177° and longitude 22.704671°, CRS WGS84, and its dimension is around 700 m $\times$ 600 m. Since 2006, four filter presses have been used to dewater the *BR*, and since 2012, all *BR* produced has been filter-pressed and stored as a "dry" (water content < 26%) by-product in an appropriate industrial landfill [31]. Producing the dry *BR* is currently known as the best technique for BR materials piling, because a lower volume of deposits is stocked, with a subsequent decrease of the risk of dam failures.

Figure 1. Location of bauxite residuals (**left**) and a high-resolution image of daily piling materials (**right**).

The samples used were collected from daily accumulated materials (daily data) including the tonnage of materials with their mean concentration value, and the area in which they were piled within the *BR* areas from June to the end of July. The samples exact locations (coordinates) were assigned where trucks discharged their daily load of materials. Figure 2 shows the daily data during two months (June and July 2019). Therefore, since, during the two months, materials were not over-accumulating, the Sentinel 2 image selected at the end of July (date: 30 July 2019) is representative of materials during June and July (accumulated in the area from the first of June until the end of July).

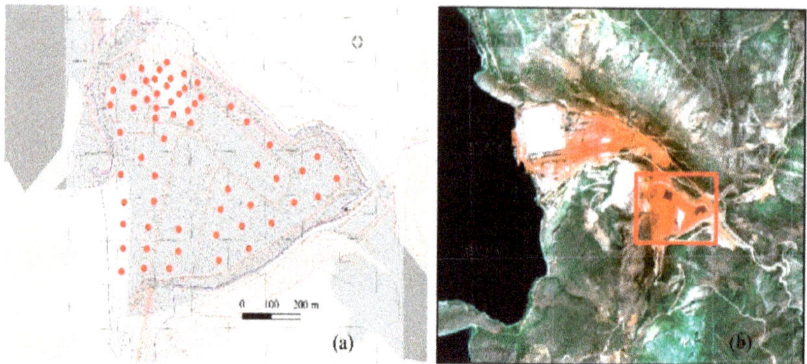

Figure 2. Samples location at the end of July 2019 (**a**) and Sentinel-2 true color composite image (RGB = 4,3,2); date: 30 July, 2019; the red line indicates the BR area (**b**).

The target RV selected is iron concentration Fe_2O_3 (%) as a strategic metal from samples obtained from BR of Greece. The daily samples were analyzed through X-ray fluorescence analysis-XRF, to obtain the iron concentration (Fe_2O_3%) information at alumina refinery of Mytilineos *S.A.* To select the secondary variable, a preliminary test has been done to choose the most relevant band ratio (with highest correlation coefficient) for iron mapping (Table 1). Since the Sentinel-2 data are used in this study, the most common iron band ratios for identifying iron [30] have been applied. It is worthwhile to note that these ratios involve only bands at the highest spatial resolution (pixel size 10 m).

Table 1. Correlation coefficients between iron grades and band ratios of Sentinel-2 [30] at the sampling points. The highest correlation is in bold.

Band Ratios	Sentinel-2A Bands with Their Central Wavelength	Correlation Coefficient with Iron Concentration (ρ)
All iron oxides	$\dfrac{4\ (664.9\ nm)}{2\ (492.1\ nm)}$	-0.130
Ferrous iron oxides	$\dfrac{4\ (664.9\ nm)}{11\ (1613.7\ nm)}$	-0.349
Ferric Iron, Fe^{3+}	$\dfrac{4\ (664.9\ nm)}{3\ (559.0\ nm)}$	-0.150
Ferrous Iron, Fe^{2+}	$\dfrac{12\ (2202.4\ nm)}{8\ (832.8\ nm)} + \dfrac{3\ (559.0\ nm)}{4\ (664.9\ nm)}$	0.194
Ferrous silicates	$\dfrac{12\ (2202.4\ nm)}{11\ (1613.7\ nm)}$	-0.125
Ferric oxides	$\dfrac{11\ (1613.7\ nm)}{8\ (832.8\ nm)}$	0.223

From the presented correlation coefficients between iron concentration and band-ratios (Table 1), the one with the highest correlation (ferrous iron oxides: 4/11) is chosen as the secondary variable to map the iron concentration variability within the BR.

The histogram of iron concentration samples and the correlation between the selected band ratios (ferrous iron oxides (4/11) band ratio) is exposed in Figure 3.

Band data are extracted from the Sentinel-2 image (see Figure 2) only inside the boundaries of the BR area. The base map of the extracted band ratio values and the histogram of data are shown in Figure 4.

Considering the iron concentration as the target variable and the ferrous iron oxides band ratio as the secondary variable, it is possible, first, to map the iron variability in BR using only iron samples. Secondly, it is possible to check if map accuracies can be improved by adding the band ratio variable. Finally, by decomposition of the band ratio variable, in the case of higher correlation, using one component can improve the iron estimation maps.

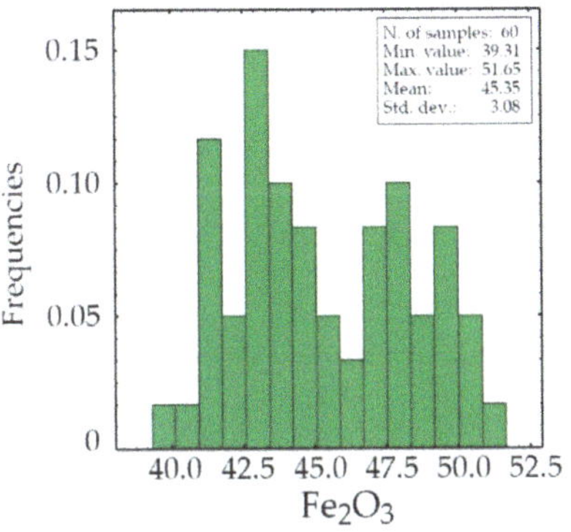

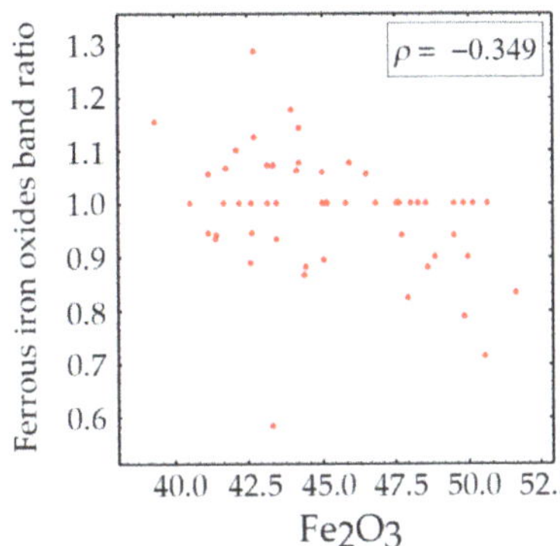

Figure 3. Histogram of Fe$_2$O$_3$ (%) of samples (**left**) and correlation with band ratio ferrous iron oxides: Band4/Band11 (**right**).

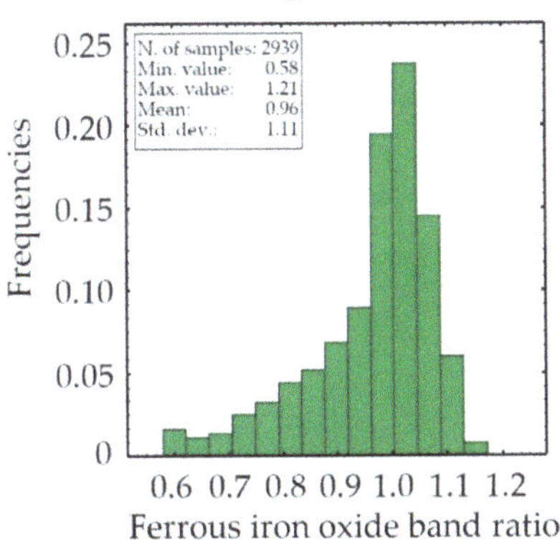

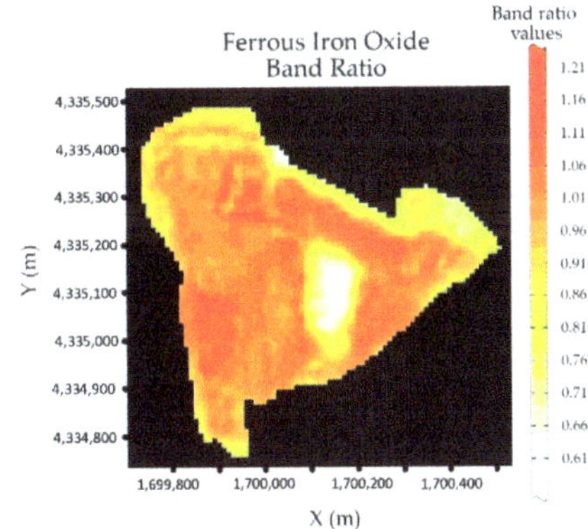

Figure 4. Histogram of ferrous iron oxides band ratio (**left**) and its base map (**right**).

3. Results

In the first step, using only iron samples, to perform *OK*, the sample variogram and variogram model is shown in Figure 5, including two spherical structures. The red bars below the variogram show the frequency of sample pairs used in variogram calculations. The variogram model parameters are presented in Table 2.

Table 2. Structures and parameters of variogram models fitted on Fe_2O_3 (%) samples variogram.

	Fe_2O_3 (%) Variogram Models				
Nugget Effect	**Spherical 1**			**Spherical 2**	
	Range (m)	**Sill**		**Range (m)**	**Sill**
2.3	70	2.9		180	3.4

Using the variogram model of Figure 5, it is possible to perform *OK*. Maps of Fe_2O_3 (%) concentration variability and estimation standard deviation are presented in Figure 6.

To improve the iron estimation results, in the second step, the presented secondary variable (the ferrous iron oxides (4/11) band ratio) can be added to map the iron concentration variability.

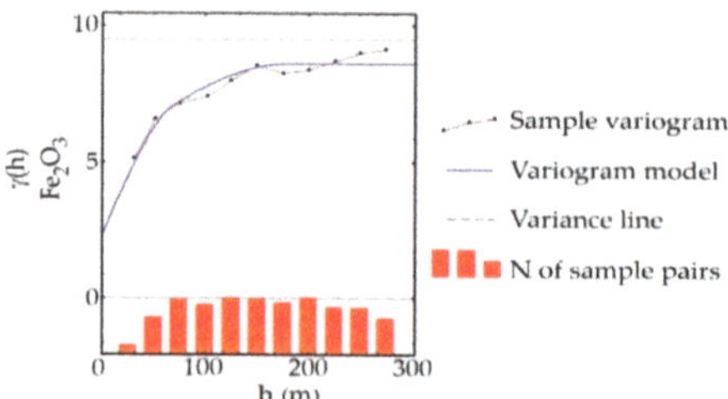

Figure 5. Sample variogram and variogram model for iron concentration obtained by samples Fe_2O_3 (%). h(m) is the distance and $\gamma(h)$ is the sample variogram.

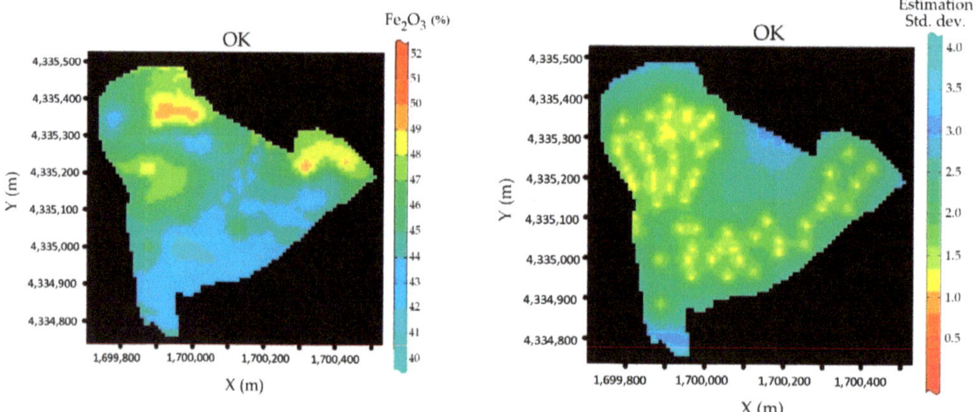

Figure 6. Fe_2O_3 (%) estimated map (*OK*) (**left**) and the estimation variance map (**right**) by performing *OK*.

To perform co-kriging, the sample variograms and variogram models considering both variables (iron as main variable and ferrous iron oxides band ratio as the secondary variable) and cross variogram are calculated and shown in Figure 7. The fitted model for iron concentration is equal to the model used in *OK*. Using the same model makes the comparisons more logical between the *OK* and *CK*. The structure and model details are presented in Table 3.

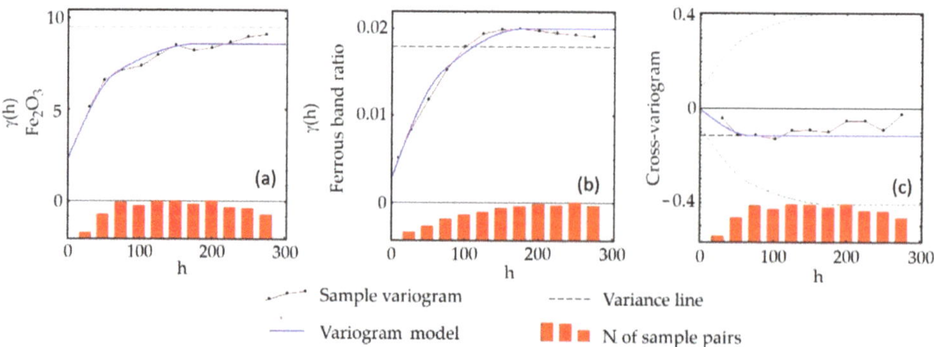

Figure 7. Sample variograms and variogram models for iron concentration obtained by samples Fe_2O_3 (%) (**a**), ferrous iron oxides band ratio (**b**) and cross-variogram (**c**). $h(m)$ is the distance and $\gamma(h)$ is the sample variogram.

Table 3. Structures and parameters of variogram models fitted on Fe_2O_3 (%) samples, ferrous iron oxides band ratio, and cross-variogram.

	Direct variable—Fe_2O_3 (%)—Variogram Models				
Nugget Effect	**Spherical 1**			**Spherical 2**	
	Range (m)	**Sill**		**Range (m)**	**Sill**
2.3	70	2.9		180	3.4
	Auxiliary Variable—Band Ratio—Variogram Models				
Nugget Effect	**Spherical 1**			**Spherical 2**	
	Range (m)	**Sill**		**Range (m)**	**Sill**
0.0027	70	0.0063		180	0.011
	Cross-Variogram Models				
Nugget Effect	**Spherical 1**			**Spherical 2**	
	Range (m)	**Sill**		**Range (m)**	**Sill**
0	70	−0.116		180	0.0001

Using the presented variogram models of Figure 7, it is possible to perform *CK*. Maps of Fe_2O_3 (%) concentration variability and estimation variance are presented in Figure 8.

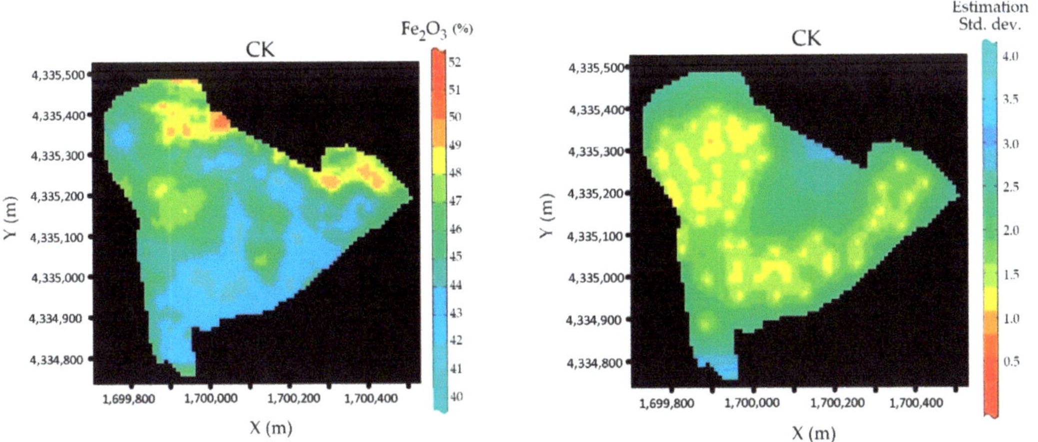

Figure 8. Fe_2O_3 (%) estimated map (*CK*) (**left**) and the estimation variance map (**right**) by performing co-kriging between samples and ferrous iron oxide band ratio.

Finally, the new approach was performed by decomposition of the secondary variable (the ferrous iron oxide band ratio).

In the first step, to choose the appropriate component, the correlation coefficients are calculated, using the variogram models' structures, and due to Equation (9):

$$\gamma_{iron} = 2.3 + 2.9 \times \gamma(R = 70) + 3.4 \times \gamma(R = 180)$$
$$\gamma_{ferrous.iron.band.ratio} = 0.0027 + 0.0063 \times \gamma(R = 70) + 0.011 \times \gamma(R = 180) \quad (17)$$
$$\gamma_{cross.variogram} = -0.116 \times \gamma(R = 70) + 0.0001 \times \gamma(R = 180)$$

Hence, the correlation coefficient between components of ferrous iron oxides band ratio and iron concentration can be calculated as bellow:

$$\begin{aligned}
\sigma_A &= 2.3 + 2.9 + 3.4 = 8.6 \\
\sigma_B &= 0.0027 + 0.0063 + 0.011 = 0.02 \\
\sigma_{AB} &= 0 - 0.116 + 0.0001 = -0.1159 \\
\sigma_{component1} &= \sigma_{Y1} = 0.0063 \\
\sigma_{component2} &= \sigma_{Y2} = 0.011
\end{aligned} \tag{18}$$

Two components are related to the small range ($R = 70$ m) and the large range ($R = 180$). Therefore, it is possible to calculate the correlation coefficients:

$$\begin{aligned}
\rho_{A/B} &= \frac{\sigma_{AB}}{\sqrt{\sigma_A^2 \times \sigma_B^2}} = \frac{-0.1159}{\sqrt{8.6 \times 0.02}} = -0.279 \\
\rho_{A/Y1} &= \frac{\sigma_{AY}}{\sqrt{\sigma_A^2 \times \sigma_Y^2}} = \frac{-0.116}{\sqrt{8.6 \times 0.0063}} = -0.498 \\
\rho_{A/Y2} &= \frac{\sigma_{AY2}}{\sqrt{\sigma_A^2 \times \sigma_{Y2}^2}} = \frac{0.0001}{\sqrt{8.6 \times 0.011}} = 0.0003 \\
\rho_{A/Y1} &= -0.498 > 0.279 = \rho_{A/B}
\end{aligned} \tag{19}$$

Since the first component has the highest correlation coefficient (negative correlation) with iron concentration, it is selected as the appropriate component to test the *CCK*.

In this step, to use the selected component in Equation (13), there is a need of estimating the component in all points of the grid. To do it, the *CK* is performed on band-ratio data, using only the first structure of the variogram model. To check the coherency of results from Equation (15), the estimated maps of both components (small range, $R = 70$ m, and large range, $R = 180$ m), plus the estimated mean are shown in Figure 9.

The sum of three maps (estimated components and mean estimated) is equal to the original values of ferrous iron oxides band ratio as Equation (15).

The small range component ($R = 70$ m) as having the higher coefficient correlation can be used as the secondary variable to perform *CCK*.

The parameters of variogram models are shown in Table 4. The same as *CK*, the fitted model on target variable (iron concentration) is equal to the previous models.

Table 4. Main parameters of the variogram model from Fe_2O_3 (%) of samples, first component of ferrous iron oxides band ratio and cross-variogram.

Nugget Effect	Fe_2O_3 (%) Variogram Models			
	Spherical 1		Spherical 2	
	Range (m)	Sill	Range (m)	Sill
2.3	70	2.9	180	3.4
	Band Ratio-Component 1 Variogram Models			
	Spherical 1			
	Range (m)		Sill	
	70		0.0063	
	Cross-Variogram Models			
	Spherical 1			
	Range (m)		Sill	
	70		−0.116	

Maps of Fe_2O_3 (%) concentration using *CCK* method and its estimation variances are presented in Figure 10.

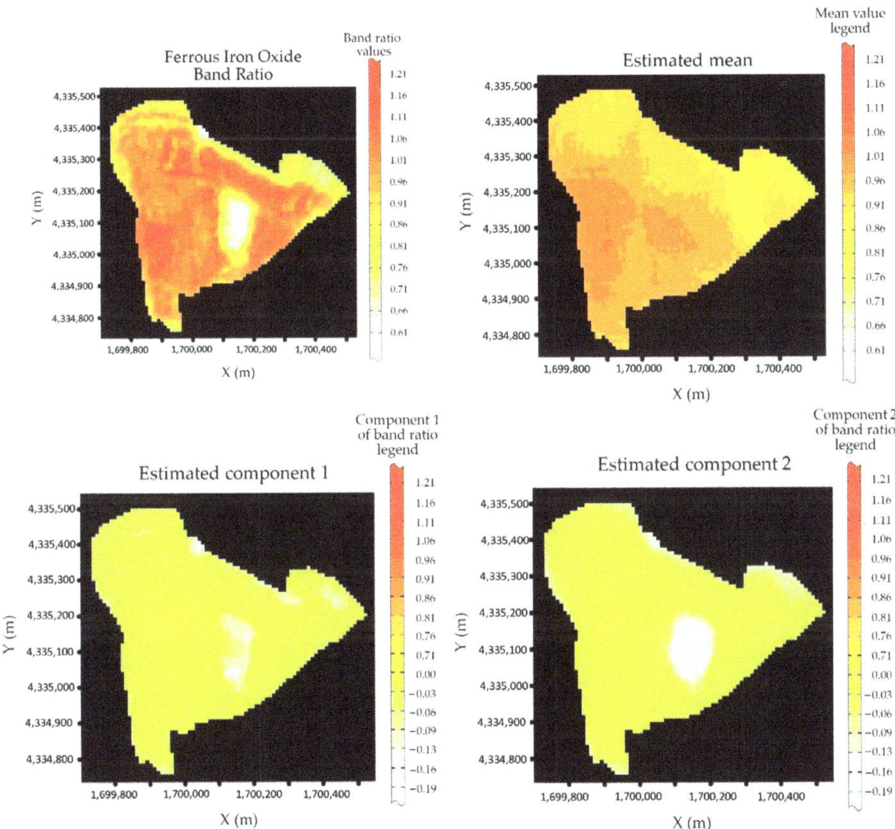

Figure 9. Base maps of real data values of ferrous iron oxides band ratios (**upper left**) and estimated components: small range component (**lower left**) and large range component (**lower right**), and the mean estimation map (**upper right**), all deposited from ferrous iron oxides band ratios.

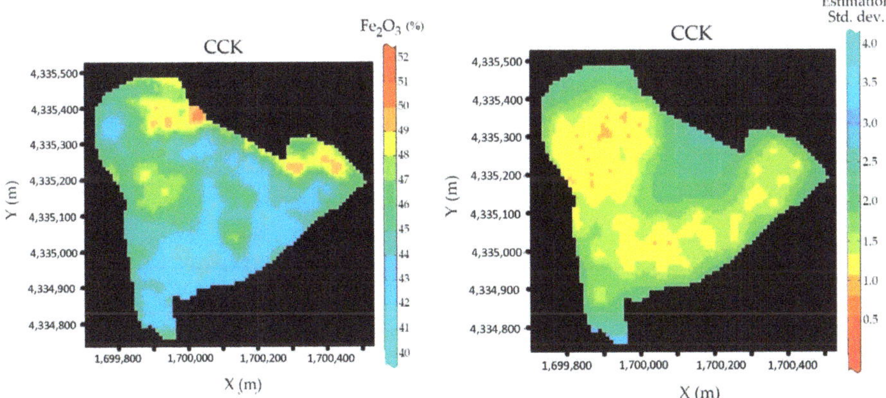

Figure 10. Fe_2O_3 (%) estimated map (*CCK*) (**left**) and the estimation variance map (**right**), by performing *CCK* between iron samples and the first component of ferrous iron oxide band ratio (Range = 70 m).

Finally, Figure 11 shows the main statistics of the cross-validation for three solutions, *OK*, *CK*, and *CCK*. Cross-validation performed by removing sample values (one-by-one) and estimating them using the selected model and neighborhood for three methods (*OK*, *CK*, and *CCK*). The scatter plots between estimated and true values of Fe_2O_3 (%) at 60 sample points, standardized estimation error, and scatter plot between error and estimated values are compared for all three solutions.

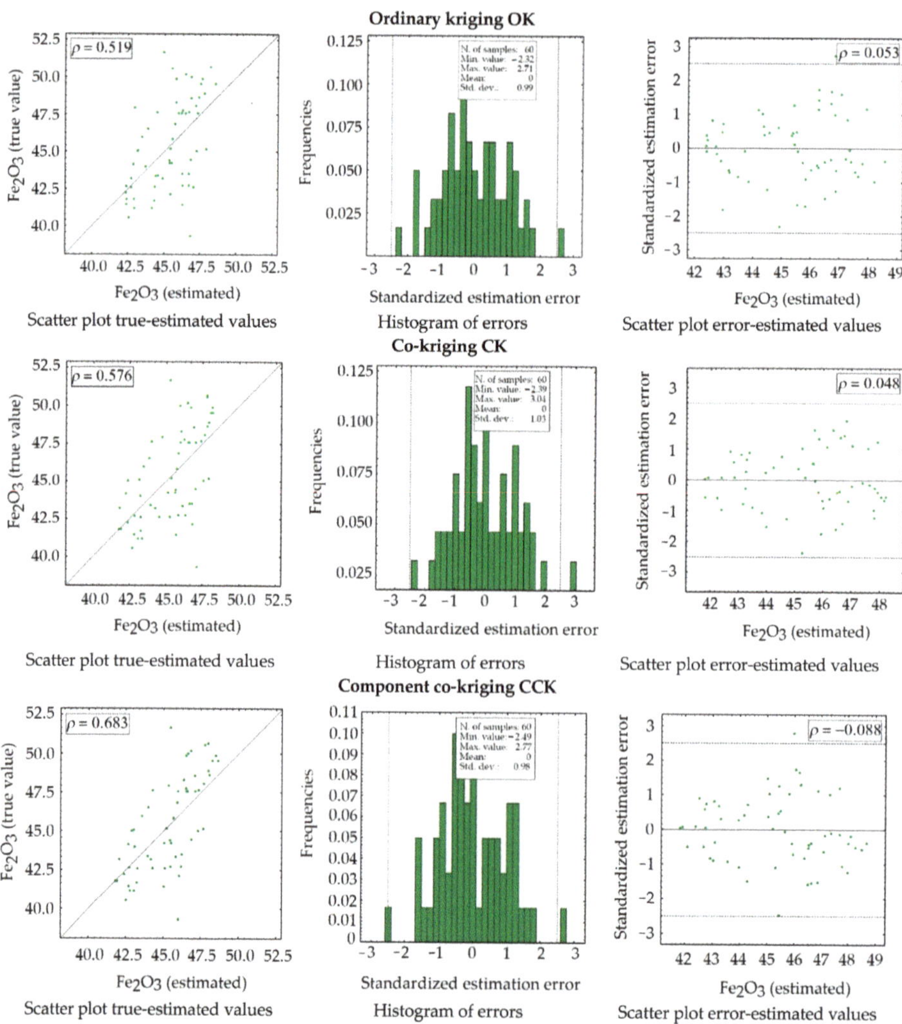

Figure 11. Statistics of cross-validation for *OK*, *CK*, and *CCK* results: scatter plot between true values of Fe_2O_3 (%) at 60 sample points and estimated values (**left**), standardized estimation error (**central**), and scatter plot between error and estimated values (**right**).

4. Discussion

Mapping a metal distribution within an artificial resource, such as a mining waste area is quite challenging and complex. Therefore, there are not many examples of using geostatistical methods for tailings characterization. Some researchers tried to map metal variability within mine tailings using in field samples and performed ordinary kriging.

However, they faced with the challenge of a small number of samples while performing geostatistical modeling [34]. Another example is characterizing the mining residues using geostatistical co-kriging estimation [35,36]. In both examples, the traditional co-kriging method is used. However, an efficient estimation of metal variability is essential, since all economic evaluations are based on metals variability maps and estimation. Therefore, the higher accuracy of the map can make the difference, when deciding whether the exploitation of a strategical resource is economically feasible and sustainable. Iron maps variability as the main target in this work is focused within a bauxite residue in Greece. Most classical estimation method of OK is performed and estimation map has shown mainly three high grade parts (more than 50% of Fe_2O_3) in north east and south east of BR. The estimation standard deviation map identified the lowest estimation variance at the samples points and a high variation where the number of samples are low (east and middle part of the BR).

By adding EO data and specifically Sentinel-2 image (a free and easily accessible image) at the date of sampling, the improvement of iron mapping was tested. The ferrous iron oxides band ratio was selected as the secondary variable to see if additional information (in a regular grid at all estimation points) can help the iron estimation mapping. Results of CK has shown a higher variability with more anomaly points (with Fe_2O_3 concentration of more than 50%) in the estimation map. Moreover, adding the band ratio data could decrease the variability and the values of the estimation variance map. Finally, the new hypothesis of using the most correlated component was tested. Due to the co-regionalization structures, it was possible to decompose the band ratio values into mean, nugget effect, and two different range components: a small range component (70 m) and a large range component (180 m).

The correlation coefficient between each component and iron was calculated and the first component with the small range (70 m) was selected due to its higher correlation with iron. To use this component in the CK system, there was a need to estimate it for all grid points, and then the estimated component was used to perform CCK. The appropriate check was done to control the equality of Equation (13). In Figure 9, by mathematically summing at each grid node, the values of all three maps (estimated component 1 and 2 and the estimated mean), the original map of the ferrous iron oxides band ratio is obtained. This mathematical check confirms that the estimated component 1 can be used in the CCK estimation based on Equation (15).

To do the comparisons among the three utilized methods (OK, CK, and CCK), in all three estimations, the iron variogram model is the same. Moreover, the neighborhoods used for estimations were equal, to have the same condition, while mapping the iron variability.

At the end, the cross-validation was performed to check the efficiency of three methods. For the scatter plots between the true and estimated values of iron (Fe_2O_3) at sample points, the higher correlation between the estimated values and true values is related to the CCK, with $\rho = 0.68$. The histogram of the standardized estimation errors provides an idea about the unbiasedness and also the quality of the estimate. It also helps locate the outliers, which are outside the two vertical lines corresponding to the threshold value. In all three methods, the mean of the histogram is close to zero and shows the acceptable estimation results. Then the scatter plot of the standardized estimation errors versus the estimated values is calculated, which should be with no preferential shape. The reason is the independency of the standardized error with the estimated values. Since the two variables are theoretically independent, this cloud should have no preferential shape, and this was confirmed by the low correlation coefficients calculated.

5. Conclusions

Mapping the strategic metals is one of the most delicate phases, and by using satellite images, an important improvement in the model quality of surface distribution can be performed. Geostatistical models offer a wide variety of powerful tools for a deep study of metals mapping and estimations. A strategic case study (a bauxite mining residue) is reported as an example to check the best method for mapping the iron concentration ($Fe_2O_3\%$). The proposed method of component co-kriging highlights not only the best secondary variable for iron estimation (with higher correlation coefficient), but also improves the classical ordinary-kriging and co-kriging estimation maps. The cross-validation results confirm the improvements of the results. Hence, to sum up:

- Remote sensing data are essential when mapping a surface feature, such as mapping the iron concentration variability;
- Band ratio can be considered an important auxiliary variable in geostatistical modeling, when there is correlation between in field samples and band ratios;
- Component co-kriging is an efficient method and, in case of high correlation coefficient between one component of the auxiliary variable and the main variable (in this work, the iron concentration), it can substantially improve the mapping results.

Author Contributions: Data curation, S.K., F.T. and E.M.; Investigation, R.B.; Methodology, R.B. and S.K.; Resources, E.B.; Writing—original draft, S.K. and F.T.; Writing—review & editing, R.B., S.K., F.T., E.M. and E.B. All authors have read and agreed to the published version of the manuscript.

Funding: This research was supported by the RawMatCop Programme (2018–2021), funded by the European Commission and EIT RawMaterials, grant agreement number 271/G/Gro/COPE/17/10036" and by the INCO-Piles 2020 Project (2020–2021), funded by EIT RawMaterials, Grant Agreement 19169.

Data Availability Statement: Not applicable.

Acknowledgments: The authors would like to thank Irene Benito Rodríguez, the RawMatCop Programme manager, Fabio Ferri, the INCO-Piles 2020 Project Officer, and Rima Dapous and Wesley Crock, all from EIT RawMaterials. Moreover, the authors would like to thank Dimitrios Panias from the National Technical University of Athens and Panagiotis Davris from Mytilineos S.A., for their support during the field trip and data access for the case study.

Conflicts of Interest: The authors declare no conflict of interest.

References

1. World Economic Forum (2005) Mining & Metals in a Sustainable World 2050, Ind Agenda. Available online: https://www.weforum.org/press/2015/09/mining-and-metals-in-a-sustainable-world-2050-report-launch/ (accessed on 20 May 2021).
2. UN General Assembly (2015) Transforming Our World: The 2030 Agenda for Sustainable Development 21 October 2015. A/RES/70/1. Available online: https://www.refworld.org/docid/57b6e3e44.html (accessed on 20 May 2021).
3. Lebre, E.; Corder, G.D.; Golev, A. The role of the mining industry in a circular economy—A framework for resource management at the mine site level. *J. Ind. Ecol.* **2017**, *21*, 662–672. [CrossRef]
4. Mancini, L.; Sala, S. Social impact assessment in the mining sector: Review and comparison of indicators frameworks. *Resour. Policy* **2018**, *57*, 98–111. [CrossRef]
5. COM (2020) 474 Final Communication from the Commission to the European Parliament, the Council, the European Economic and Social Committee and the Committee of the Regions Critical Raw Materials Resilience: Charting a Path towards Greater Security and Sustainability. Available online: https://eur-lex.europa.eu/legal-content/EN/TXT/?uri=CELEX%3A52020DC0474 (accessed on 20 May 2021).
6. Ferro, P.; Bonollo, F. Materials selection in a critical raw materials perspective. *Mater. Des.* **2019**, *177*, 107848. [CrossRef]
7. Jutz, S.; Milagro-Perez, M. Copernicus Program. *Compr. Remote Sens.* **2017**, *1*, 150–191. [CrossRef]
8. Follador, M. Using remote sensing for mineral characterization in tropical forest areas of Brazil, GIS and Spatial Analysis. In *2005 Annual Conference of the International Association for Mathematical Geology*; IAMG: Fortaleza, Brazil, 2005; pp. 127–132, ISBN 0973422017.
9. Ferrier, G. Application of imaging spectrometer data in identifying environmental pollution caused by mining at Rodaquilar, Spain. *Remote Sens. Environ.* **1999**, *68*, 125–137. [CrossRef]
10. Mars, J.C.; Crowley, J.K. Mapping mine wastes and analyzing areas affected by selenium-rich water runoff in southeast Idaho using AVIRIS imagery and digital elevation data. *Remote Sens. Environ.* **2003**, *84*, 422–436. [CrossRef]
11. Choe, E.; Van der Meer, F.; Van Ruitenbeek, F.; Van der Werff, H.; Boudewijn de Smeth, B.; Kim, K.W. Mapping of heavy metal pollution in stream sediments using combined geochemistry, field spectroscopy, and hyperspectral remote sensing: A case study of the Rodalquilar mining area, SE Spain. *Remote Sens. Environ.* **2008**, *112*, 3222–3233. [CrossRef]
12. Pascucci, S.; Belviso, C.; Cavalli, R.M.; Palombo, A.; Pignatti, S.; Santini, F. Using imaging spectroscopy to map red mud dust waste: The Podgorica aluminum complex case study. *Remote Sens. Environ.* **2012**, *123*, 139–154. [CrossRef]

13. Werner, T.T.; Bebbington, A.; Gregory, G. Assessing impacts of mining: Recent contributions from GIS and remote sensing. *Extr. Ind. Soc.* **2019**, *6*, 993–1012. [CrossRef]

14. Lopez-Granados, F.; Jurado-Exposito, M.; Pena-Barragan, J.M.; Garcia-Torres, L. Using geostatistical and remote sensing approaches for mapping soil properties. *Eur. J. Agron.* **2005**. [CrossRef]

15. Liu, Y.; Cao, G.; Zhao, N.; Mulligan, K.; Ye, X. Improve ground-level PM2.5 concentration mapping using a random forests-based geostatistical approach. *Environ. Pollut.* **2018**. [CrossRef]

16. Bzdęga, K.; Zarychta, A.; Urbisz, A.; Szporak-Wasilewska, S.; Ludynia, M.; Fojcik, B.; Tokarska-Guzik, B. Geostatistical models with the use of hyperspectral data and seasonal variation—A new approach for evaluating the risk posed by invasive plants. *Ecol. Indic.* **2021**. [CrossRef]

17. Matheron, G. The intrinsic random functions and their applications. *Adv. Appl. Probab.* **1973**, *5*, 439–468. [CrossRef]

18. Wackernagel, H. Multivariate Geostatistics. In *An Introduction with Applications*; Springer: Heidelberg/Berlin, Germany, 2003; pp. 121–208. [CrossRef]

19. Chiles, J.P.; Delfiner, P. *Geostatistics Modeling Spatial Uncertainty*, 2th ed.; Wiley: Hoboken, NJ, USA, 2012; pp. 118–177, ISBN 978-0-470-18315-1.

20. Van der Meer, F. Extraction of mineral absorption features from high-spectral resolution data using non-parametric geostatistical techniques. *Int. J. Remote Sens.* **1994**, *15*, 2193–2214. [CrossRef]

21. Yamaguchi, Y.; Kahle, A.B.; Tsu, H.; Kawakami, T.; Pniel, M. Overview of Advanced Spaceborne Thermal Emission and Reflection Radiometer (ASTER). *IEEE Trans. Geosci. Remote Sens.* **1998**, *36*, 1062–1071. [CrossRef]

22. Yamaguchi, Y.; Fujisada, H.; Tsu, H.; Sato, I.; Watanabe, H.; Kato, M.; Kudoh, M.; Kahle, A.B.; Pniel, M. ASTER early image evaluation. *Adv. Space Res.* **2001**, *28*, 69–76. [CrossRef]

23. Rouskov, K.; Popov, K.; Stanislav Stoykov, S.; Yamaguchi, Y. Some applications of the remote sensing in geology by using of ASTER images. In Proceedings of the Scientific Conference "SPACE ECOLOGY SAFETY" with International Participation, Sofia, Bulgaria, 10–13 June 2005.

24. de Morais, M.C.; Martins Junior, P.P.; Paradella, W.R. Multi-scale approach using remote sensing images to characterize the iron deposit N1 influence areas in Carajás Mineral Province (Brazilian Amazon). *Environ. Earth Sci.* **2012**, *66*, 2085–2096. [CrossRef]

25. Van der Meer, F.D.; Van der Werff, H.M.A.; Van Ruitenbeek, F.J.A.; Hecker, C.A.; Bakker, W.H.; Noomen, M.F.; Van der Meijde, M.; Carranza, E.J.M.; de Smeth, J.B.; Woldai, T. Multi- and hyperspectral geologic remote sensing: A review. *Int. J. Appl. Earth Obs.* **2012**, *14*, 112–128. [CrossRef]

26. Gopinathan, P.; Parthiban, S.; Magendran, T.; Al-Quraishi, A.M.; Singh, A.K.; Singh, P.K. Mapping of ferric (Fe^{3+}) and ferrous (Fe^{2+}) iron oxides distribution using band ratio techniques with ASTER data and geochemistry of Kanjamalai and Godumalai, Tamil Nadu, south India. *Remote Sens. Appl. Soc. Environ.* **2020**, *18*, 100306. [CrossRef]

27. Guha, A.; Singh, V.K.; Parveen, R.; Vinod Kumar, K.; Jeyaseelan, A.T.; Dhanamjaya Rao, E.N. Analysis of ASTER data for mapping bauxite rich pockets within high altitude lateritic bauxite, Jharkhand, India. *Int. J. Appl. Earth Obs. Geoinf.* **2013**, *21*, 184–194. [CrossRef]

28. Krishnamurthy, Y.V.N.; Sreenivasan, G. Remote Sensing Technology for Exploration of Mineral Deposits with Special Reference to Bauxite and Related Minerals. In Proceedings of the 16th International Symposium of ICSOBA, "Status of Bauxite Alumina, Aluminium, Downstream Products and Future Prospects", Nagpur, India, 28–30 November 2005; pp. 68–83.

29. Ben Dor, E.; Irons, J.R.; Epema, G.F. Soil Reflectance. In *Remote Sensing for the Earth Sciences: Manual of Remote Sensing*, 3rd ed.; Rencz, A.N., Ryerson, R.A., Eds.; John Wiley & Sons: New York, NY, USA, 1999; Volume 3, pp. 111–173.

30. Drusch, M.; Del Bello, U.; Carlier, S.; Colin, O.; Fernandez, V.; Gascon, F.; Hoersch, B.; Isola, C.; Laberinti, P.; Martimort, P.; et al. Sentinel-2: ESA's Optical High-Resolution Mission for GMES Operational Services. *Remote Sens. Environ.* **2012**, *120*, 25–36. [CrossRef]

31. Van der Werff, H.; Van der Meer, F. Sentinel-2A MSI and Landsat 8 OLI Provide Data Continuity for Geological Remote Sensing. *Remote Sens.* **2016**, *8*, 883. [CrossRef]

32. Balomenos, E.; Davris, P.; Pontikes, Y.; Panias, D.; Delipaltas, A. Bauxite residue handling practice and valorisation research in Aluminium of Greece. In: Pontikes Y (ed) Proceedings of Bauxite residue valorization and best practices conference. *Athens* **2018**, *7*, 27–36.

33. Davris, P.; Balomenos, E.; Panias, D.; Paspaliaris, I. Selective leaching of rare earth elements from bauxite residue (red mud), using a functionalized hydrophobic ionic liquid. *Hydrometallurgy* **2016**, *164*, 125–135. [CrossRef]

34. Candeias, C.; F Ávila, P.F.; Ferreira da Silva, E.; Paulo Teixeira, J. Integrated approach to assess the environmental impact of mining activities: Estimation of the spatial distribution of soil contamination (Panasqueira mining area, Central Portugal). *Environ. Monit. Assess.* **2015**, *187*, 135. [CrossRef]

35. Kasmaee, S.; Tinti, F.; Bruno, R. Characterization of metal grades in a stockpile of an iron mine (case study- Choghart iron mine, Iran). *Rud. Geol. Naft. Zb.* **2018**, *33*, 51–59. [CrossRef]

36. Kasmaeeyazdi, S.; Mandanici, E.; Balomenos, E.; Tinti, F.; Bonduà, S.; Bruno, R. Mapping of Aluminum Concentration in Bauxite Mining Residues Using Sentinel-2 Imagery. *Remote Sens.* **2021**, *13*, 1517. [CrossRef]

Article

Shear-Related Gold Ores in the Wadi Hodein Shear Belt, South Eastern Desert of Egypt: Analysis of Remote Sensing, Field and Structural Data

Mohamed Abd El-Wahed [1], Basem Zoheir [2,3], Amin Beiranvand Pour [4,*] and Samir Kamh [1]

1   Geology Department, Faculty of Science, Tanta University, Tanta 31527, Egypt;
    mohamed.abdelwahad@science.tanta.edu.eg (M.A.E.-W.); skamh@science.tanta.edu.eg (S.K.)
2   Geology Department, Faculty of Science, Benha University, Benha 13518, Egypt; basem.zoheir@ifg.uni-kiel.de
3   Institute of Geosciences, University of Kiel, 24118 Kiel, Germany
4   Institute of Oceanography and Environment (INOS), University Malaysia Terengganu (UMT),
    Kuala Nerus 21030, Terengganu, Malaysia
*   Correspondence: beiranvand.pour@umt.edu.my; Tel.: +60-9-6683824; Fax: +60-9-6692166

Citation: Abd El-Wahed, M.; Zoheir, B.; Pour, A.B.; Kamh, S. Shear-Related Gold Ores in the Wadi Hodein Shear Belt, South Eastern Desert of Egypt: Analysis of Remote Sensing, Field and Structural Data. *Minerals* **2021**, *11*, 474. https://doi.org/10.3390/min11050474

Academic Editors: Paul Alexandre

Received: 23 February 2021
Accepted: 27 April 2021
Published: 30 April 2021

Abstract: Space-borne multispectral and radar data were used to comprehensively map geological contacts, lithologies and structural elements controlling gold-bearing quartz veins in the Wadi Hodein area in Egypt. In this study, enhancement algorithms, band combinations, band math (BM), Principal Component Analysis (PCA), decorrelation stretch and mineralogical indices were applied to Landsat-8 OLI, ASTER and ALOS PALSAR following a pre-designed flow chart. Together with the field observations, the results of the image processing techniques were exported to the GIS environment and subsequently fused to generate a potentiality map. The Wadi Hodein shear belt is a ductile shear corridor developed in response to non-coaxial convergence and northward escape tectonics that accompanied the final stages of terrane accretion and cratonization (~680–600 Ma) in the northern part of the Arabian–Nubian Shield. The evolution of this shear belt encompassed a protracted ~E–W shortening and recurrent sinistral transpression as manifested by east-dipping thrusts and high-angle reverse shear zones. Gold-mineralized shear zones cut heterogeneously deformed ophiolites and metavolcaniclastic rocks and attenuate in and around granodioritic intrusions. The gold mineralization event was evidently epigenetic in the metamorphic rocks and was likely attributed to rejuvenated tectonism and circulation of hot fluids during transpressional deformation. The superposition of the NW–SE folds by NNW-trending, kilometer scale tight and reclined folds shaped the overall framework of the Wadi Hodein belt. Shallow NNW- or SSE-plunging mineral and stretching lineations on steeply dipping shear planes depict a considerable simple shear component. The results of image processing complying with field observations and structural analysis suggest that the coincidence of shear zones, hydrothermal alteration and crosscutting dikes in the study area could be considered as a model criterion in exploration for new gold targets.

Keywords: multispectral satellite data; PALSAR; gold exploration; Wadi Hodein shear belt; transpressional deformation; gold-bearing quartz veins

1. Introduction

Remote sensing data are commonly used to identify geological structures associated with extensive shear belts, particularly in vast and rugged terrains. The innovative technologies in the remote sensing promoted the application of space-borne imagery data for detailed structural geology mapping [1–6]. The application of multisensor satellite imagery can be considered as a cost-efficient exploration strategy for prospecting orogenic gold mineralization in transpression and transtension zones, which are located in harsh regions around the world [7,8]. In Egypt, structural analysis aided by the interpretations of space-borne data has been efficiently employed to constrain the major structures in the South

Eastern Desert [9–12]. Zoheir and Emam [7], Zoheir et al. [11,12] tested the effectiveness of processing space-borne multispectral and radar imagery data for the geological mapping, particularly to highlight elements controlling the distribution of gold occurrences in the south Eastern Desert of Egypt. Iron oxides, clay and carbonate $\pm$ sulfate mineral phases in the hydrothermal alteration zones have specific spectral signatures in the visible, near infrared, and shortwave infrared radiation regions [13]. The hydrous mineral phases with the OH groups (Mg-O-H, Al-O-H, Si-O-H) and CO_3 acid group have diagnostic absorption features in the shortwave infrared region (SWIR) (2.0–2.50 µm) [14].

Landsat-8 OLI with high radiometric resolution (16 bits) is an effective remote sensing sensor for detailed lithological mapping [15]. Landsat-8 OLI imagery consists of nine spectral bands, from which seven bands measure the reflected VNIR and SWIR radiation with 30-m spatial resolution for bands 1–7 and 9, while the panchromatic band 8 has 15-m resolution. The ultra-blue band 1 is operative in coastal and aerosol targets, whereas band 9 is valued for cloud detection. The TIR bands collects two thermal bands (10 and 11) that measure the emitted radiation through the 10.6–12.5 µm wavelength region (TIR) with 100-m spatial resolution [16,17]. The Advanced Spaceborne Thermal Emission and Reflection Radiometer (ASTER) data showed high capabilities in discriminating lithological units and alteration zones associated with hydrothermal ore deposits [18–22]. The ASTER data cover a wide spectral range of 14 bands, measuring reflected radiation in three bands between 0.52 and 0.86 µm (visible-near infrared, VNIR) with 15-m resolution, and six bands from 1.6 to 2.43 µm (shortwave Infrared, SWIR) with 30-m resolution. The emitted radiation is measured at 90-m resolution in five bands through the 8.125-µm–11.65-µm wavelength region (thermal infrared; TIR) [23,24]. Synthetic Aperture Radar (SAR) remote sensing technology allows the application of space-borne imagery data for detailed structural geology mapping [1–6,11,12]. Phased Array type L-band Synthetic Aperture Radar (PALSAR) is a L-band synthetic aperture radar, which can penetrate sand and vegetation because of the radar's longer wavelengths (15.0–30.0 cm) [5]. L-band SAR data can provide detailed geological structure information for desert and tropical environments [25]. PALSAR has multimode observation functions such as Fine mode, Direct downlink, ScanSar mode, and Polarimetric mode. Multipolarization configuration (HH, HV, VH, and VV), variable off-nadir angle (9.9–50.8°), and spatial resolution of 10 m for Fine mode, 30 m for Polarimetric mode, 100 m for ScanSar mode are designed. The swath width observation is 30 km for Polarimetric mode, 70 km for Fine mode, and 250–350 km for ScanSar modes [26]. The fusion of these multisensor imagery data can offer a wide-ranging means for tracing along the strike of structural elements and to investigate controls of the orogenic gold occurrences in well exposed and arid regions such as the Wadi Hodein shear belt, South Eastern Desert of Egypt.

The Eastern Desert of Egypt is part of the Arabian–Nubian Shield (ANS) [27]. The latter is made up mainly of three distinctive tectonostratigraphic units: (1) infracrustal gneisses, locally exposed as discrete core complexes [28–31]; (2) supracrustal dismembered ophiolites and island arc volcanic/ volcanosedimentary assemblages; and (3) intramountainous molasse sediments (Hammamat Group) and Cordilleran calc-alkaline volcanics (Dokhan Volcanics) [32]. Widespread granitoid intrusions of different ages cut both the infracrustal and supracrustal rocks. The South Eastern Desert experienced a history of superimposing compressional and transpressional deformation, manifested by extensive NW–SE trending high-angle thrusts, folds and transcurrent faults [12,28,30,31]. The Wadi Hodein shear belt is a major NW-oriented transcurrent shear zone in the southern Eastern Desert (Figure 1). It is suspected to have accommodated up to 300 km of sinistral displacement during the late stages of transpression [29–34]. Opinions differ about the tectonic character of this shear zone, where some authors consider it as the youngest major Najd-related shear in the Egyptian Eastern Desert [29–31,33] or even a transpressional corridor [34,35].

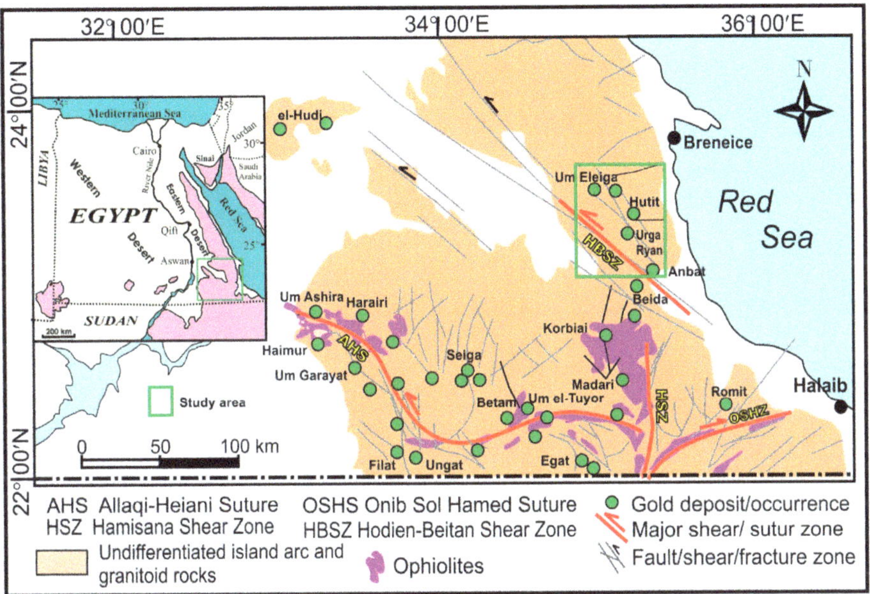

Figure 1. Distribution of the known gold mining sites and gold quartz vein occurrences relative to the major fault/shear structures and ophiolitic masses in the South Eastern Desert terrane (modified after [11]). Inset shows a location map of the study area.

Gold bearing quartz veins and their alteration zones occur in different geologic and structural settings in the southern Eastern Desert [10] (Figure 1). These different settings include: (1) Au–quartz veins hosted by ductile shear zones at contacts between ophiolitic and island arc terranes (e.g., Um El-Tuyor El-Foqani, Betam, Seiga, Shashoba, Um Garayat, and Haimur deposits); (2) Au–quartz veins occurring along steeply dipping anastomosing shear zones wrapped around or cutting syn- or late-orogenic granitoid intrusions (e.g., Korbiai, Madari, Romite, Egat deposits); (3) Au–quartz-carbonate veins in association with brittle–ductile shear zones separating listvenized serpentinite from the underlying successions of intercalated metavolcanic and volcaniclastic metasediments (e.g., Hutit, El-Beida, and El-Anbat deposits). In the present study, structural analysis supported by remote sensing studies was aimed at revealing what controls the distribution of the scattered gold–quartz vein occurrences in the Wadi Hodein area. Multisensor satellite data were used to resolve the spatial relationship between gold mineralization and district-scale structures. The relative timing of the successive structural events was also envisaged to place the gold mineralization in the broader evolutionary framework of the South Eastern Desert.

2. Geological Setting

The Neoproterozoic rocks exposed in Hodein shear belt comprises variably deformed ophiolitic mélange rocks, island-arc metavolcanic and metavolcaniclastic successions (Figure 2). These rocks are unconformably overlain by Cretaceous sandstone formations (Nubian Sandstone) and locally by Miocene marine turbidites and carbonate. Dioritic and granitic gneisses and migmatites are exposed around Wadi Beitan and Wadi Khuda, in the western part of the study area. Wadi Khuda gneisses occur as an ENE–WSW trending belt forming a major doubly plunging anticline (Figure S1a,b), amphibolite and migmatites [35]. It is intruded by syn-tectonic tonalite and diorite as well as late-to post-tectonic syenogranites and leucogranites. The gneissic rocks of Wadi Beitan form a major NW-trending belt in the northwestern part of the Wadi Hodein shear system [27] (Figure 2). This belt comprises fine-to coarse-grained biotite gneisses, biotite-hornblende gneisses and hornblende gneisses together with subordinate varieties of garnet–biotite–hornblende gneisses, augen gneiss, amphibolites (Figure S1c,d), migmatites and mylonites [11,27]. The contact between the Wadi Beitan gneisses belt and the volcaniclastic rocks is defined by a SW-verging thrust fault associated with discrete zones of mylonitized granites (Figure S1d).

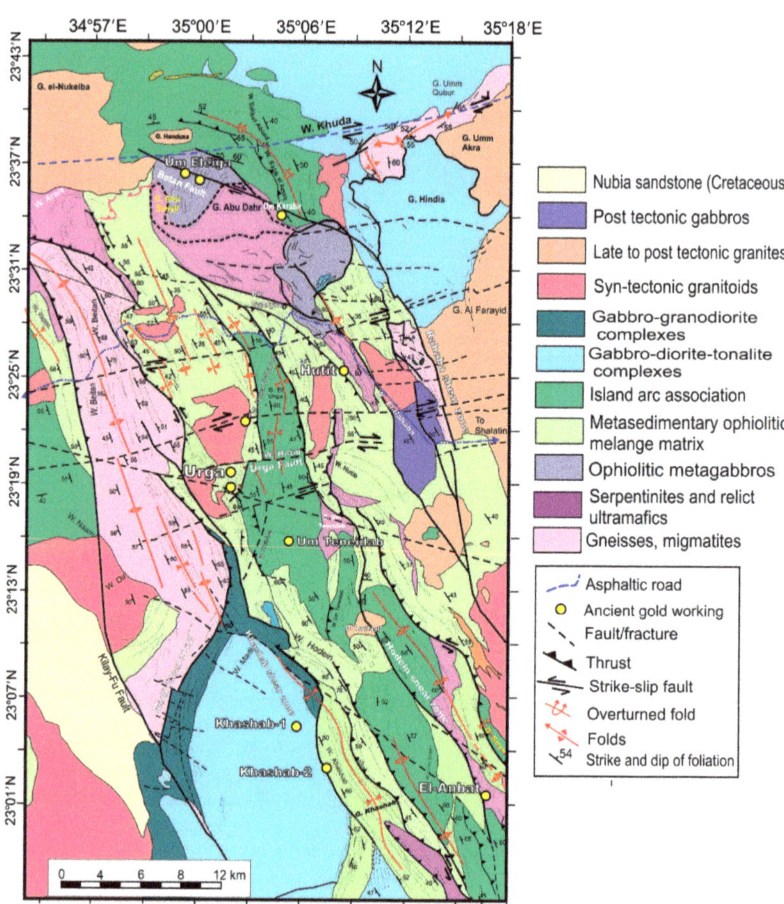

Figure 2. Geological map of Wadi Hodein–Beitan shear belt (complied and modified from [36–39]).

The Wadi Khashab–Gabal Sirsir ophiolites form two discontinuous NNW–SSE oriented belts of mainly serpentinite, metabasalt, ophiolitic and metagabbro rocks [12,36,38,39]. Serpentinite and associated talc carbonate occur as masses elongated in a NW–SE direction (e.g., Gabal Sirsir and Wadi El-Beida) and tectonically emplaced within the foliated metavolcanics and volcaniclastic metasediments. Serpentinites and foliated metabasalts in the Gabal Sirsir–El-Anbat belt occur as elongate lenses along NW to NNW–SSE shear zones [37–40]. The mafic metavolcanics form NW-elongated belts bound by highly sheared rocks and mylonites. The metavolcanics rocks are intruded by metagabbro–diorite, granodiorite and granitoid intrusions. They are strongly foliated and exhibit well developed pencil cleavage and boudinage patterns. They comprise metaandesite/basaltic andesite and tuffaceous intercalations. The metasedimentary rocks include metagreywacke, bedded metamudstone, metaconglomerate and schists with minor calcareous beds. The Gabal Abu Dahr ophiolitic mélange in the central and eastern parts of the study area occur as elongate, fault-bounded blocks, tectonically admixtured with island arc metavolcanic and metavolcaniclastic rocks along Wadi Rahaba and around Gabal Abu Dahr. The ophiolitic and island arc rocks are cut by a metagabbro/diorite complex and granitic intrusions [41–43]. The Wadi Rahaba serpentinite occurs as NW-elongate masses and small isolated irregular lenses in the volcaniclastic metasediments (Figure S1e). The metagabbro–diorite intrusions of the Wadi Rahaba occur as isolated, irregular masses, cutting ophiolites, metavolcanic and metavolcaniclastic rocks. Contacts of the serpentinized ultramafic rocks with metagabbro–diorite complex are sharp and irregular. The serpentinized ultramafic rocks are dominated by a harzburgite, dunite veins, chromitite pods and talc-carbonate rocks [42]. The serpentinized ultramafic rocks contain large masses of dunite and concordant sills of wehrlite, pyroxenite and gabbro [43]. Ophiolitic metagabbro occurs as moderate relief masses of pyroxene metagabbro, hornblende metagabbro, and meta-anorthosite (Figure S1f).

The ophiolitic mafic metavolcanics to northeast of Gabal Abu Dahr are massive or pillowed (Figure S2a) metabasalt and metabasaltic andesite. Pillowed metabasalt bodies are exposed as ellipsoidal, globular and tabular shaped bodies, ranging between 10 cm and 1 m in diameter (Figure S2a,b). The mafic metavolcanics along Wadi Hutib and Wadi Urga El Rayan form a NNW-elongated belt and show variable degrees of shearing. The Wadi Arais ophiolites include masses of serpentinite, serpentinized peridotites (Figure S2c), metagabbro, dunite, pyroxenite, chlorite schist, metavolcanics rocks and listwaenite along the thrust and shear zones [43,44]. The serpentinized peridotite occurs as steep mountains along NW–SE striking thrust faults (Figure S2c). Irregular serpentinite masses are incorporated in the volcaniclastic metasediments. The ophiolitic metagabbro rocks form hills of moderate relief, which have tectonic contacts against the serpentinite and metavolcanic blocks. The Gabal Arais mass is composed chiefly of actinolite hornblende metagabbro and hornblende metagabbro and less commonly of appinitic metagabbro [45]. The arc metavolcanic rocks along Wadi Um Araka form an E–W elongate belt comprised mainly meta-andesite, metadacite, dacitic tuff and andesitic tuff with minor amounts of metabasalt [45]. The foliated varieties of the Um Araka metavolcanics are mainly schistose metatuffs. The ophiolitic mélange of Wadi Arais comprises volcaniclastic metasediments and exotic clasts of metagabbro and amphibolite.

The metagabbro–diorite–tonalite–granodiorite complex forms a large elongated belt in the Wadi Khuda area. These rocks intrude Wadi Khuda gneisses (Figure S2d) and the eastern parts of Wadi Rahaba–Gabal Abu Dahr ophiolitic mélange and intruded by late to post-tectonic biotite and muscovite granites (Figure S2e). Their contacts with Wadi Khuda gneisses are characterized by the strongly foliated diorite and amphibolite [46,47]. The gabbro–diorite complex forms variable degrees of deformation, from massive to strongly sheared or foliated. The metagabbro–diorite rocks of the Um Eleiga exhibit circular zonation where gabbro forms the core and the granodiorite occupies the margin [41].

The tonalite of Wadi Urga El Rayan form N–S trending, strongly weathered and weakly foliated plutons in the central part of the study area. Granodiorite forms a N–S trending pluton along the Wadi Hutib and E–W oriented small intrusions along the Wadi Beitan (Figure S2f). The granitoid intrusions that cut the ophiolitic mélange of the Wadi Khashab–Wadi Um Karaba range in composition from tonalite to granodiorite. The granitic rocks of Gabal El Maarafay and Gabal Al Farayid include monzogranite, granodiorite, and tonalite and are locally foliated [48]. At Wadi Khuda, leucogranite forms small stocklike and dikelike bodies intruding on the core of the Khuda gneisses and the granite pluton [46]. The ophiolitic rocks of Wadi Arais are cut by monzogranite and syenogranite of the Gabal El-Nukeiba and Gabal Handusa. A small circular intrusion of younger gabbro in the southwestern part of the study area (G. Homraii) cuts metavolcanic rocks and the older granitoid terranes. The gold-bearing quartz veins in the study area cut sheared metavolcanics, volcaniclastic metasediments and gabbro–diorite complex. The mineralized quartz veins are controlled consistently by NW- and NNW-trending shear zones, and are associated with extensive pervasive hydrothermal alteration. Gold occurrences in the study area include El-Beida, Khashab, El-Anbat, Um Teneidab, Urga Ryan, Hutit and Um Eleiga.

3. Remote Sensing Data Characteristics and Analysis

Iron oxides, clay and carbonate ± sulfate in the hydrothermal alteration zones have specific spectral signatures in the visible near infrared (VNIR), and shortwave infrared (SWIR) spectral regions, respectively [13,14]. False color composite (FCC) and band combination images are used in geological applications based on known spectral signatures of mineral phases in specific wavelength regions. Mapping the hydrothermal alteration zones is a prime focus of mineral exploration programs using the remote-sensing data [1–8,11,12]. Most of the hydrothermal alteration mineral species have distinctive features in the shortwave infrared (SWIR) region, making the Landsat-8 OLI and ASTER sensors suitable data sources for mineral mapping, particularly in arid regions. Moreover, considerable numbers of image processing techniques, such as band-rationing (BR), principal component analysis (PCA), and the spectral mineralogical indices, have been proven effective in lithological and hydrothermal alteration mapping if verified by the fieldwork [1,7,8,11,12]. The present study integrates Landsat-8 OLI/TIRS, ASTER and ALOS PALSAR data for comprehensive mapping of the lithological units and geological structures. Additionally, we aimed to detect the mineralized zones in the Wadi Hodein area. The characteristics of the used remote sensing data are presented in Tables 1 and 2.

Table 1. Characteristics of the Landsat-8 OLI/TIRS and ASTER data [16,17].

Landsat-8 (OLI/TIRS)		Wavelength (μm)	ASTER	Wavelength (μm)	Band
Band	**Resolution (m)**		**Resolution (m)**		
Band 1	30 Costal/Aerosol	0.435–0.451	15	0.52–0.6	Band1
Band 2	30 Blue	0.452–0.512	15	0.63–0.69	Band 2
Band 3	30 Green	0.533–0.590	15	0.76–0.86	Band 3
Band 4	30 Red	0.636–0.673	30	1.60–1.70	Band 4
Band 5	30 NIR	0.851–0.879	30	2.145–2.185	Band 5
Band 6	30 SWIR-1	1.566–1.651	30	2.185–2.225	Band 6
Band 7	30 SWIR-2	2.107–2.294	30	2.235–2.285	Band 7
Band 8	15 Pan	0.503–0.676	30	2.295–2.365	Band 8
Band 9	30 Circus	1.363–1.384	30	2.360–2.430	Band 9
band10	100 TIR-1	10.60–11.19	90	8.125–8.475	Band 10
band11	100 TIR-2	11.50–12.51	90	8.475–8.825	Band 11
-	-	-	90	8.925–8.275	Band 12
-	-	-	90	10.25–10.95	Band 13
-	-	-	90	10.95–11.65	Band 14

Table 2. Characteristics of the ALOS PALSAR data [26].

	Fine Resolution		ScanSAR	Polarimetric
Beam Mode	FBS, DSN	FBD	WB1, WB2	PLR
Center Frequency		L-Band (1.27 GHz)		
Polarization	HH or VV	HH + HV or VV + VH	HH or VV	HH + HV + VV + VH
Spatial Resolution	10 m	20 m	100 m	30 m
Swath Width	70 km	70 km	250–350 km	30 km
Off-Nadir Angle		34.3° (default)	27.1° (default)	21.5° (default)

Abbreviation: DSN = Direct Downlink, FBD = Fine Resolution Mode, Dual polarization, PLR = Polarimetry, HH, VV, HV, VH = Polarization types.

3.1. Data and Processing Techniques

Two cloud-free ASTER scenes' AST_L1T00312252006082430 and 003122520060082422' were acquired on 24 August 2006. The Landsat-8 OLIL/TIRS scene 'LC08_L1TP1730442019073001_T1' was acquired on 30 July 2019 with path/raw 173/44. Four ALOS PALSAR scenes, ALP-

SRP075090450, ALPSRP075090460, ALPSRP077570450 and ALPSRP077570460, of L-Band level 1.5 images. For detailed information on the preprocessing techniques of the remote sensing data, adopted software and methodology see the supplementary data file (Section S1 in supplementary material). The remote sensing data have been processed for lithological mapping, highlighting structural elements and delineating the alteration zones. Image processing techniques, i.e., enhancement algorithms, band combinations (FCC), band math (BM), Principal Component Analysis (PCA), decorrelation stretch and mineralogical indices were applied for outlining geological mapping and the hydrothermal alteration zones in the study area. Finally, multicriteria approach was applied to produce the potentiality gold mineralization map of the study area. Figure S3 shows the flow chart of the methodology adopted for the present study.

3.2. Landsat-8 OLI and ASTER-Based Lithological Mapping

The Landsat-8 OLI data have higher radiometric resolutions (16 bits) and lower spectral resolution compared with ASTER data. Several Landsat-8 OLI and ASTER band combinations images are employed for mapping lithological units and structural elements in the study area. The Landsat-8 band combination images (RGB-753; Figure 3a) and ASTER (RGB-431; Figure S4a) discriminate well between different lithologic units and highlight the structural elements. The ophiolitic rocks appear as dark green pixels, metavolcanics and metasediments are represented by dark to medium brown pixels while the highly weathered granitoids and gneisses exhibit light brown signature. The E–W, NW-, NE-trending faults/fractures that cut these rock outcrops are conspicuous. The band math technique, which is used for reducing the effects of topography and enhancing the spectral differences between bands, divides the digital number value of one band by the digital number value of another band [49]. Band math is a useful for highlighting hydrothermal alteration minerals, as it minimizes the illumination differences caused by the topographic features. Landsat-8 band math image (6/7) was generated to map CO_3 and OH-bearing minerals. This image ratio highlights clay minerals, serpentine, and many alteration zones as bright pixels, where the metavolcanic–metasedimentary rocks are manifested by gray pixels (Figure 3b). On the other hand, the grey-scale image of ASTER band math (7 + 9/8) presented as carbonate/chlorite index. It highlights the highly sheared and carbonatized rocks (serpentinite and talc-carbonate) and metavolcanics with bright image signature (Figure S4b). Ophiolitic gabbros and metasedimentary rocks appear as medium grey pixels, while the gneisses and granitoid rocks have black spectral signatures. Hassan and Sadek [15] used this ratio to differentiate between talc-carbonate and ophiolitic basic metavolcanic rocks in the Gerf area south of the present study area.

Segal [50] used band ratio images for the enhancement of spectral contrasts among the bands considered in the ratio operation, and they have successfully been used in the mapping of alteration zones. Therefore, Landsat-8 FCC of the Abrams ratio (RGB-6/7, 4/3, 5/4), Chica-Olma ratio (RGB-6/7, 6/5, 4/2) and Kaufmann ratio (RGB-7/5, 5/4, 6/7) (Figure 4a–c, respectively) were generated to characterize iron-oxide, hydroxyl minerals, clay minerals and containing rocks. The Abrams ratio (RGB-6/7, 4/3, 5/4) image successfully delineated mafic mineral-rich rocks, including the ophiolitic serpentinite, and highly tectonized ophiolites, which are clearly demarcated as bright red color, and the highly tectonized mafic metavolcanic rocks appear as purple and dark blue pixels (Figure 4a). The Chica-Olma ratio (RGB-6/7, 6/5, 4/2; Figure 4b) image clearly delineates the serpentinite rocks and their sharp contacts with the surrounding rocks of metavolcanic–metasedimentary association and granitoids, where theses contacts are considered the favorable sites for gold mineralization. Serpentinite appear as bloody red pixels, metasedimentary rocks as dark green, whereas the metavolcanics and quartz–carbonate (listvenite) have a bright green signature. The altered materials and clay minerals appeared as olive green pixels. The ophiolite assemblage is highlighted successfully as bright blue and quartz-carbonate (listvenite) as violet pixels in the Kaufmann ratio (RGB-7/5, 5/4, 6/7; Figure 4c) image. The metasedimentary rocks are characterized by dark cyan color, whereas the metavolcanics appeared by reddish brown pixels. These distinctive spectral signatures helped in heightening the foliation trajectories of the highly deformed and tectonized rocks.

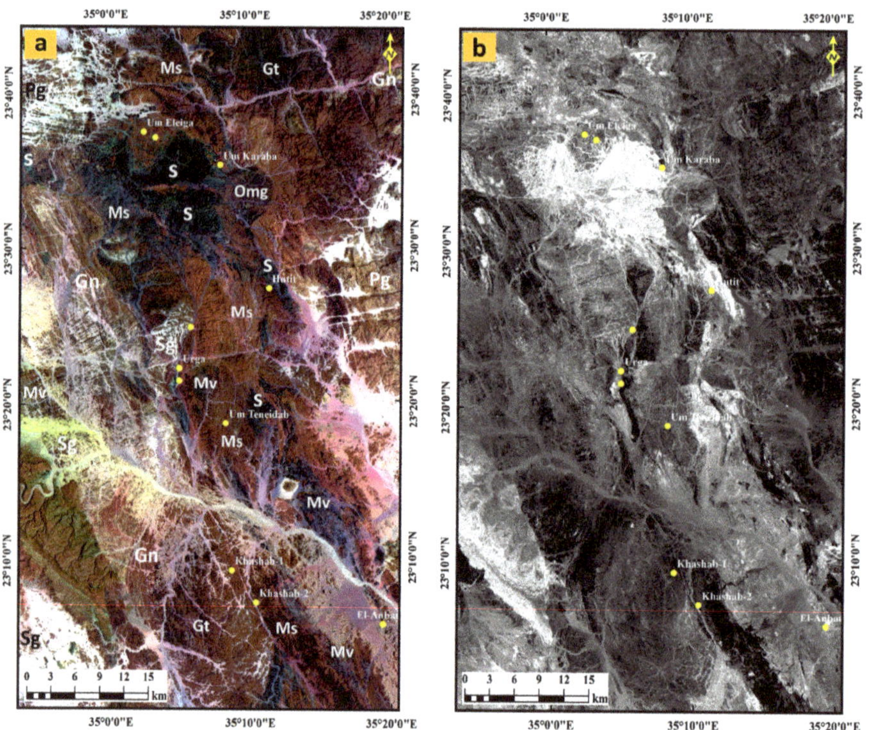

Figure 3. Lithological discrimination using (**a**) Landsat-8 (RGB-753) and (**b**) grey scale Landsat-8 band ratio (6/7). Abbreviation: Gn = Gneisses and migmatites, S = Serpentinites, Omg = Ophiolitic metagabbros, Ms = Metasedimentary ophiolitic mélange matrix, Mv = Island arc association, Gt = gabbro-tonalite-granodiorite complex, Sg = Syn-tectonic granitoids, Pg = Post-tectonic granitoids.

The NW-thrusted contacts and E–W strike slip faults can be detected from the three image ratios (Figure 4a–c). Moreover, decorrelation stretching is an important image enhancement method that is used to improve visual interpretation of satellite images [51]. For bands having the maximum variance, decorrelation stretching was applied to 7-6-5 of Landsat-8 bands (Figure 4d). The dark blue pixels of serpentinite can be differentiated from the greenish brown pixels of ophiolitc metagabbros. The metasedimentary rocks and metavolcanics attained greenish red and bright green colors, respectively. Kaolinite- and illite-bearing rocks show bright red signature, whereas the gneissic and granitoid rocks appear as light violet pixels (Figure 4d).

Principal Component Analysis (PCA), as one of the spectral enhancement techniques, has been used for lithological discrimination [15]. Principal component analysis (PCA) is a multivariate statistical technique that selects uncorrelated linear combinations (eigenvector loadings) of variables in such a way that each successively tracted linear combination, or principal component (PC), has a smaller variance [52]. The PCA transformation was carried out for the VNIR and SWIR bands of Landsat-8 and ASTER to extract lithological and alteration zones information, which related to gold mineralization. The eigenvector matrices of the Landsat-8 and ASTER data derived from the PCA are given in Tables S1 and S2, respectively). The three first PCA images (PC1, PC2 and PC3), containing the highest topographical and spectral information, are found suitable for lithological discrimination.

The PCA images can bear crucial information related to the alteration minerals, which could be reflected in the eigenvector loading of the absorption and reflection bands [12]. Gupta et al. [53] suggested that when the eigenvector loading is strong (as positive or negative sign) in the reflection and absorption bands of the target mineral or mineral group, the enhanced pixels related to the mineral or mineral group will manifest as bright pixels when this loading is positive in the reflection band; conversely, the pixels will manifest as dark when the loading is negative in the reflection band.

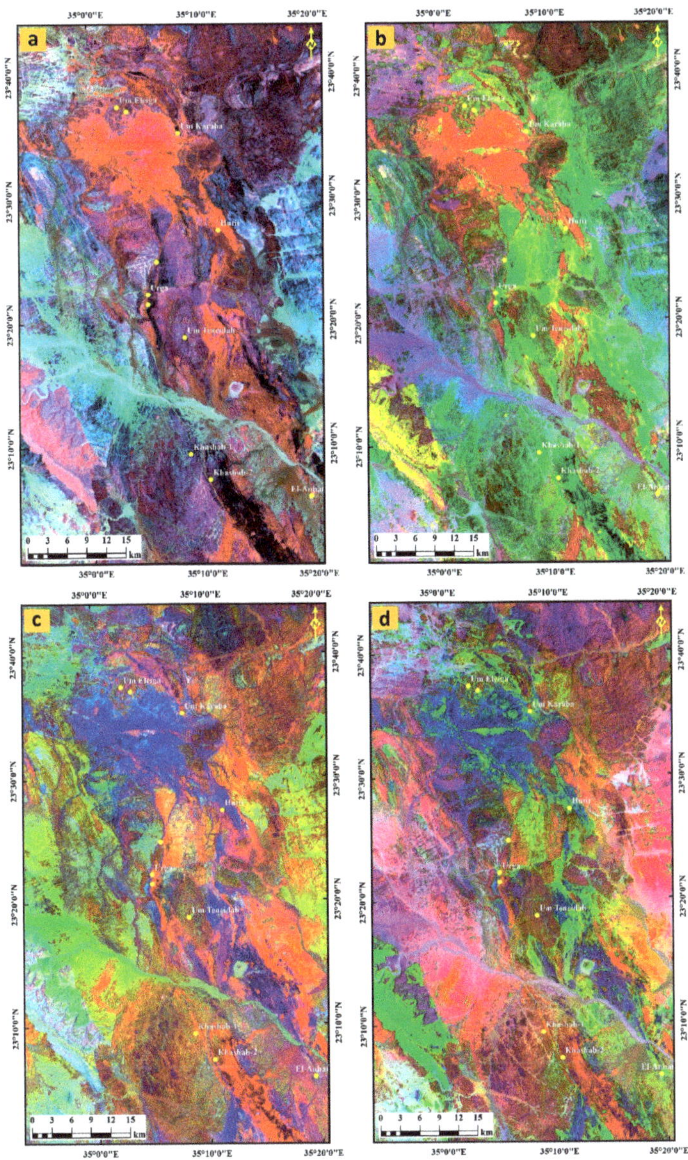

Figure 4. Spectral discrimination of ophiolites and related rocks using processed Landsat-8 (**a**) Abrams ratio (RGB-6/7, 4/3, 5/4), (**b**) Chica-Olma ratio (RGB-6/7, 6/5, 4/2) and (**c**) Kaufmann ratio (RGB-7/5, 5/4, 6/7). (**d**) Decorrelation stretching to (7-6-5).

According to the analysis of the eigenvector loadings of the seven bands of Landsat-8 PCA (Table S1), we produced the color composite images (RGB-PC2, PC1, PC4 and RGB-PC2, PC4, PC5; Figure S5a,b, respectively). Bands 2, 4, 5, 6, 7 of Landsat-8 contain valuable information for iron oxides and hydroxyl-bearing minerals' mapping [54]. The rocks consisting of high contents of Al and/or Mg-OH-bearing minerals (ophiolites) were best delineated by bright lemon-yellow pixel signatures in Figure S5a and yellowish orange pixels in Figure 5a. The two images emphasize the contact between felsic and basic metavolcanics and volcaniclastic metasediments and gneisses along Wadi Hodein and Wadi Beitan. The areas with rich altered materials show cyan color tones. Based on the eigenvector loadings of the nine bands of ASTER PCA (Table S2), the color composite images (RGB-PC1, PC2, PC3 and RGB-PC6, PC3, PC1; Figure S5a,b, respectively) were prepared. Figure S5b shows bright magenta pixels for iron oxide/hydroxide minerals, which are mostly associated with ophiolitic serpentinite and metagabbros. The metasedimentary rocks are manifested by light magenta pixels, metavolcanics as bright green, and gneisses and granitoids as cyan pixels. Figure 5b shows the metavolcanic/metavolcaniclastic rocks appearing as reddish color pixels. The light cyan color domains within the ophiolitic mélange terranes whilst metasediments represent grass green and gneisses and granitoids are presented by yellow green pixels. The four images enhance the visualization of the E–W strike slip faults and foliation trajectories.

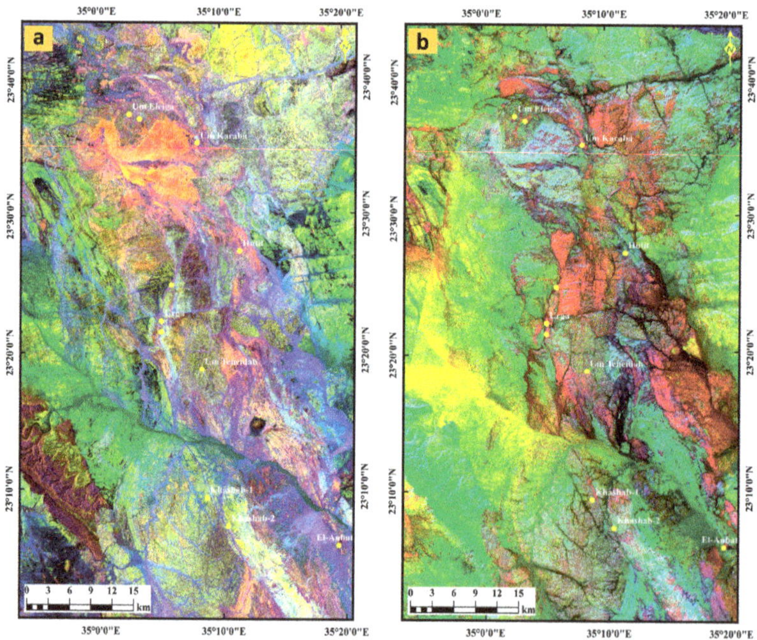

Figure 5. False-color composite of principal component analysis (PCA) of (**a**) Landsat-8 RGB-PC2, PC4, PC5 and (**b**) ASTER RGB-PC6, PC3, PC1.

3.3. Hydrothermal Alteration Zones Detection

Shortwave infrared (SWIR) channels of ASTER can increase the accuracy of the spectral identification of minerals and rock units [55–61]. ASTER indices including kaolinite, clay and muscovite and carbonate alteration zones [12]. The mineralogical indices are reflectance combinations of two or more spectral bands signifying the relative abundance of the target objects. The application of mineralogical indices for the ASTER bands is efficient for lithological and hydrothermal alteration discrimination [62]. The six ASTER–SWIR bands were used to map the hydrothermal alteration minerals. Therefore, four spectral mineralogical indices (OH-bearing mineral index, OHI; kaolinite index, KLI; calcite index, CLI and alunite index, ALI) were used to characterize the alteration zone in the study area. The OHI is calculated as (band 7/band 6) × (band 4/band 6), the KLI is developed by way of (band 4/band 5) × (band 8/band 6), CLI is calculated as (band 6/band 8) × (band 9/band 8), while ALI is expressed as (band 7/band 5) × (band 7/band 8) [62]. The hydrothermal alteration zones are traced as bright pixels in the highly tectonized ophiolites, metasedimentary and island arc metavolcanic and metavolcaniclastic rocks (Figure 6a–d). The island arc rocks appear as bright pixels likely due to the abundant OH-minerals such as mica, amphiboles, chlorite and epidote (Figure 6a). It is noticed that the Urga, Um Teneidab, Khashab-1, Khashab-2 and El-Anbat gold mines overlap these alteration zones. Kaolinitic and clay minerals rich rocks such as granitoids, wadi deposits and altered ophiolitic rocks show bright pixels in Figure 6b, where Um Eleiga, Um Karaba and Hutit gold mines are seen. The calcite index grey image discriminates perfectly the calcite rich rocks zone of serpentinite, talc carbonate rocks, island arc metavolcanics and highly sheared metasediments as bright and gray image pixels.

The false-color composite (FCC) ratio image (OHI, KLI, CLI) was elaborated by combination of the three gray scale mineralogical indices images in the RGB channels (Figure 6d). This image characterizes the OH-bearing rocks (meta-ultramafic) as light blue pixels, metavolcanics and metasediments as dark blue and reddish blue pixels, respectively. The quartz-rich rocks have a reddish yellow image signature, whereas the clay minerals bearing rocks and stream sediments have lemon and bright green image signatures. The hydrothermal alteration zones in the study area are associated with the highly deformed ophiolitic and the ductile deformation zone. The results of alteration mapping derived from ASTER datasets reveal the spatial association of highly sheared rocks with gold occurrences.

3.4. PALSAR-Based Lineaments Extraction

Lineaments' extraction and analysis is considered a fast tool in geologic mapping and mineral exploration [63]. Hydrothermal ore deposits occur within or nearby fracture zones. Thus, lineament mapping is useful to infer locations of unexposed mineralization or dilatation sites [64]. High-resolution PALSAR data with full polarization and variable off-nadir angle have significantly enhanced the structural mapping and lineaments extraction. The adaptive Lee and Local Sigma filters can be applied to PALSAR to eliminate speckles and to enhance structural lineaments [6,65]. In the present study, four PALSAR scenes were orthorectified, mosaicked and enhanced by Lee filter to trace the linear features and major structures. The HH, HV, HH + HV bands were transformed by PCA method to PCA RGB images (PC1, PC2, PC3; Figure 7a). The automated lineament extraction was carried out using PC1 image (Figure 7b) by using LINE algorithm PCI Geomatica software. This algorithm is adopted with parameters, including radius of filter in pixels (RADI), threshold for edge gradient (GTHR), threshold for curve length (LTHR), threshold for line fitting error (FTHR), threshold for angular difference (ATHR), and threshold for linking distance (DTHR).

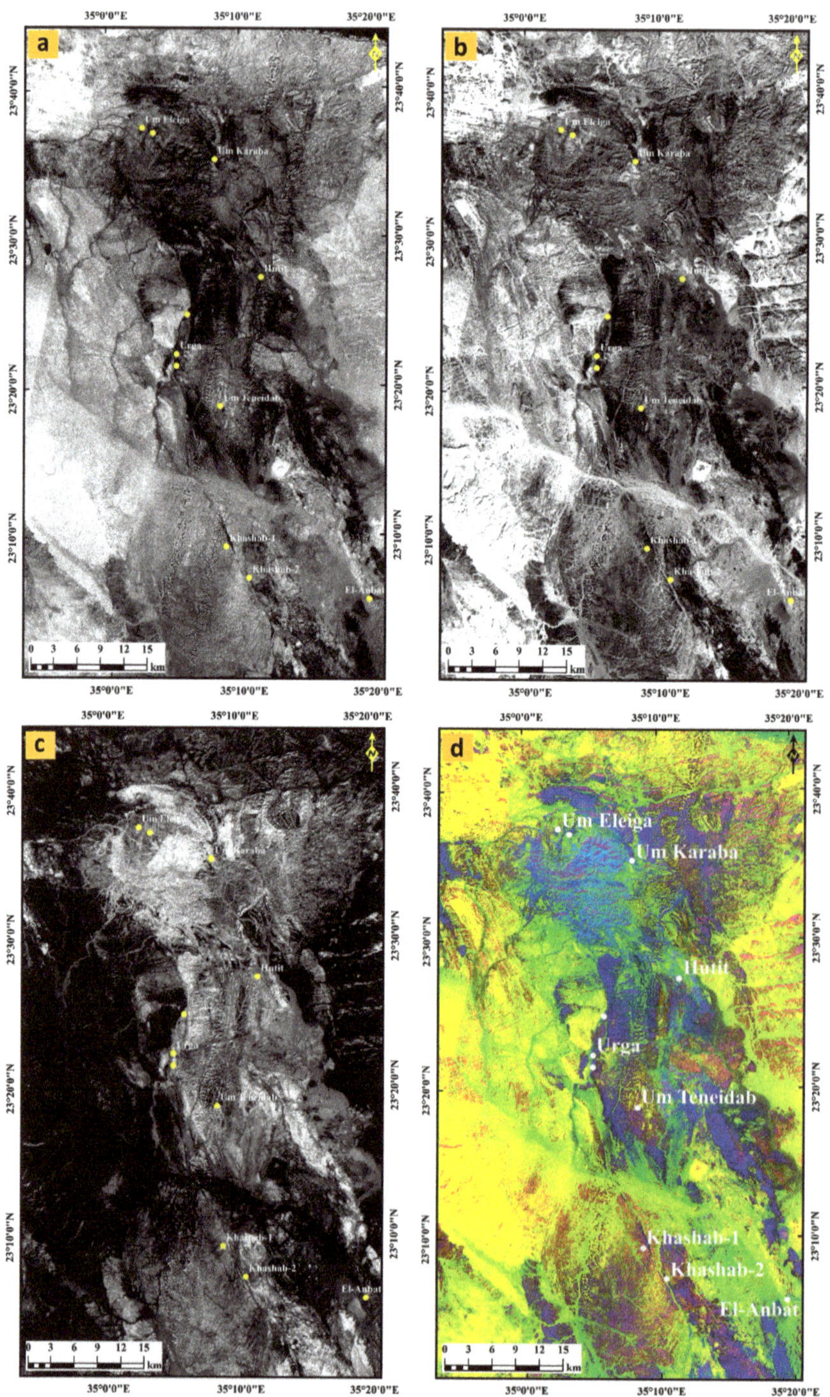

Figure 6. Grey scale ASTER band ratio images representing different mineral indices (**a**) OHI, (**b**) KLI (**c**) CLI and (**d**) Fused FCC RGB-OHI, KLI, CLI highlights the alteration zones.

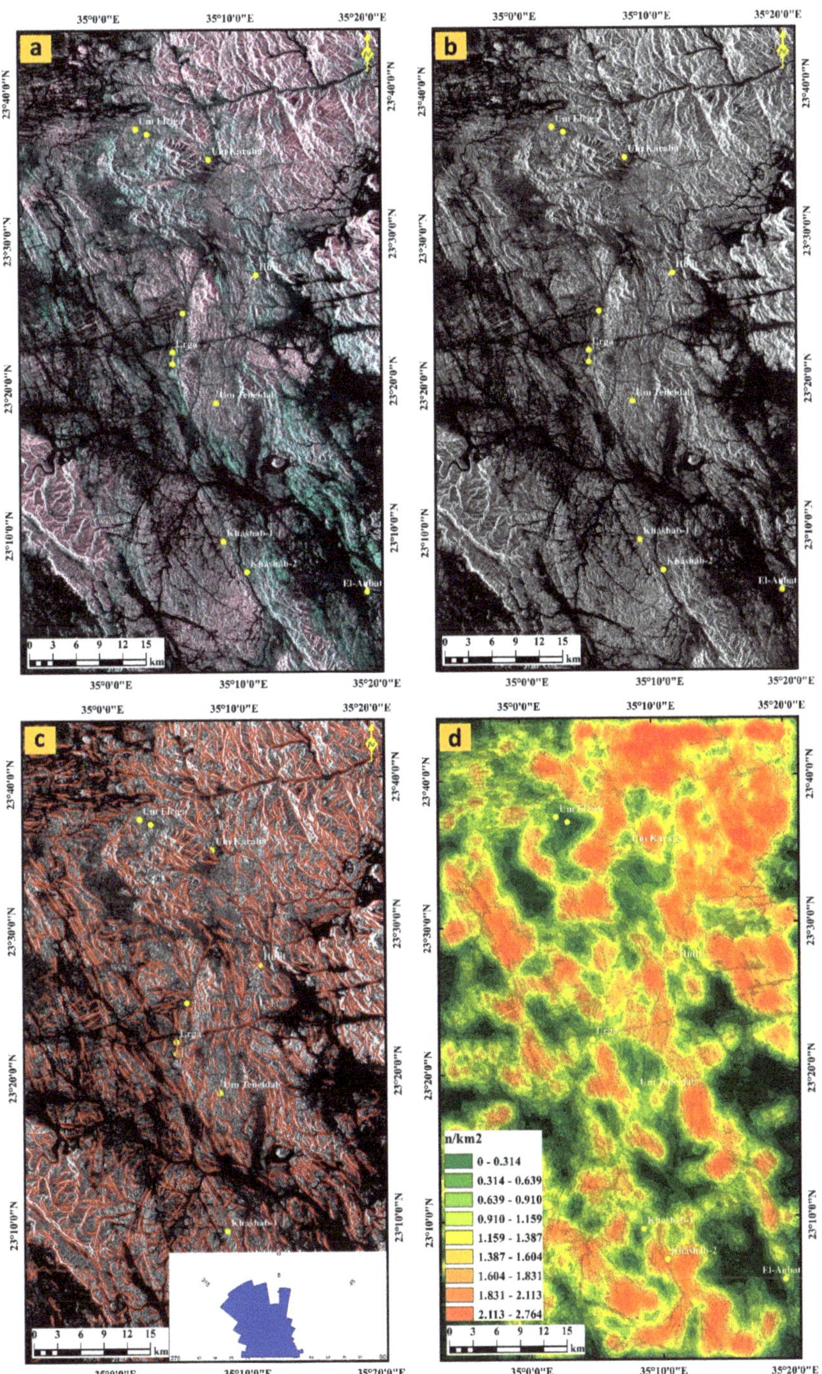

Figure 7. Lineaments extraction using processed PALSAR data (**a**) PCA of enhanced Lee filter of HH, HV, HH + HV in RGB-PC1, PC2, PC3, (**b**) PC1 highlights lineaments features, (**c**) the extracted lineaments dropped over PC1 with inset azimuth-frequency diagram and (**d**) lineament density map.

The extracted lineaments were overlaid on the PC1 image (Figure 7c) and exported to Rockworks software to generate the azimuth-frequency diagram (inset, Figure 7c). The lineaments layer was imported in the ArcMap (version 10.5, ESRI (Environmental Systems Research Institute, Redlands, CA, USA) environment to generate the lineament density map using the line density module in the spatial analyst toolbox (Figure 7d). The resultant lineaments layer shows predominantly NW–SE, N–S and E–W trending structural grain, which agrees mostly with the major thrusts and faults in Figure 2. Well-developed foliations, shear cleavages and strike slip faults offsets are also recognized (Figure 7a,b). The gold mining sites occur close to the high-density zones of the lineaments (Figure 7d), similar to conclusions reached by Zoheir and Emam [7], and Zoheir et al. [11,12] in the south Eastern Desert of Egypt.

3.5. Geospatial Modelling for Gold Mineralization Zones

Geographic Information System (GIS) is capable of manipulating, overlaying, integrating, and storing digital data in the geo-database as thematic layers. This approach plays an essential role in creating and integrating numerous geo-datasets from different geospatial data such as geology, structural density, etc. [66]. Remote sensing and (GIS) were integrated to identify new gold mineralization sites. For geospatial modelling of these mineralization zones, the evidence layers were integrated according to their relative contribution to the gold mineralization [67]. Where three main processes are recommended, including ranking processes for the evidence layers, weighting process are used for the classes of each layer, and data integration is used to assess gold mineralization potential zones [67,68]. Multi-source datasets were obtained, analyzed and integrated to develop a geospatial model. Seven thematic maps for the study area were prepared and then converted into raster or vector form to be easily integrated with the GIS tools. A weight was assigned to each factor layer according to the different ranking methods [69–73] (Table 3). The weighted sum of all the factor layers was calculated (Figure S6). The layers were then processed using ArcGIS analysis tools to place the assigned gold occurrences on a knowledge-based hierarchy, utilizing the Spatial Analyst tool of ArcGIS (Figure S6). To perform this process three basic steps were followed: spatial database building, data analysis, and data integration. The created layers for the geospatial model were subjected to reclassification and assigned suitable weights (Figure S6 and Table 3). Moreover, the geospatial thematic maps of the gold mineralization model were ranked from 1 to 5 (where 5 is the most favorable and 1 is the least) based on their suitability to host gold mineralization.

Gold mineralization is associated with the ophiolitic rocks and their contacts, close to contacts, and mostly along shear zones. A mineralization potential map was created and classified the study area into five relatively zones, varying from very low to very high potential zones (Figure 8). The high potential zones are distributed mainly around ophiolitic rocks contacts (serpentinite, talc carbonate and ophiolitic metagabbros) and metavolcanic and metasediments association rocks, respectively. The new promising sites occur close to the already known gold mining sites (Figure 8). The integrated approach demarcates several potential zones along the NW-Hodein shear zone. Validation of this model was accomplished by testing ancient mining localities in the study area. Figure 8 shows that Um Eleiga, Um Karaba, Hutit and Urga occur in the very high potential zone, while Um Teneidab, Khashab 1, 2 and El-Anbat match the high potential zone. Additionally, as constrained by new fieldwork and the available literature [11,12], occurrences of gold-bearing quartz veins are mostly confined to thrust ophiolite/metavolcanic-metasediments contacts.

Table 3. Weights assigned for different gold mineralization parameters in the study area.

Thematic Layer	Class Ranges	Layer Weight	Influence (%)	Class Rank
Lithology map	Gn/Pgb/NSS/WD	0.38	38	1
	Sg/Pg			2
	Gt			3
	Ms/Mv			4
	S/Omg			5
Alteration Zone image (threshold)	Clay minerals	0.19	19	5
	OH-bearing rocks			4
	Calcite-bearing rocks			3
	Quartz-bearings rocks			2
	Sedimentary rocks			1
Proximity to gold mines (km)	<10	0.12	12	3
	10–20			2
	>20			1
Proximity to favorable contacts (km)	<2	0.10	10	3
	2–10			2
	>10			1
Proximity to major faults (km)	<1	0.08	8	3
	1–3			2
	>3			1
Major faults density (km/km^2)	0–0.28	0.07	7	1
	0.29–0.56			2
	0.57–0.84			3
	0.85–1.12			4
	1.13–1.4			5
Lineaments density (km/km^2)	0–0.55	0.06	6	1
	0.56–1.11			2
	1.12–1.66			3
	1.67–2.21			4
	2.22–2.76			5

Abbreviation: Gn = Gneisses and migmatites, S = Serpentinites, Omg = Ophiolitic metagabbros, Ms = Metasedimentary ophiolitic mélange matrix, Mv = Island arc association, Gt = gabbro-tonalite-granodiorite complex, Sg = Syn-tectonic granitoids, Pg = Post-tectonic granitoids, Pgb = Post-tectonic gabbros, NSS = Nubian Sandstone, WD = Wadi deposits.

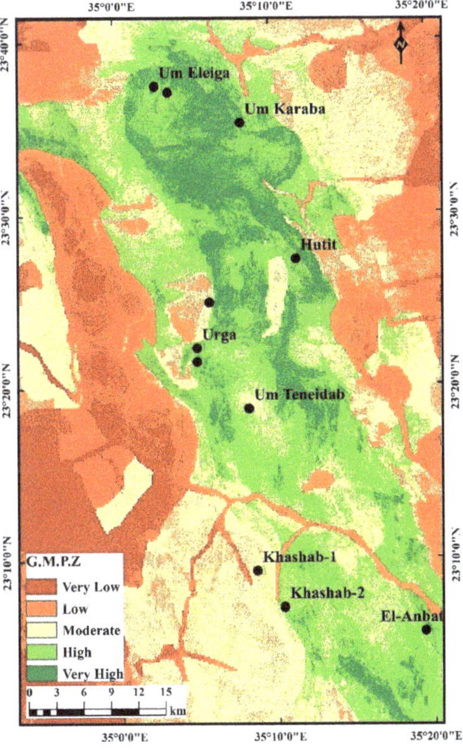

Figure 8. Gold mineralization potential zones (G.M.P.Z) map of the study area.

4. Structural Setting and Analysis

The Wadi Hodein shear belt (Figure 9) is a high angle NW-oriented transcurrent shear zone in the South Eastern Desert. It exhibits a sinistral sense of shear manifested by various kinematic indicators such as biotite fishes, asymmetric porphyroclasts, S-C structures and deflected markers [34,40,74]. Evidence for a dextral sense of shear superimposed on the main sinistral sense of shear was also reported within the W. Kharit–W. Hodein shear zone [74]. Abdel-Karim et al. [75] described high-grade regional metamorphism contemporaneous with the regional foliation and intrafolial folds. A later phase was mainly a dynamic metamorphism associated with the emplacement of ophiolites onto the gneissic rocks. The Wadi Rahaba–Gabal Abu Dahr and Gabal Arais ophiolites include imbricate thrust-bounded sheets and slices of serpentinized ultramafic rocks, amphibolite and metabasalt, embedded in a highly tectonized matrix of talc carbonate, schists and metasiltstones. According to several authors, the Wadi Hodein area has evolved throughout four phases of deformation, D1–D4 [11,37–40,61,76]. The D1 deformation phase was an early NNE–SSW crustal shortening related to arc–arc collision and is manifested by S1 axial planar foliations, WNW-trending tight intrafolial and overturned folds (F1) and SSE-verging thrusts [37,61]. D2 is expressed in NNW–SSE folds and NNW–SSE crenulation cleavage (S2). F2 folds have en-echelon geometry and verge towards the WSW [37–39,61]. The NNW–SSE crenulations and kink folds (F2) might have been developed through oblique non-coaxial deformation of cleaved rocks. D3 was a sinistral transpression along the NNW–SSE ductile shear zones due to NW-ward nappe stacking and thrusting [37]. D4 was a brittle deformation event that led to formation of the ENE–WSW dextral strike–slip faults that deformed the preexisting rocks and dislocated the earlier structures [37,61].

4.1. The Gold-Mineralized Shear Zones

4.1.1. Wadi Khashab Shear Zone

The NNW-trending Wadi Khashab shear zone (Figure 9) and related splays cut across the southwest part of the area, and it dips moderately or steeply to the east. This shear zone is cut by WNW–ESE and ENE–WSW strike–slip faults. The Wadi Khashab shear zone hosts several occurrences of gold-bearing quartz veins that are commonly associated with sericitized and silicified rocks. Of these occurrences, the El-Beida and Wadi Khashab occurrences are heavily worked out by artisanal miners at present. The Wadi Urga occurrence is confined to highly sheared metavolcanic rocks with signs of hydrothermal alteration along their contacts with a large granodiorite–tonalite intrusion. The ~E–W Urga fault is a major, fault that dissects the Wadi Khashab shear zone and apparently displaces its trace ~1–2 km right-laterally. Similar parallel structures are abundant north and south of this major fault. A number of prospective targets, including Wadi Beitan, Wadi Urga Rayan, Um Teneidab and north Wadi Hutib, have been identified along similar, ~E–W trending faults, or where the main NNW structural trend is cut by WNW–ESE strike slip faults. To the north, it seems that this structure may host other unexplored occurrences, based on the widespread carbonate alteration in the footwalls of the ophiolitic blocks.

4.1.2. Wadi Rahaba Shear Zone

The Wadi Rahaba shear zone (Figure 9) represents an intense NNW-trending, steeply dipping, brittle-ductile shear zone cutting a generally carbonated mafic–ultramafic rocks and volcaniclastic successions in the central and northern parts of the study area. The shear zone also extends to the south, where it hosts the Anbat mine and other scattered surface mining sites. Towards the east of the Wadi Rahaba shear zone, a major thrust fault boundary marks the base of a hanging wall block of metagabbro and meta-granodiorite; and varies significantly from NW-trending and moderately steeply east-dipping in the central-east to shallow south-dipping and easterly trending in its northeastern part. A number of parallel, NW-trending, shallow to moderately easterly dipping thrust faults (or splays) occur immediately west of the Wadi Rahaba shear zone. These thrust faults mark the western boundary of the gneissic domains and are spatially associated with a number of prospective targets. Numerous significant, E–W to ENE trending late cross-faults appear to offset the major thrust-faults within this area.

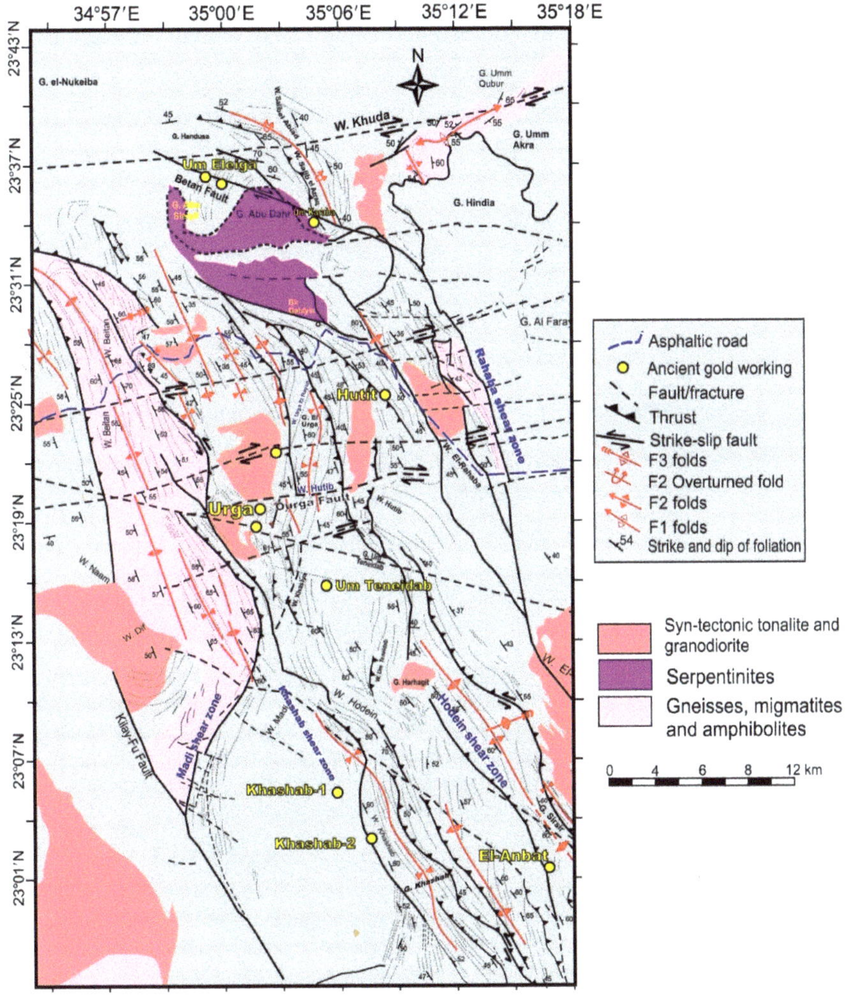

Figure 9. Structural map of the Wadi Hodein shear belt.

4.1.3. Wadi Beitan Fault Zone

The Wadi Beitan fault zone (Figure 9) is a major, steeply dipping, WNW-trending silicified breccia zone with a strike length of some +25 km and is traced across the Um Eleiga mine area and eastwards. This structure cuts rocks of variable types in the area, including the Um Eleiga gabbro, Abu Dahr ophiolites and Rahaba gabbro–diorite complex. This zone converges with a NNW-SSE shear zone to the east and delineates the eastern structure contact of the island arc and ophiolitic rocks. Hydrothermal breccia and silica alteration of the host rocks are common along this structure, and abundant zones of sulfide mineralization with local malachite occurrences are observed. In the vicinity of the Um Eleiga gold mine, albitization of the gabbroic host rocks are pervasive. Iron oxides and copper hydro-carbonate are observed in some localities within the Um Eleiga mine area. Noteworthy, the Wadi Beitan fault zone could have been genetically linked with the extensive dike swarms spread over the large are between G. Handusa and G. Nukeiba northwest of the Abu Dahr ophiolitic massif. These dike swarms overprint the earlier NNW-SSE structural trend. Numerous basic dikes along the WNW–ESE direction cut the Um Eleiga gabbro–diorite intrusion and extend eastward to the Abu Dahr ophiolites.

4.2. Deformation Events

The Wadi Hodein shear belt (Figure 9) is a NNW-striking belt (~105 km-long), deformed greenschist metamorphosed ophiolites and island arc rocks. It was developed via four phases of deformation; (i) D1, NNE–SSW crustal shortening, (ii) D2: NE-SW oblique convergence and transpression, (iii) D3, E–W compressional regime and (iv) D4: extension tectonics. Lithological layering represents the primary bedding surface (S0) which is still locally recognizable in the ophiolitic metagabbros and bedded metamudstone as cm-scale alternating lighter and darker layers.

4.2.1. D1: NNE–SSW Crustal Shortening

N–S shortening and terrane accretion occurred as a result of a collision between the Gabgaba and Gerf terranes over a ~N-dipping subduction zone [77]. D1 structures are mainly preserved in the Wadi Khuda gneisses and in the Wadi Khashab–Gabal Sirsir ophiolitic mélange. Wadi Khuda gneisses represent a small structural window later cut by tonalite-granodiorite and biotite granite intrusions. In the Wadi Khuda gneisses, quartz–feldspar-micas gneissosity (S1) strikes ENE–WSW, and dips toward the NNW and SSE (Figure 10a). The axial plane gneissosity is only evident in the hinges of F1 folds which represented by metrewide, high-amplitude isoclinal, near recumbent to shallowly dipping folds. Boudins (B1) are developed on the quartzo-feldspathic layers in banded amphibolites. They have a symmetric shape and their long axes are parallel to the gneissosity (S1), suggesting a coaxial strain. The largest F1 fold is an asymmetric, doubly plunging anticline extending for up to 8 km in the Wadi Khuda gneisses (Figure 9). The axis of F1 major anticline is displaced by a NNW-striking sinistral strike–slip fault and the southern limb is superimposed by a NNW-tending anticline (F2). The early thrusts are SSE verging and are folded during the D2 event. These thrusts are preserved in the Arais ophiolitic mélange, marking the contacts between Beitan gneisses belt and the ophiolite blocks, as well as the contact between serpentinites and volcaniclastic metasediments (Figure 9). These thrusts converge downwards onto a basal S-dipping decollement.

4.2.2. D2: NE–SW Oblique Convergence and Transpression

The D2 deformation phase is manifested by regional foliation (S2), mylonitic foliation (S2m), major and minor folds (F2), stretching and intersection lineations (L2) and major thrusts (T2). The structures include small-scale, rootless intrafolial folds (Figure 10b), shearband boudins (Figure 10b), recumbent (Figure 10c,d), tight isoclinal and sheath folds along the first-generation ductile shear zone. These structures are largely preserved as decimetre-scale folds in a transposing and intrafolial foliation (S2) in the Beitan gneisses belt, schistose metavolcanics and volcaniclastic metasediments. The S2 in foliated metavolcanics and volcaniclastic metasediments is axial plane schistosity for the F2 folds and is defined by a strong preferred alignment of actinolite, chlorite and biotite. In the bedding-parallel quartz veins, asymmetrical "shearband-type" boudins (B2; Figure 10e) and asymmetrical "domino-boudinaged" quartz veins are common. The shearband boudins (B2) were only observed in shear zone close to the gold occurrences. Ptygmatic folding is widespread in the Beitan gneisses (Figure 10f). S2 planes have moderate to high dips predominantly toward the SW and NE. Around the synorogenic gneissose granites, ophioltic metagabbros and serpentinite mountainous blocks, the S2 foliations show important changes along strike or define different geometric patterns. In the central segment of Hodein shear belt, S1 traces strike NW–SE to NNW–SSE to NNE–SSW, whereas they curve back into a NW–SE orientation to the southwest of Gabal Dahr serpentinite's peridotites.

Figure 10. Field photographs of structural elements in the study area: (**a**) gneissosity (S1) in Khuda gneisses defined by the graiN–Shape preferred orientation of quartz–feldspar aggregates, hornblende and biotite; (**b–d**) NNW-plunging, reclined and overturned and recumbent F2 folds and refolded quartz vein in volcaniclastic metasediments of Wadi Rahaba; (**e**) asymmetrical "shearband-type" boudinaged quartz veins in volcaniclastic metasediments around Urga gold occurrence; (**f**) Ptygmatic folds in Beitan gneisses; (**g**) Major F2 asymmteric anticline in volcaniclastic metasediments, Wadi Rahaba; (**h**) SW-verging volcaniclastic metasediments thrust over Beitan gneisses belt, south of Gabal Arais; (**i**) NE-verging serpentinite thrust over volcaniclastic metasediments, Wadi Rahaba.

The L2 lineation plunges have a N25–35° W or N40° E within the S2 gneissosity and schistosity and show all the characters of a mineral and stretching lineation. The hinge lines of microfolds in gneisses, schistose metavolcanics and volcaniclastic metasediments define a vertical crenulation lineation trending mainly NNW–SSW. A number of large-scale F2 folds have been mapped, and most of their axes are NNW- or NW-oriented (Figures 10g–i,11a–i). Other F2 axes may have resulted from later reworking by F3. Axial planar to these folds is a spaced crenulation cleavage (S2), which is commonly well developed. F2 folds range from centimeter- to kilometer-scale and are common in Beitan gneisses belt and the ophiloitic mélanges (Figure 11a–i). These folds are open, asymmetrical to isoclinal and their axes are nearly upright or plunge NNW or SSW. Some of the F2 fold axes plunge NNE or SSW, which may have resulted from later slight reorientation by the D3 open folding and/or the second-generation low-angle ductile shear zone. The axial traces of F2 folds strike mainly NW–SE or NNW–SSE. A doubly plunging asymmetric F2 anticline in the Wadi Khuda gneisses dips moderately to NW or SE. A series of WSW-verging thrust faults (T2) (Figure 9), parallel to the hinges of the largest F2 folds could have formed by progressive F2 folding in a compressional setting.

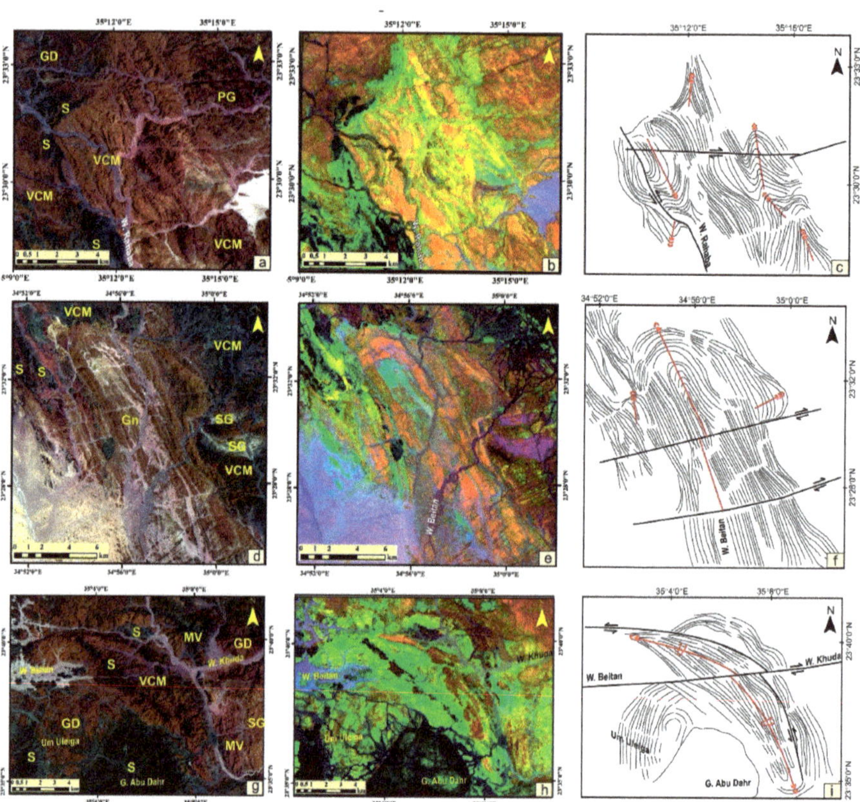

Figure 11. (**a,d,g**) Landsat-8 OLI band composites RGB-753, (**b,e,h**) Landsat-8 OLI band ratio RGB-7/6, 6/5, 4/2 highlighted the different folding types and trends, and (**c,f,i**) the resultant interpretation maps of the different fold generations (thin lines for foliation trajectories and red arrows lines are interpreted fold axial traces and major faults (black lines). Gn; gneisses, S; serpentinites, VCM; volcaniclastic metasediments, MV; metavolcanics, SG; syn-orogenic granites, GD; metagabbro–diorite complex, PG, post-tectonic granite.

The ophiolitic mélanges in the Wadi Hodein–Wadi Beitan shear belt are thrusted over each other and over the Beitan gneisses belt (infracrustal rocks) along three major imbricate thrusts (T2). Three imbricate thrusts are named in this study: the Sirsir–Rahaba thrust, the Arais imbricate thrust and the Hodein–Beitan imbricate thrusts (Figure 9). The geometry of Wadi Khashab–Gabal Sirsir ophiolitic mélange is controlled by the oppositely dipping Sirsir–Rahaba and the southern extension of Wadi Hodein–Wadi Beitan sinistral reverse shear zone, respectively (Figure 12a–c). Gently to moderately ENE- and WSW-dipping foliation (S2) forms the dominant fabric that intensifies into belts of mylonites associated with D2 ductile thrusts. The Sirsir–Rahaba thrust dips ENE and separates ophiolitic blocks of Gabal Sirsir and the volcaniclastic metasediments of Gabal Anbat (Figures 9 and 12a–c). The Wadi Hodein–Wadi Beitan imbricate thrust forms an imbricated fan (Figure 12a–c). Axial planar to tight-to-isoclinal, reclined F2 sheath folds have axes concentrated around the L2 mineral lineation. To the north of Gebel of Abu Dahr the thrusts mark the tectonic contacts between the ophiolitic slabs (serpentinites and talc carbonates) and the overlying metavolcanics and volcaniclastic metasediments. Additionally, Arais imbricate thrusts control the contact between Arais opoilitic mélange and the Beitan gneisses belt (Figure 9) and also mark the tectonic contacts between the ophiolitic slices. These NE-dipping thrusts constitute a backthrust to the WSW-dipping Hodein–Beitan imbricate thrusts.

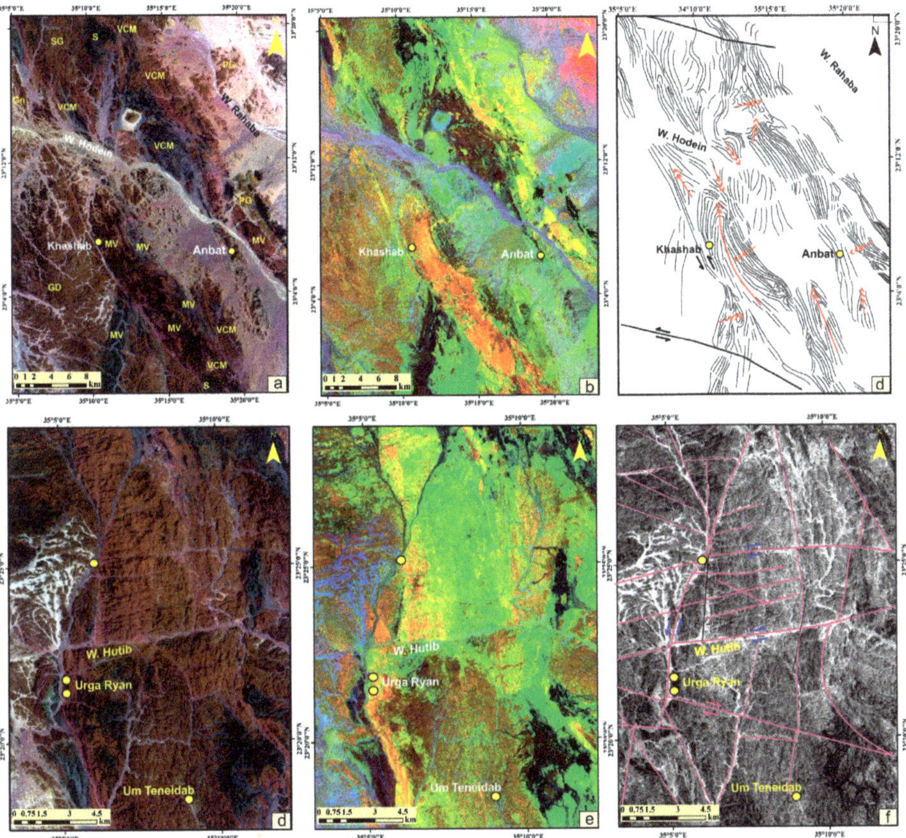

Figure 12. (**a**,**d**) Landsat-8 OLI band composites RGB-753, (**b**,**e**) Landsat-8 OLI band ratio RGB-7/6, 6/5, 4/2 highlighted ENE–WSW strike slip faults dissected the island arc metavolcanics, (**c**) the resultant interpretation maps of the different fold generations (thin lines for foliation trajectories and red arrows lines are interpreted fold axial traces and major faults (black lines) and (**f**) Landsat-8 PC3 shows the ENE–WSW dextral strike–slip faults cutting across pre-existing structures and controlling the drainage system. S; serpentinites, VCM; volcaniclastic metasediments, MV; metavolcanics, SG; syn-orogenic granites, GD; metagabbro–diorite complex, PG, post-tectonic granite.

Most of the small-scale, tight to isoclinal, asymmetric F2 folds in the Wadi Hodein-Wadi Beitan imbricate thrusts show top-to-the-ENE thrusting. Many shear sense indicators, including mineral lineations, asymmetric shear folds, asymmetric porphyroblasts, shearband boudins, S-C fabrics and shear bands indicate that the ophiolite blocks along Sirsir–Rahaba thrust and Arais thrusts are moved top to WSW and those along Hodein–Beitan thrusts are moved up to ENE (Figure 13a–c). Each thrust zone comprises several thrust segments showing anastomosing morphologies and form WSW-dipping tectonic duplexes. S2 gneissosity in the Wadi Beitan gneisses occur as asymmetrical folds around these boudins. Migmatitization is common in the shadows of the rotated boudins. The boudins do not display any foliations but are surrounded by the S2 gneissosity, suggesting that they formed during the D2 deformation. Shearband boudins show a top-to-the-WSW movement.

Figure 13. Field photographs of structural elements in the study area: (**a**) Foliation-parallel shearband boudins of quartz along S2 foliation in volcaniclastic metasediments,Wadi Rahaba, indicating sinistral transpressive shearing along thrust planes, (**b**) Metagabbro prophyroclast in volcaniclastic metasediments indicating top-to NE sense of movement along thrust planes, Wadi Beitan, (**c**) Asymmetrical folds and shearband boudins of quartz veins along S2 foliation in volcaniclastic metasediments, Wadi Rahaba, indicating top-to NE sense of movement along thrust planes, (**d**) Strike–slip shear zones appear as steep ductile zones in sheared metavolcanics which sheared the S2 foliation. S3 foliation strike N20–30° W with steep dip toward the SW, Wadi Beitan, (**e**) Crenulation lineation in metasedisment with a subhorizontal plunge to the NNW, (**f**) Asymmetric tight and reclined folds in volcaniclastic metasediments, south of Gabal Abu Darh, indicating sinistral sense of shearing, (**g**,**h**) Quartz and metagabbro asymmetric prophyroclast in volcaniclastic metasediments, Wadi Um Teneidab and Wadi Rahaba, (**i**) en-echelon folded quartz veins in schistose metavolcanics, Wadi Hutib, indicating sinistral sense of shearing.

4.2.3. D3: E–W Shortening and Sinistral Shearing

D3 structures are related to transposition and reorientation of the D2 structures. S3 schistosity overprinted the pre-existing S2 schistosity. S3 foliation strikes N20–30° W and dips 60–70° to the NE or SW (Figure 12d). Mylonitic foliation in C3 shear zones is superimposed on the S2 foliation (Figure 13d). The spacing of the strike slip shear zones varies from a few centimeters to some meters. Kilometer scale NNW-oriented strike–slip shear zones are commonly developed in the Wadi Khashab–Gabal Sirsir ophiolitic mélange, especially along Wadi Khashab, along the main planes of Hodein–Beitan shear zone and Sirsir–Rahaba shear zone (Figure 9). The ophiolitic and metavolcanic rocks on both sides of the shear zone show S-shaped fault–drag folds (F3), consistent with the sinistral movement. Quartz veins showing signs of ductile shearing are sulfide-bearing and are associated with green malachite alteration in places.

The strike–slip shear zones in Hodein shear belt separate the Wadi Beitan gneisses from the overlying ophiolitic mélange (Figure 9). The mylonitized rocks are characterized by stretched quartz ribbons with an anastomosing pattern of intervening variably deformed lenses. The angle between S3 and C3 is often clearly visible and can reach 5°, indicating strong shear values. Asymmetric boudins are developed in the Wadi Beitan banded gneisses and amphibolite. Mylonitic rocks exhibit subvertical to steeply dipping foliations (N20° W and N40° W) and dip steeply to the SW-dipping high-angle S3 foliations in the high strain zones along the main planes of the Beitan imbricate thrusts and Sirsir–Rahaba thrust. The S3 foliation is defined by kink bands or spaced cleavage that are parallel to the F3 axial planes. S3 kink bands are sharp deflections of S2. The high-strain zones with subvertical foliation planes are subparallel and are oriented NNW–SSW except in the central part of the Wadi Hodein–Wadi Beitan shear belt where they are deflected into NNE-SSW. In these high-strain zones, isoclinal and intrafolial folds with shallow plunging fold hinges and a subvertical axial planar foliation are abundant. Shallow NNW- or SSE-plunging mineral and stretching lineations are present on the steep foliation planes in the NNW–SSE striking high-strain zones.

Two distinct types of mineral lineations L3 on the S3 foliations have been recognized by the measurement of the minerals that grew during D3. The first type, here defined as L3a, is a SW and SE-plunging mineral stretching lineation along the ENE-verging thrust (T3). L3a is marked by platy quartz, amphibole fish and feldspar rods. The second type, here referred to as L3b, is a b-type lineation orientated subparallel to the F3 fold axes. The L3b is a crenulation lineation with a subhorizontal plunge to the NNW or SSE (Figure 13e). The D3 deformation produced regional-scale NNW-trending folds (F3) defined by open folds with nearly vertical axial planes. Superposition of F3 on F2 folds resulted in extensive NNW-trending tight and reclined folds and NE-trending open folds. The most common macroscopic shear sense indicators include asymmetric tight and reclined folds (Figure 13f), sigmoidal porphyroclasts (Figure 13g,h), S-C fabrics, folded and en-echelon quartz lenses (Figure 13i), shearband quartz boduins and pop up structures. The lineation indicates that the tectonic transport took place while far-field compression progressed and switched to oblique thrusting.

4.2.4. D4: Extension and Terrane Exhumation

D4 was a prolonged phase of brittle deformation associated with intrusion and emplacement of late-orogenic granitoid intrusions and terrane exhumation. During D4, the Wadi Hodein–Wadi Beitan shear belt was dissected by a number of dextral strike–slip faults, deformed the post-orogenic granites and the D1–D3 structures. These ENE–WSW right-lateral strike–slip faults occur in the central part of the Wadi Hodein–Wadi Beitan shear belt and along Wadi Khuda (Figures 9 and 12d–f). Another major sinistral strike–slip fault striking NNE-SSW controls the course of Wadi Urga El-Rayan. NW–SE oriented gold-bearing quartz veins were originated during D2 and were subsequently deformed by D3–D4 events.

5. Gold Occurrences in the Wadi Hodein Shear Belt

A number of gold occurrences in the Wadi Hodein shear belt are mainly associated with the high strain zones. The latter are confined to fault-bounded imbricated thrust sheets of allochthonous ophiolitic blocks. The Roman and Ptolemaic gold occurrences are confined to deformed granitoid and metagabbro–diorite complexes in the northern and central parts of the Hodein shear belt. In the following sections, we describe the internal structures of the mineralized quartz veins and the structural controls of several occurrences, namely: Wadi Khashab, the El-Anbat mine, the Um Teneidab mine, the Urga El–Ryan occurrence, the Hutit mine and the Um Eleiga deposit.

5.1. The Wadi Khashab Occurrence

In the Wadi Khashab area (Figure 14a), the ophiolite slices are overthrust on the arc's metavolcanic rocks and volcaniclastic metasediments. The serpentinites of Gabal El Beida form a NW-elongated large belt and slices of different sizes tectonically incorporated in the volcaniclastic metasediments. The serpentinites form dissected elongated massive masses with sheared and foliated peripheries. These serpentinite bodies are altered to talc–

carbonate rocks along the shear zones. Blocks and masses of ophiolitic metagabbros usually occur within the volcaniclastic metasediments. They locally exhibit weakly developed foliation and compositional layering. Ultramafic rocks are commonly associated with pillow and amygdaloid metabasalts. The metavolcanic rocks are represented by strongly foliated metaandesite, metabasaltic andesite and intermediate to acidic metatuffs [39].

Gold-sulfide mineralization is confined to discrete shear zones of highly silicified, ferruginated metavolcanics and volcaniclastic metasediments commonly associated with sulfide-bearing granophyric dikes and quartz± carbonate veins [39]. The Wadi Khashab zone is a major sinistral strike–slip shear zone, striking NNW–SSE and dipping steeply (65–75°) towards the ENE. Rocks along this shear zone are carbonatized and consist mainly of highly sheared metavolcanics, serpentinites and volcaniclastic metasedimentary rocks. The wall rock alteration is present as strongly foliated flakes in quartz veins and is dominated by the silicification and chloritization associated with carbonate [39,78]. The silicified shear zone (SiO_2 > 80%) of the pillow metabasalts is relatively enriched in Au (25–45 ppm) whereas the carbonatized shear zone (SiO_2 < 50%) within the obducted area of the serpentinite rocks has a very low Au content (0.24–0.45 ppm) [78]. The samples analyzed by Zoheir [39] yielded 2.1–6.6 ppm Au, and the sulfidized granophyric dikes are also gold-bearing (0.2–2.1 ppm Au).

5.2. The El-Anbat Deposit

The Gabal El-Anbat–Gabal Sirsir belt consists mainly of heterogeneously sheared talc carbonate, carbonatized, serpentinite, ophiolitic metagabbro, pillow metabasalts, and arc metavolcanic and volcaniclastic rocks (Figure 14b). Syn-orogenic metagabbro–diorite, granodiorite and post-orogenic granitoid intrusions are widespread in the study area. The serpentinite occurs as large masses, forming the main body of Gabal Sirsir and small slices incorporated in the volcaniclastic metasedimentary rocks. The serpentinite is variably altered to talc-carbonate and its contacts with the neighboring volcaniclastic metasediments and ophiolitic metagabbro are highly sheared. The gold grade in the quartz-carbonate veins and hydrothermal alteration zones is ~0.3–46 g/t Au [79]. The metavolocanic succession of the Gabal El-Anbat–Gabal Sirsir area comprises variably sheared metabasalt, metabasaltic andesite, meta-andesite, metadacite, and metapyroclastic rocks [80].

In the El-Anbat mine area (23°17′01″ N, 35°09′21″ E), the alteration zones occur as sheets or lenses along the NW- and NNW-striking thrusts, between the carbonatized serpentinite and foliated intermediate metavolcanics (Figure 14b). S2 is a regional foliation in El-Anbat area striking NNW–SSW and dipping moderately or steeply to the ENE and less commonly to the WSW. The F1 folds are still preserved in ophiolitic metagabbros and mafic metavolocanics. The F2 folds are major NNW-trending asymmetric left-stepping folds in the volcaniclastic metasediments. Different styles of minor folds are recognized— open concentric folds, closed tight folds, symmetric and asymmetric folds, isoclinal folds and chevron folds. During D3, Conjugate sinistral and dextral strike–slip faults and shear zones are apparently developed as a component of a major transpression system.

The mineralized shear zones and the bleached sheared host rocks are confined to the steeply dipping thrusts between carbonatized serpentinites and the underlying volcaniclastic metasediments. These zones are characterized by arrays of recrystallized quartz pods, fracture-filling ankerite, sulfidized granitic offshoots, boudinage of the granitic bodies, silicified listvenite masses and intermingling carbonate sericite and malachite veinlets [38], suggesting significant gold grades in the hydrothermally altered, sheared host rocks. The analyses indicate that gold is related to mineralized quartz veins (with up to 7.5 g/t Au) and to pervasively altered wall rocks (with up to 5 g/t Au) [38].

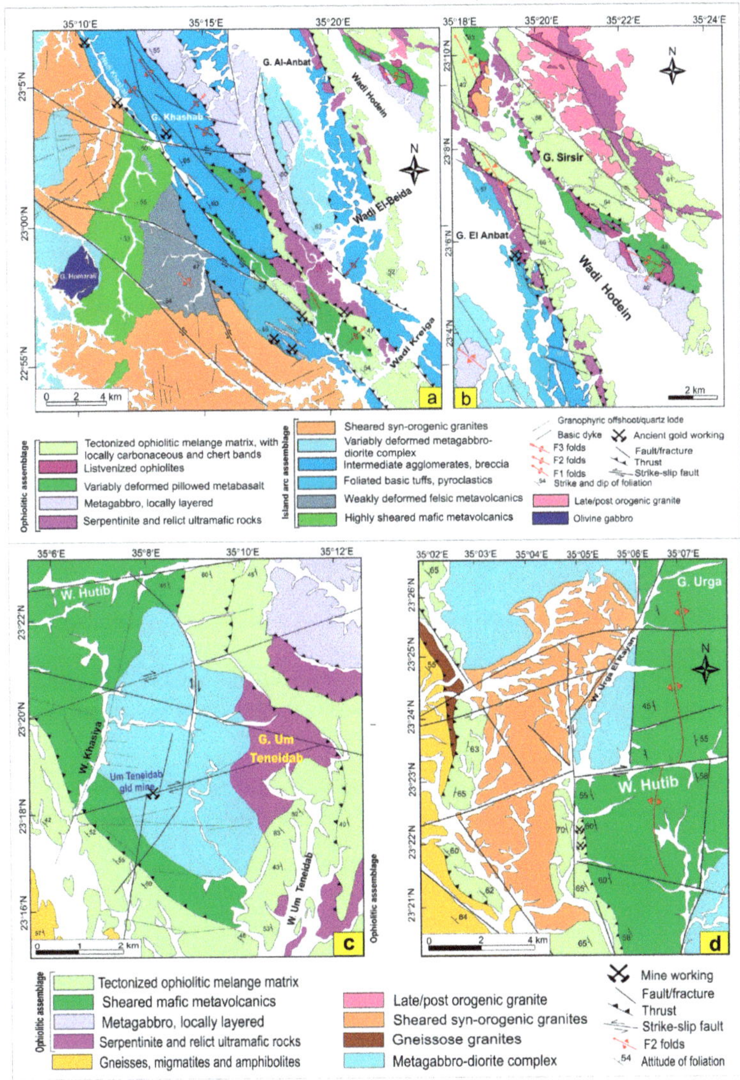

Figure 14. (**a**) Geological map of the Wadi El Beida–Wadi Khashab area (modified from Zoheir [39]); (**b**) geological map of Gabal El-Anbat–Wadi Hodein area (modified from [38]); (**c**) geological map of Um Teneidab area (modified from Hamimi and Sakran, [81]); and (**d**) geological map of Wadi Urga El Rayan area (modified from El Baraga, [48]).

5.3. The Um Teneidab Deposit

The Um Teneidab mine (23°17′01″ N, 35°09′21″ E) is located ~13 km to the southwest of the Hutit mine and to the west of Gabal Um Teneidab peak. The Um Teneidab area is underlain by gneisses and migmatites, serpentinite, volcaniclastic metasediments, and deformed mafic metavolcanic rocks. These rocks are cut by metagabbro–diorite complex (Figure 14c), as well as unmappable masses and offshoots of post-orogenic leucogranites [12,81]. Serpentinite constitutes the main mass of Gabal Um Teneidab and forms blocks and slices of variable sizes incorporated in the volcaniclastic metasediments. The mélange blocks are randomly distributed, varying from cobbles to mountain size and usually oriented in the NWSE or NNW–SSE directions. The gabbroic rocks in the Um Teneidab mine area were mapped as island arc, synorogenic metagabbro–diorite complex [12,82] and as post-orogenic younger gabbros [81]. In the present work, we agree with Hassan and El-Manakhly [82] and Zoheir et al. [12] and consider the gabbroic rocks in the Um Teneidab mine area as a variably foliated, metagabbro–diorite complex.

The Um Teneidab metagabbro–diorite complex is dissected by a number of ENE–WSW and N–S dextral strike–slip faults and WNW–ESE major fractures with no obvious lateral displacement. The zones in which the leucogranites send offshoots into the gabbroic rocks (Figure 14c) are characterized by intense shearing and alteration as well as the presence of abundant quartz veins and felsic dikes. Shearing in the area is predominantly brittle, with little ductile deformation being experienced, with alteration zones and quartz veins that are generally controlled by NW–SE shear/fault sets. The subvertical NW- and NNW-striking milky quartz veins extend up to 250 m with 8 cm to 45 cm thickness and lensoid morphology. These veins are in part or completely recrystallized. The main lode is a zone of stockwork of veinlets (70 cm wide) bordered by hydrothermally altered wall rocks forming together a ~2-m-wide mineralization zone [12]. The gold in these veinlets and the hydrothermally altered wall rocks occurs as fillings in the microfractures and is scattered as sliver in altered pyrite and galena. The gold content ranges from 1 to 30 g/t in quartz veins, whereas the altered wall rocks locally contain ~8 g/t Au [12].

5.4. The Urga El-Ryan Occurrence

The Urga Ryan gold deposit (35°5′0″ E and 23°21′23″ N) is located at ~3 km to the south of the intersection between Wadi Urga Ryan and Wadi Hutib. The area around Wadi Urga El-Ryan (Figure 14d) is underlain by gneisses, volcaniclastic metasediments and mafic to intermediate metavolcanic rocks. The volcaniclastic metasediments include wide varieties of metagreywacke, metamudstone, slate, pelitic schists and metaconglomerate in decreasing order of abundance. They are strongly NW–SE foliated and less commonly NNE–SSW cleaved [48].

The metavolcanics are composed of repeatedly alternating and interfingered successions of basaltic flows together with layered andesite and subordinate dacitic flows and their corresponding metatuffs [48] as well as chlorite schists. The metavolcanic rocks are variably deformed by km-scale shear zones that led to intense shearing in the NNW–SSE direction (S2 and S3), overprinting the WNW–ESE schistosity (S1). The metagabbro is emplaced into both metavolcanics from the northern and western parts and volcaniclastic metasediments from the southern part [48]. The metagabbroic rocks are foliated and sheared, particularly along the shear zones. The syn-orogenic granites are slightly foliated and form an elongated intrusion (10 m × 2.5 m width) trending roughly N–S, which forms discontinuous hills. It is mainly tonalite in composition and is bounded from the west by the Wadi Beitan gneisses, and by metagabbro from the north. Volcaniclastic metasediments and metavolcanics border these rocks from the southeastern side. Xenoliths and roof pendants from metagabbro are enclosed within the syn-orogenic tonalite.

In the Urga El-Ryan occurrence (Figure 14d), the old mining houses spread along the main Wadi Urga Ryan and their tributaries, possibly reflecting significant mine activities. The mineralized regions are limited only to NNW–SSE striking and westward dipping zones of intense shearing in metavolcanic rocks. The quartz vein strikes N–S (355°) concordantly with the shear zone system and dips at 80° E [83]. It was exploited in a deep mine down to a level of 15 m below the surface. These shear zones (up to 40 m in length and 5–30 cm in thickness) are strongly mylonitized and marked by recrystallized quartz lenses and asymmetric boudinaged quartz veins [12]. The narrow zones of sheared wall rocks are characterized variable degrees of hydrothermal alteration. The gold grade is 1 to 7g/t Au in the quartz veins from the Urga El-Ryan occurrence [84].

5.5. Hutit Deposit

The Hutit gold deposit occurs in the Wadi Huzama, which is a small tributary of the larger Wadi Rahaba (Figure 15a). The Hutit mine area is underlain by serpentinite, mafic metavolcanics, volcaniclastic metasediments, ophiolitic metagabbros, syn-orogenic metagabbros and post-orogenic granites (Figure 15a,b) [12,48,82,84,85]. Structurally, the Hutit deposit is controlled by the Rahaba–Sirsir sinistral shear zone and their kilometer-scale imbricate thrust system (Figure 15b). The SW-directed thrusts bound the ophioltic blocks and dip moderately or steeply to the NE. Imbricate slices and sheets of serpentinite are tectonically overlying metavolcanics and volcaniclastic metasediments. The oblique thrust planes accommodate sinistral displacement and asymmetrical anastomosing and kinked fabrics. Asymmetric quartz lenses and drag folds indicate a sinistral sense of shear, whereas subvertical slickensides along the quartz vein walls document the reverse slip of the hanging wall block. Joints and fractures are abundant in the rock units mainly with NE–SW and NW–SE directions. Mafic to felsic dikes (basalt, basaltic andesite, andesite, rhyodacite and dacite) as well as granodiorite dikelike bodies strike NW–SE and less commonly NE–SW.

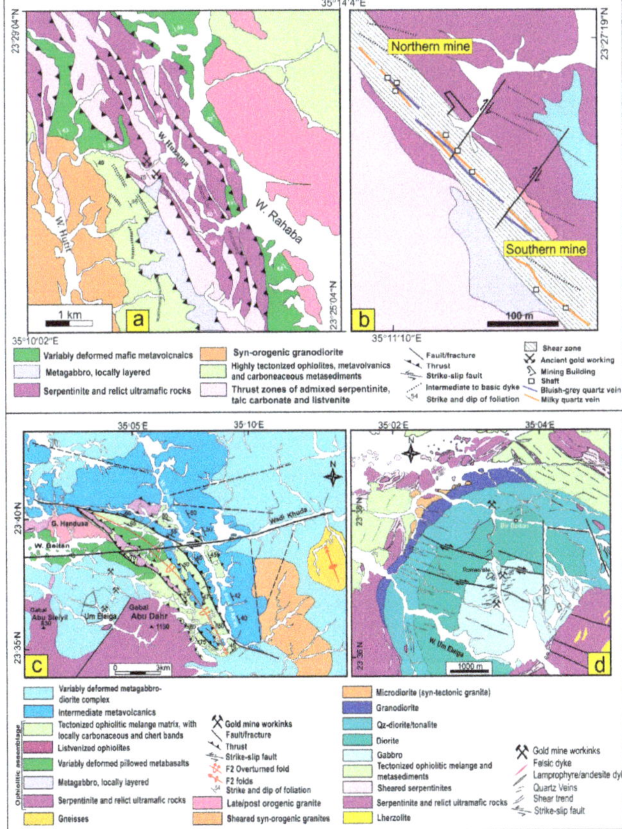

Figure 15. (**a**) Geological map of the Hutit gold mine and surroundings, (modified after from Hassan and El-Manakhly [82]; Zoheir et al. [12]), (**b**) Detailed geological map of the main lode in the northern mine and southern mine (modified after Zoheir et al. [12]). (**c**) Geological map of Abu Dahr ophiolite (modified from Ashmawy, [86]), and (**d**) Geological map of the Um Eleiga gold mine (modified from Zoheir et al. [41]).

Two main old mine workings (northern; 23°27′26″ N, 35°11′34″ E and southern; 23°27′16″ N, 35°11′41″ E) (Figure 15b) were mapped at the 1:1000 scale by Hassan and El-Manakhly [82]. The ruins of the separation and grinding plants are still observed in the northern old mine, whereas the loading station, leaching basins and crusher stages were observed in the southern mine. In both mines, a main entrance through a horizontal ~E–W adit leads to the veins at a distance of 20 m or 35 m [12]. The mineralized quartz veins occur along the thrust contact between the serpentinite masses and the mafic metavolcanics and volcaniclastic metasediments. Two types of anastomosing and undulating gold-bearing quartz veins were reported in the mine area, including bluish-gray (common in the northern mine) and milky quartz veins (dominant in the southern mine). Concerning the ore grade, Gabra [84] reported 1–36 g/t in samples from quartz veins the southern mine and 1–40 g/t in quartz veins from the northern mine. Takla et al. [85] reported an average of 20 g/t in samples from the two different types of quartz veins and 8 g/t Au on average in altered wall rocks.

5.6. The Um Eleiga Deposit

The Um Eleiga mine (24°36′44″ N, 35°03′19″ E) occurs ~50 km west of the Red Sea coast. Traces of placer gold workings during Roman–Byzantine and early Islamic times include hundreds of shallow pits, dumps, ancient mining camps, stone anvils, hammers, and grinding mills [12,41,83]. The geology of Um Eleiga area is dominated by a 7 km wide intrusive complex with a gabbroid core cutting the serpentinite-chromitite of Gabal Abu Dahr and the northern part of Wadi Rahaba–Gabal Abu Dahr ophiolite mélange. The latter composed masses and blocks of serpentinized ultramafics, pillow metabasalts and mafic metavolcanics incorporated within highly tectonised matrix of pelitic and carbonaceous metasediments. The ophiolite fragments occur as NE-dipping slabs and sheets thrusted southwestward forming imbricate thrust system. Contacts of the serpentinized ultamfics with Um Eleiga metagabbros-diorite complex are sharp and irregular. In the upper part of the Abu Dahr massif, serpentinite contains large masses of serpentinized dunite, sills or layers of wehrlite, pyroxenite and gabbro [42,86].

The mafic metavolcanic rocks comprise blocks and thrusted slices of metabasalt and basaltic andesite [42,86]. These rocks structurally overlay the Um Eleiga intrusive metagabbros-diorite complex and Abu Dahr serpentinized ultramafics (Figure 15c). Pillow metabasalt occurs in the core of a major fault-propagation fold (Figure 15c) within the highly tectonized matrix with schists and sheared serpentinite. Pillow metabasalt forms ellipsoidal, globular and tabular shaped bodies, ranging between 10 cm to 1 m in diameter. The Um Eleiga intrusive metagabbro–diorite complex is approximately 7 km long and 4 km across, and encompasses a compositional continuum from gabbro to granodiorite through diorite and subordinate tonalite (Figure 15d). This complex exhibits a circular zonation, in which gabbro occupies the core and subordinate granodiorite forms the margin [41]. The Um Eleiga metagabbro and metabasalt have a tholeiitic to calc-alkaline affinity and are interpreted to have been formed in a forearc setting [42].

In the Um Eleiga mine, the mineralized quartz veins trend mainly NE–SW or ENE–WSW and cut the gabbroic rocks in the central part of the complex and extend beyond the gabbro–diorite boundary [12]. The fault/joint intersections are the main structural control of intensely hydrothermal alteration zones and high gold contents in the central part of the Um Eleiga complex [41]. Late barren quartz veins are confined to intersections of fault and tension gashes in the brecciated gabbro core and trend mainly N–S, NW–SE and E–W (Figure 15d). Sulfide-bearing quartz veins (5–40 cm thick) cut the metagabbro–diorite complex and are partially stopped out [12]. Analyses of quartz dumps in several pits gave gold values up to 28 g/t [82]. According to Takla et al. [85], most of the ancient workings in the mine area were confined to the old terraces (lithified wadi alluvium) within and adjacent to the intensely kaolinitised gabbro. Atomic absorption analyses of heavy concentrates of these old terraces indicated that gold contents were variable from barren to 10 g/t [85].

6. Discussion

6.1. Remote Sensing Targeting of New Gold Occurrences

The Nubian Shield is a typical example of well-exposed crystalline basement rocks, and mafic–carbonate–hydrous mineralogical indices were used to extract the representative pixels in the satellite images. Zoheir et al. [12] used Landsat-8 OLI and ASTER band ratio images and successfully benefited from the effective absorption features of the mafic rock-forming minerals and their metasomatic products. Therefore, Landsat-8 OLI and ASTER data with comprehensive fieldwork were used for mapping geological structures, lithological units and alteration zones associated with orogenic gold mineralization in the study area. The image processing techniques of these remotely sensed data maximize the relatively small differences in the spectral responses of rock forming minerals within a specific wavelength range [87]. The applied approach includes false band combination (FCC), band rationing, principal component and mineralogical indices to the Landsat-8 OLI and ASTER data. It is evident that the band math of Landsat-8 (6/7) distinguished clay minerals, serpentine, and many alteration zones, whereas ASTER (7 + 9/8) demarcated carbonatized rocks, metavolcanics and metasedimentary rocks.

Landsat-8 FCC of Abrams ratio (RGB-6/7, 4/3, 5/4), Kaufmann ratio (RGB-7/5, 5/4, 6/7) and Chica-Olma ratio (RGB-6/7, 6/5, 4/2) characterized the contacts between the different rock units in the study area and upgraded the exciting geologic maps (see Figure 4). Moreover, they deciphered the structural elements especially folding and thrusting in the Wadi Hodein–Beitan shear belt as well as the ENE-dextral strike slip faulting. The RGB of the principal component analysis of Landsat-8 (PC2, PC1, PC4 and PC2, PC4, PC5) and ASTER (PC1, PC2, PC3 and PC6, PC3, PC1) discriminated the rocks consisting of high contents of Al and/or Mg-OH-bearing minerals and emphasized the contact between felsic and basic metavolcanics and volcaniclastic metasediments and gneisses along Wadis Hodein and Beitan (see Figure 5).

The mineralogical indices of Ninomiya [62] (OHI, KLI, CLI and ALI) were used to characterize the alteration zone in the study area using ASTER data. The resultant images highlighted effectively the zone of the hydrothermal alterations. In addition, they stressed the fact that the alteration minerals zones are found in both ductile and brittle deformation zones where the integration between the structural segments (density and direction) associated with the presence of the altered minerals may help to localize and predict new occurrences of gold mineralization. A false-color composite (FCC) ratio image (OHI, KLI, CLI) was generated. This RGB FCC image shows the OH-bearing rocks (meta-ultramafic–mafic ophiolites and island arc meta-basic rocks) and the clay-minerals (see Figure 6). The evaluation of the remote sensing alteration mapping results with field data using an error matrix approach and Kappa Coefficient shows a very good match, which indicates the overall accuracy of 88.33% and the Kappa Coefficient of 0.77 for Landsat-8 data and the overall accuracy of 74.10% and the Kappa Coefficient of 0.65 for ASTER data, respectively (Table 4A,B). It is noticed that the old working gold occurrences are associated with the high deformation and fracturing zones bounding the sheared ophiolitic belt and faulted contacts such as El-Beida, Urga Ryan, Hutit and Um Eleiga gold mines. Multi-source datasets were selected to control factor layers indicating the distribution of gold mineralization potentiality zones such lithology, alteration zones, proximity to existing gold mines, proximity to favorable contacts, proximity to major structures (especially thrusts), faults density maps and lineaments density maps. The seven evidence layers were derived, analyzed, and integrated in a GIS platform to develop the gold mineralization model (see Figure 8).

Table 4. Error matrix for alteration mapping derived from remote sensing data versus field data. (**A**) Landsat-8 data; (**B**) ASTER data.

(A) Landsat-8 Alteration Map	Field Data		
Classes	Iron Oxide/Hydroxides	OH-Bearing and Carbonate Minerals	Totals User's Accuracy
Iron oxide/hydroxides	52	8	60–87%
OH-bearing and carbonate minerals	6	54	60–90%
Totals	58	62	120
Producer's accuracy	89.65%	87.10%	-
Overall accuracy = 88.33%	Kappa Coefficient = 0.77		

(B)ASTER Alteration Map	Field Data					
	Iron Oxide/Hydroxides	OHI	KLI	CLI	Totals	User's Accuracy
Iron oxide/hydroxides	22	6	1	1	30	73%
OHI	3	20	6	1	30	67%
KLI	1	4	24	1	30	80%
CLI	1	4	2	23	30	77%
Totals	27	34	33	26	120	
Producer's accuracy	81%	59%	73%	88%	-	-
Overall accuracy = 74.10%	Kappa Coefficient = 0.65					

6.2. Transpressional Tectonics in the Evolution of the Wadi Hodein Shear Belt

The Wadi Hodein shear belt has evolved throughout a multistage deformation history (D1–D4), in which D2 and D3 were the most dominant. NNE–SSW crustal shortening (D1) led to development of early ENE–WSW striking S1 gneissosity and F1 doubly plunging anticline in the Wadi Khuda gneisses. D1 structures in the Wadi Khashab, Wadi El Beida and Gabal Anbat and Gabal Arais areas exhibit WNW-striking S1 schistosity, F1 minor tight, recumbent and overturned folds and T1 thrusts. NE–SW oblique convergence and transpression (D2) led to the development of NW–SE striking S2 regional foliation, S2m mylonitic foliation, F2 major kilometer scale anticlines and synclines (see Figures 10–13,16a–d) and minor open, asymmetrical to isoclinal folds, NW-plunging stretching and mineral lineations (L2) and WSW-verging major thrusts (T2). The major D2 fabrics are represented by a series of large-scale folds (F2) associated with northwest-vergent foliation (S2), indicating that the F2 folds formed in response to a NW–SE-oriented stress field.

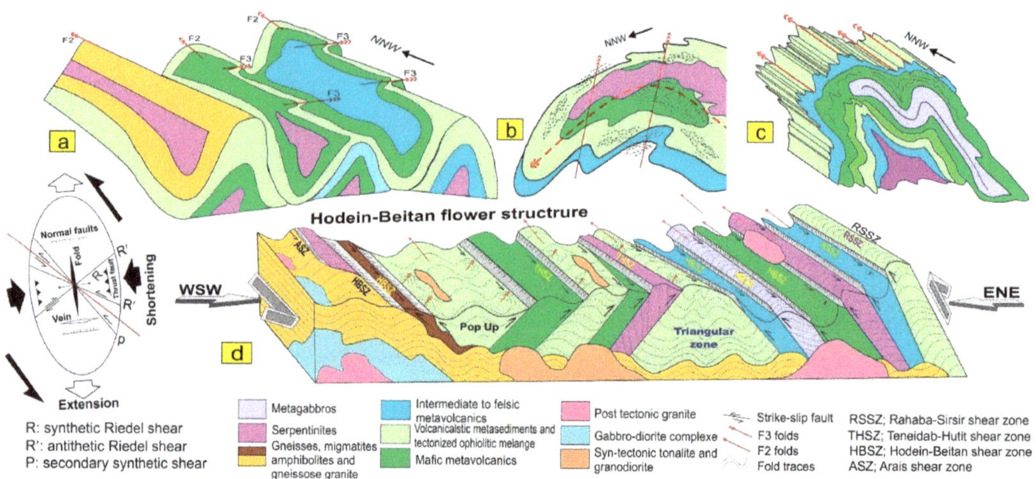

Figure 16. (**a–c**) F2 folds superimposed by the upright open F3 folds in the Wadi Hodein shear belt (Wadi Hodein, Wadi El Beida and Wadi Khashab areas). (**d**) Sketch block diagram illustrating the structural evolution of the Wadi Hodein shear belt.

The Wadi Hodein–Beitan shear belt is cut by a suite of variably deformed granitic intrusions. Many of these intrusions occur in the vicinity of the major thrusts and are considered to have emplaced during the regional D2 deformation event. E–W shortening (D3) produced subvertical NNW-Striking foliation (S3), NNW–SSE thrust dominated strike slip shear zones (C3), NNW- and ENE-trending folds (F3). L3a is a SW and SE-plunging stretching lineation along the NNE–SSW T3 minor thrusts. L3b lineation is a crenulation lineation with a subhorizontal plunge. During D3, the major NNW-oriented shear zones rooted into a low-angle, deep crustal detachment, delineating the point at which oblique thrusting is partitioned into separate transpressional strike–slip and reverse-sense shear zones, i.e., Sirsir–Rahaba, Hodien–Beitan and Um Teneidab–Hutit shear zones. The sinistral porphyroclasts and NNW-plunging reclined and vertical folds as well as NE-trending open folds in sheared metavolcanics and volcaniclastic metasediments indicate a transpressive stress regime including both pure and simple compressional components. The latest extensional structures (D4) in Hodein shear belt include major brittle strike slip faults and microfaults and dikes associated with the intrusion of late to post-tectonic granites. Structures assigned to this deformation phase include ENE–WSW dextral and NNE–SSW sinistral strike–slip faults.

The newly produced structures map (see Figure 9) shows that the overall structural setting of the Wadi Hodein shear belt bears a resemblance to an asymmetric flower structures in which the ophiolitic mélange and the ophiolitie blocks are cut and bounded on both sides by reverse-slip faults, and strike–slip shear zones. The eastern (Sirsir–Rahaba and Um Teneidab–Hutit shear zones) and western boundary faults (Hodein–Beitan and Arais shear zones) correspond to a reverse-sinistral thrusts. The transpressional imbricate thrusts forming the Sirsir–Rahaba, Hodein–Beitan and Arais shear zones dip toward ENE, whereas those pertaining to the Um Teneidab-Hutit shear zones dip to WSW (see Figures 9 and 16d). All the shear zones show a moderately sinuous trace in map-view, up to 2 km-thick band of fine-laminated mylonites, developed in the footwall gneisses and volcaniclastic metasediments. The high-angle, NNW-oriented thrust-dominated, strike slip shear zones marked a near-vertical ductile deformation zone rather than an imbricate structure [88–90]. The thrusts show a typical sequence of propagation in the footwall of the previous thrust in a piggy-back model. Many kinematic indicators, including asymmetric porphyroblasts such as "σ-type" augens, and pressure shadows, indicate a sinistral sense of shearing. The predominance of reverse slip movement over the along strike shearing suggests a thrust-dominated transpression along the Wadi Hodein shear zone. The NW–SE to NNW–SSE subparallel Sirsir–Rahaba, Hodein–Beitan, Um Teneidab-Hutit and Arais strike–slip shear zones exhibit a consistently sinistral sense of shear.

6.3. Structures Controlling Gold Occurrences

The distribution of the gold occurrences is confined to the margins of the Wadi Hodein shear belt (see Figure 9) especially along the thrust-dominated transpressional major zones and their splays formed during D3. Conjugate sinistral (N) NW–(S) SE and dextral (E)NE–(W)SW strike–slip faults and shear zones apparently developed During D3 as a component of a major transpression system. These conjugate shear zones control the occurrence of gold in Hodein shear belt. Twisting and anastomosing S2 and T2 thrusts led to NE-trending F3 open folds and oblique sinistral shear zones on the pre-existing NW-striking thrusts. The S-C fabrics and asymmetric quartz lenses replicate left-lateral shearing, which resulted in widespread NE–SW-striking dilation bents within and around the shear zones. At the Hutit and El-Anbat mine areas, the mineralized quartz veins and the alteration zones occur along the NW- and NNW-striking thrusts between the carbonatized serpentinite and foliated metavolcanics and the metavolcanics and volcaniclastic metasediments. In the Um Eleiga mine, the mineralized quartz veins trend mainly NE–SW or ENE–WSW mainly along the fault/joint intersections. In the Urga El-Ryan occurrence, the mineralized regions are limited only to NNW–SSE striking and westward dipping zones of intense shearing in metavolcanic rocks. In the Um Teneidab deposit alteration zones and quartz veins are generally controlled by NW–SE shear/fault sets. In the Wadi Khashab occurrence, the quartz veins occur as closely spaced swarms running parallel and subparallel to NNW-trending sinistral strike–slip shear zone of Wadi Khashab.

7. Conclusions

Detailed geological mapping of the Wadi Hodein area was successfully accomplished by employing different processing techniques, i.e., band combinations, band math (BM), Principal Component Analysis (PCA), decorrelation stretch and mineralogical indices, to Landsat-8 OLI, ASTER and ALOS PALSAR data. Field data were integrated with results of the remote sensing studies, which enabled an updated understanding of the structural evolution of the Wadi Hodein–Wadi Beitan shear belt. The structural framework of the Wadi Hodein–Wadi Beitan shear belt was shaped by the superposition of the NW–SE folds (F2) by the NNW-trending, km-scale tight and reclined folds (F3). The overall configuration of the investigated shear belt is identical to that of an asymmetrical flower structure in which the ophiolitic mélange and the ophiolitic blocks are cut and bounded on both sides by reverse slip and strike–slip shear zones. The field studies showed that most of the gold mineralization is mainly associated with quartz veins within extensive sinistral transpressional shear zones assigned to the last ductile deformation event (D3) in the evolution of the entire shear belt. This event was characterized by strong stretching and mylonitic lineation, abundant undulatory shear zones and shear foliation disturbance. The gold-bearing quartz veins are hosted by narrow, steeply SW- and NE-dipping NNW- and NW-trending mylonitic zones that are characterized by a steeply NNW-plunging mineral lineation. This may imply the important role of oblique convergence in the formation of the auriferous quartz veins.

The processing flow chart introduced in this study has identified areas with the potential for unexplored gold occurrences and helped to illustrate the controls of the distribution of gold mineralization sites. The processing results were classified into five categories, from very high to very low potentiality zones. Gold mineralization is here interpreted to show preferential distribution along the Wadi Hodein–Wadi Beitan shear belt. The most important conclusion of this study is that gold mineralization in the Wadi Hodein–Wadi Beitan shear belt was mostly confined to steeply dipping strike–slip shear zones in the marginal parts of the shear belt. The occurrence of gold-bearing quartz veins in central shear zone is not evident so far. A conceptual model for testing new and unexplored targets of gold–quartz veins in the region translates the coincidence of shear zones, alteration and acid dikes to highly potential zones for new and possibly important gold occurrences in the study area.

Supplementary Materials: The following are available online at https://www.mdpi.com/article/10.3390/min11050474/s1: Figure S1: Field photographs showing the different rock units in the study area; (a) and (b) hornblende gneiss alter-nating with bands of biotite gneiss forming ENE-trending ridges along Wadi Khuda, (c) banded amphibolite from Beitan gneisses belt, (d) contact between hornblende gneisses and gneissose granite from Beitan gneisses belt (e) car-bonated ophiolitic serpentinite enclosing small bodies of listivenitized ultramafites (dark ridges) from Wadi Rahaba, (f) small prophyroclast of ophiolitic metagabbro between volcaniclastic metasediment (lower) and serpentinite (up-per) from Wadi Rahaba; Figure S2: Field photographs showing the different rock units in the study area: (a) pillowed metabasalts from Wadi Khuda, northeast of Gabal Abu Dahr; (b) ellipsoidal Pillowed metabasalt from Wadi Hutib; (c) serpentinized peridotites of Gabal Arais thrusted over the Beitan gneisses belt and are intruded by a small mass of alkali feldspar granite; (d) syn-orogneic gabbro-diorite intruded the gneisses of Wadi Khuda; (e) post-orogenic biotite granite intrudes volcaniclastic metasediments along Wadi Rahaba; (f) syn-orogonic granodiorite intrudes volcaniclastic metasediments along Wadi Beitan, 3.1. Data and Processing Techniques; Figure S3: Flowchart of the remote sensing and GIS methodologies adopted in the present study; Figure S4: Lithological discrimination using (a) ASTER (RGB-431) and (b) grey scale ASTER band ratio (7 + 9/8); Figure S5: False-color composite of principal component analysis (PCA) of (a) Landsat-8 RGB-PC2, PC1, PC4 and (b) ASTER RGB-PC1, PC2, PC3; Figure S6: Model Builder of the geospatial thematic maps used to locate the high potential zones for gold mineralization in the study area; Table S1: Eigenvector loadings of principal component analysis for Landsat-8 OLI data; Table S2: Eigenvector loadings of principal component analysis for ASTER data; Section S1: Data and Processing Techniques

Author Contributions: Conceptualization, M.A.E.-W., S.K. and B.Z.; methodology, S.K. and A.B.P.; software, S.K.; validation, S.K., M.A.E.-W. and B.Z.; data curation, S.K.; writing—original draft preparation, M.A.E.-W. and S.K.; writing—review and editing, B.Z. and A.B.P.; supervision, M.A.E.-W.; fieldwork, S.K. and M.A.E.-W., shaping and fine-tuning the conclusions, B.Z. All authors have read and agreed to the published version of the manuscript.

Funding: The present work has received no funding or grant.

Data Availability Statement: The data presented in this study are available on request from the corresponding author.

Acknowledgments: The authors appreciate the help of Mohamed Hamdy and Ismail Thabet (Tanta University, Egypt) during the fieldwork. Additionally, the Institute of Oceanography and Environment (INOS), Universiti Malaysia Terengganu (UMT) is acknowledged.

Conflicts of Interest: The authors declare no conflict of interest.

References

1. Pour, B.A.; Hashim, M.; van Genderen, J. Detection of hydrothermal alteration zones in a tropical region using satellite remote sensing data: Bau gold field, Sarawak, Malaysia. *Ore Geol. Rev.* **2013**, *54*, 181–196. [CrossRef]
2. Pour, B.A.; Hashim, M. Structural geology mapping using PALSAR data in the Bau gold mining district, Sarawak, Malaysia. *Adv. Space Res.* **2014**, *54*, 644–654. [CrossRef]
3. Pour, B.A.; Hashim, M.; Marghany, M. Exploration of gold mineralization in a tropical region using Earth Observing-1 (EO1) and JERS-1 SAR data: A case study from Bau gold field, Sarawak, Malaysia. *Arab. J. Geosci.* **2014**, *7*, 2393–2406. [CrossRef]
4. Pour, A.B.; Hashim, M.; Makoundi, C.; Zaw, K. Structural Mapping of the Bentong-Raub Suture Zone Using PALSAR Remote Sensing Data, Peninsular Malaysia: Implications for Sedimenthosted/Orogenic Gold Mineral Systems Exploration. *Resour. Geol.* **2016**, *66*, 368–385. [CrossRef]
5. Pour, B.A.; Hashim, M. Structural mapping using PALSAR data in the Central Gold Belt, Peninsular Malaysia. *Ore Geol. Rev.* **2015**, *64*, 13–22. [CrossRef]
6. Pour, A.B.; Hashim, M. Application of Landsat-8 and ALOS-2 data for structural and landslide hazard mapping in Kelantan, Malaysia. *Nat. Hazards Earth Syst. Sci.* **2017**, *17*, 1285. [CrossRef]
7. Zoheir, B.; Emam, A. Field and ASTER imagery data for the setting of gold mineralization in Western Allaqi–Heiani belt, Egypt: A case study from the Haimur deposit. *J. Afr. Earth Sci.* **2014**, *99*, 150–164. [CrossRef]
8. Pour, A.B.; Park, T.-Y.; Park, Y.; Hong, J.K.; Muslim, A.M.; Läufer, A.; Crispini, L.; Pradhan, B.; Zoheir, B.; Rahmani, O.; et al. Landsat-8, Advanced Spaceborne Thermal Emission and Reflection Radiometer, and WorldView-3 Multispectral Satellite Imagery for Prospecting Copper-Gold Mineralization in the Northeastern Inglefield Mobile Belt (IMB), Northwest Greenland. *Remote Sens.* **2019**, *11*, 2430. [CrossRef]
9. Kusky, T.M.; Ramadan, T.M. Structural controls on Neoproterozoic mineralization in the South Eastern Desert, Egypt: An integrated field, Landsat TM and SIR-C/X SAR approach. *J. Afr. Earth Sci.* **2002**, *35*, 107–121. [CrossRef]
10. Zoheir, B.; Emam, A.; El-Amawy, M.; Abu-Alam, T. Auriferous shear zones in the central Allaqi-Heiani belt: Orogenic gold in post-accretionary structures, SE Egypt. *J. Afr. Earth Sci.* **2018**, *146*, 118–131. [CrossRef]
11. Zoheir, B.A.; Emam, A.; Abd El-Wahed, M.; Soliman, N. Gold endowment in the evolution of the Allaqi-Heiani suture, Egypt: A synthesis of geological, structural, and space-borne imagery data. *Ore Geol. Rev.* **2019**, *110*, 102938. [CrossRef]
12. Zoheir, B.; Emam, A.; Abdel-Wahed, M.; Soliman, N. Multispectral and Radar Data for the Setting of Gold Mineralization in the South Eastern Desert, Egypt. *Remote Sens.* **2019**, *11*, 1450. [CrossRef]
13. Hunt, G.R. Spectral Signatures of Particulate Minerals in The Visible and Near Infrared. *Geophysics* **1977**, *42*, 501–513. [CrossRef]
14. Clark, R.N.; King, T.V.V.; Klejwa, M.; Swayze, G.A.; Vergo, N. High spectral resolution reflectance spectroscopy of minerals. *J. Geophys. Res.* **1990**, *95*, 12653–12680. [CrossRef]
15. Hassan, S.M.; Sadek, M.F. Geological mapping and spectral based classification of basement rocks using remote sensing data analysis: The Korbiai-Gerf nappe complex, South Eastern Desert, Egypt. *J. Afr. Earth Sci.* **2017**, *134*, 404–418. [CrossRef]
16. Irons, J.R.; Dwyer, J.L.; Barsi, J.A. The next Landsat satellite: The Landsat Data Continuity Mission. *Remote Sens. Environ.* **2012**, *145*, 154–172. [CrossRef]
17. Roy, D.P.; Wulder, M.A.; Loveland, T.A.; Woodcock, C.E.; Allen, R.G.; Anderson, M.C.; Helder, D.; Irons, J.R.; Johnson, D.M.; Kennedy, R.; et al. Landsat-8: Science and product vision for terrestrial global change research. *Remote Seins. Environ.* **2014**, *145*, 154–172. [CrossRef]

18. Pour, A.B.; Sekandari, M.; Rahmani, O.; Crispini, L.; Läufer, A.; Park, Y.; Hong, J.K.; Pradhan, B.; Hashim, M.; Hossain, M.S.; et al. Identification of Phyllosilicates in the Antarctic Environment Using ASTER Satellite Data: Case Study from the Mesa Range, Campbell and Priestley Glaciers, Northern Victoria Land. *Remote Sens.* **2021**, *13*, 38. [CrossRef]

19. Sekandari, M.; Masoumi, I.; Beiranvand Pour, A.; Muslim, A.M.; Rahmani, O.; Hashim, M.; Zoheir, B.; Pradhan, B.; Misra, A.; Aminpour, S.M. Application of Landsat-8, Sentinel-2, ASTER and WorldView-3 Spectral Imagery for Exploration of Carbonate-Hosted Pb-Zn Deposits in the Central Iranian Terrane (CIT). *Remote Sens.* **2020**, *12*, 1239. [CrossRef]

20. Shirmard, H.; Farahbakhsh, E.; Beiranvand Pour, A.; Muslim, A.M.; Müller, R.D.; Chandra, R. Integration of Selective Dimensionality Reduction Techniques for Mineral Exploration Using ASTER Satellite Data. *Remote Sens.* **2020**, *12*, 1261. [CrossRef]

21. Takodjou Wambo, J.D.; Pour, A.B.; Ganno, S.; Asimow, P.D.; Zoheir, B.; Reis Salles, R.D.; Nzenti, J.P.; Pradhan, B.; Muslim, A.M. Identifying high potential zones of gold mineralization in a sub-tropical region using Landsat-8 and ASTER remote sensing data: A case study of the Ngoura-Colomines goldfield, eastern Cameroon. *Ore Geol. Rev.* **2020**, *122*, 103530. [CrossRef]

22. Pour, A.B.; Park, Y.; Crispini, L.; Läufer, A.; Kuk Hong, J.; Park, T.-Y.S.; Zoheir, B.; Pradhan, B.; Muslim, A.M.; Hossain, M.S.; et al. Mapping Listvenite Occurrences in the Damage Zones of Northern Victoria Land, Antarctica Using ASTER Satellite Remote Sensing Data. *Remote Sens.* **2019**, *11*, 1408. [CrossRef]

23. Abrams, M.; Yamaguchi, Y. Twenty Years of ASTER Contributions to Lithologic Mapping and Mineral Exploration. *Remote Sens.* **2019**, *11*, 1394. [CrossRef]

24. Abrams, M.; Tsu, H.; Hulley, G.; Iwao, K.; Pieri, D.; Cudahy, T.; Kargel, J. The Advanced Spaceborne Thermal Emission and Reflection Radiometer (ASTER) after Fifteen Years: Review of Global Products. *Int. J. Appl. Earth Obs. Geoinf.* **2015**, *38*, 292–301. [CrossRef]

25. Paillou, P.; Lopez, S.; Farr, T.; Rosenqvist, A. Mapping subsurface geology in Sahara using L-Band SAR: First results from the ALOS/PALSAR imaging Radar. *IEEE J. Sel. Top. Appl. Earth Obs. Remote Sens.* **2010**, *3*, 632–636. [CrossRef]

26. Rosenqvist, A.; Shimada, M.; Watanabe, M.; Tadono, T.; Yamauchi, K. Implementation of systematic data observation strategies for ALOS PALSAR, PRISM andAVNIR-2. In Proceedings of the 2004 IEEE International Geoscience and Remote Sensing Symposium, Anchorage, AK, USA, 20–24 September 2004.

27. Abd El-Wahed, M.; Hamimi, Z. The Infracrustal Rocks in the Egyptian Nubian Shield: An Overview and Synthesis. In *The Geology of the Egyptian Nubian Shield*; Regional Geology, Reviews; Hamimi, Z., Arai, S., Fowler, A.R., El-Bialy, M.Z., Eds.; Springer: Berlin/Heidelberg, Germany, 2021. [CrossRef]

28. Abd El-Wahed, M.A. Thrusting and transpressional shearing in the Pan-African nappe southwest El-Sibai core complex, Central Eastern Desert, Egypt. *J. Afr. Earth Sci.* **2008**, *50*, 16–36. [CrossRef]

29. Abd El-Wahed, M.A.; Harraz, H.Z.; El-Behairy, M.H. Transpressional imbricate thrust zones controlling gold mineralization in the Central Eastern Desert of Egypt. *Ore Geol. Rev.* **2016**, *78*, 424–446. [CrossRef]

30. Abd El-Wahed, M.A.; Kamh, S.Z.; Ashmway, M.; Shebl, A. Transpressive Structures in the Ghadir Shear Belt, Eastern Desert, Egypt: Evidence for Partitioning of Oblique Convergence in the Arabian Nubian Shield during Gondwana Agglutination. *Acta Geol. Sin.* **2019**. [CrossRef]

31. Hamimi, Z.; Abd El-Wahed, M.A. Suture(s) and Major Shear Zones in the Neoproterozoic Basement of Egypt. In *The Geology of Egypt*; Regional Geology Reviews; Hamimi, Z., El-Barkooky, A., Martínez Frías, J., Fritz, H., Abd El-Rahman, Y., Eds.; Springer: Berlin/Heidelberg, Germany, 2020; pp. 153–189. [CrossRef]

32. Abd El-Wahed, M.A.; Kamh, S.Z. Pan-African dextral transpressive duplex and flower structure in the Central Eastern Desert of Egypt. *Gondwana Res.* **2010**, *18*, 315–336. [CrossRef]

33. Fritz, H.; Wallbrecher, E.; Khudeir, A.A.; Abu El Ela, F.; Dallmeyer, D.R. Formation of Neoproterozoic core complexes during oblique convergence (Eastern Desert, Egypt). *J. Afr. Earth Sci.* **1996**, *23*, 311–329. [CrossRef]

34. Greiling, R.O.; Abdeen, M.M.; Dardir, A.A.; El Akhal, H.; El Ramly, M.F.; Kamal El Din, G.M.; Osman, A.F.; Rashwan, A.A.; Rice, A.H.; Sadek, M.F. A structural synthesis of the Proteorozoic Arabian-Nubian Shield in Egypt. *Geol. Rundsch.* **1994**, *83*, 484–501. [CrossRef]

35. Asran, A.M.H.; Kamal El-Din, G.M.; Akawy, A. Petrochemistry and tectonic significance of gneiss amphibolite-migmatite association of Khuda metamorphic core complex, Southeastern Desert, Egypt. In Proceedings of the 5th International Conference of the Arab World, Cairo University, Cairo, Egypt, 21–24 February 2000; pp. 15–34.

36. El Amawy, M.A.; Wetait, M.A.; El Alfy, Z.S.; Shweel, A.S. Geology, geochemistry and structural evolution of Wadi Beida area, south Eastern Desert, Egypt. *Egypt J. Geol.* **2000**, *44*, 65–84.

37. Abdeen, M.M.; Sadek, M.F.; Greiling, R.O. Thrusting and multiple folding in the Neoproterozoic Pan-African basement of Wadi Hodein area, south Eastern Desert, Egypt. *J. Afr. Earth Sci.* **2008**, *52*, 21–29. [CrossRef]

38. Zoheir, B. Transpressional zones in ophiolitic mélange terranes: Potential exploration targets for gold in the South Eastern Desert, Egypt. *J. Geochem. Explor.* **2011**, *111*, 23–38. [CrossRef]

39. Zoheir, B.A. Lode-gold mineralization in convergent wrench structures: Examples from South Eastern Desert, Egypt. *J. Geochem. Explor.* **2012**, *114*, 82–97. [CrossRef]

40. Nano, L.; Kontny, A.; Sadek, M.F.; Greiling, R.O. Structural evolution of metavolcanics in the surrounding of the gold mineralization at El Beida, South Eastern Desert, Egypt. *Ann. Geol. Surv. Egypt* **2002**, *25*, 1122.

41. Zoheir, B.A.; Qaoud, N.N. Hydrothermal alteration geochemistry of the Betam gold deposit, south Eastern Desert, Egypt: Mass-volume-mineralogical changes and stable isotope systematics. *Appl. Earth Sci.* **2008**, *117*, 55–76. [CrossRef]

42. Gahlan, H.A.; Mokhles, K.; Azer, M.K.; Khalil, A.E.S. The Neoproterozoic Abu Dahr ophiolite, South Eastern Desert, Egypt: Petrological characteristics and tectonomagmatic evolution. *Min. Pet.* **2015**, *109*, 611–630. [CrossRef]
43. Khedr, M.Z.; Arai, S. Chemical variations of mineral inclusions in Neoproterozoic high-Cr chromitites from Egypt: Evidence of fluids during chromitite genesis. *Lithos* **2016**, *240–243*, 309–326. [CrossRef]
44. Ahmed, A.H. Highly depleted harzburgite-dunite-chromitite complexes from the Neoproterozoic ophiolite, south Eastern Desert, Egypt: A possible recycled upper mantle lithosphere. *Precambrian Res.* **2013**, *233*, 173–192. [CrossRef]
45. Ghoneim, M.F.; Lebda, E.M.; Nasr, B.B.; Khedr, M.Z. Geology and Tectonic Evolution of the Area around Wadi Arais, Southern Eastern Desert, Egypt. In Proceedings of the Six International Conference on Geology of the Arab World, Cairo University, Cairo, Egypt, 11–14 February 2002; Volume 1, pp. 45–66.
46. Raslan, M.F.; Ali, M.A.; El-Feky, M.G. Mineralogy and radioactivity of pegmatites from South Wadi Khuda area, Eastern Desert, Egypt. *Chin. J. Geochem.* **2010**, *29*, 343–354. [CrossRef]
47. Asran, A.M.; Hassan, S. Remote sensing-based geological mapping and petrogenesis of Wadi Khuda Precambrian rocks, South Eastern Desert of Egypt with emphasis on leucogranite. *Egypt. J. Remote Sens. Space Sci.* **2021**, *24*, 15–27. [CrossRef]
48. El-Baraga, M.H. Geological and Mineralogical and Geochemical Studies on the Precambrian Rocks around Wadi Rahaba, South Eastern Desert, Egypt. Ph.D. Thesis, Geology Department, Faculty of Science, Tanta University, Tanta, Egypt, 1993; 247p.
49. Sabins, F.F. *Remote Sensing Principles and Interpretation*; W. H. Freeman Company: New York, NY, USA, 1997; pp. 366–371.
50. Segal, D.B. Use of Landsat Multispectral Scanner Data for Definition of Limonitic Exposures in Heavily Vegetated Areas. *Econ. Geol.* **1983**, *78*, 711–722. [CrossRef]
51. Canbaz, O.; Gürsoy, Ö.; Gökce, A. Detecting Clay Minerals in Hydrothermal Alteration Areas with Integration of ASTER Image and Spectral Data in Kösedag-Zara (Sivas), Turkey. *J. Geol. Soc. India* **2018**, *91*, 483–488. [CrossRef]
52. Singh, A.; Harrison, A. Standardized principal components. *Int. J. Remote Sens.* **1985**, *6*, 883–896. [CrossRef]
53. Gupta, R.P.; Tiwari, R.K.; Saini, V.; Srivastava, N.A. Simplified approach for interpreting principal component images. *Adv. Remote Sens.* **2013**, *2*, 111–119. [CrossRef]
54. Loughlin, W.P. Principal components analysis for alteration mapping. *Photogramm. Eng. Remote Sens.* **1991**, *57*, 1163–1169.
55. Crosta, A.P.; Souza Filho, C.R.; Azevedo, F.; Brodie, C. Targeting key alteration 648 minerals in epithermal deposits in Patagonia, Argentina, using ASTER imagery and principal component analysis. *Inter. J. Remote Sens.* **2003**, *24*, 4233–4240. [CrossRef]
56. Ninomiya, Y.; Fu, B.; Cudhy, T.J. Detecting lithology with Advanced Spaceborne Thermal Emission and Refection Radiometer (ASTER) multispectral thermal infrared "radiance-at-sensor" data. *Remote Sens. Environ.* **2005**, *99*, 127–135. [CrossRef]
57. Gad, S.; Kusky, T.M. ASTER spectral ratioing for lithological mapping in the Arabian-Nubian Shield, the Neoproterozoic Wadi Kid area, Sinai, Egypt. *Gondwana Res.* **2007**, *11*, 326–335. [CrossRef]
58. Gad, S.; Kusky, T.M. Lithological mapping in the Eastern Desert of Egypt, the Barramiya area, using Landsat thematic mapper (TM). *J. Afr. Earth Sci.* **2006**, *44*, 196–202. [CrossRef]
59. Hassan, S.M.; Sadek, M.F.; Greiling, R.O. Spectral analyses of basement rocks in El-Sibai-Umm Shaddad area, Central Eastern Desert, Egypt using ASTER thermal infrared data. *Arab. J. Geosci.* **2014**, *8*, 6853–6865. [CrossRef]
60. Gabr, S.S.; Hassan, S.M.; Sadek, M.F. Prospecting for new gold-bearing alteration zones at El-Hoteib area, South Eastern Desert, Egypt, using remote sensing data analysis. *Ore Geol. Rev.* **2015**, *71*, 1–13. [CrossRef]
61. Sadek, M.F.; Ali-Bik, M.W.; Hassan, S.M. Late Neoproterozoic basement rocks of Kadabora-Suwayqat area, Central Eastern Desert, Egypt: Geochemical and remote sensing characterization. *Arab. J. Geosci.* **2015**, *8*, 10459–10479. [CrossRef]
62. Ninomiya, Y. A Stabilized Vegetation Index and Several Mineralogic Indices Defined for ASTER VNIR and SWIR Data. In Proceedings of the IEEE 2003 International Geosciences and Remote Sensing Symposium (IGARSS '03), Toulouse, France, 21–25 July 2003; Volume 3, pp. 1552–1554.
63. Bannari, A.; El-Battay, A.; Saquaque, A.; Miri, A. PALSAR-FBS L-HH Mode and Landsat-TM Data Fusion for Geological Mapping. *Adv. Remote Sens.* **2016**, *5*, 246–268. [CrossRef]
64. Mars, J.C.; Rowan, L.C. Mapping Phyllic and Argillic-Altered Rocks in Southeastern Afghanistan Using Advanced Spaceborne Thermal Emission and Reflection Radiometer (ASTER) Data. *USGS Open-File Rep.* **2007**. [CrossRef]
65. Lang, F.; Yang, J.; Li, D. Adaptive-window polarimetric SAR image speckle filtering based on a homogeneity measurement. *IEEE Trans. Geosci. Remote Sens.* **2015**, *53*, 5435–5446. [CrossRef]
66. Arnous, M.; El-Rayes, A.; Geriesh, M.; Ghodeif, K.; Al-Oshari, F. Groundwater potentiality mapping of tertiary volcanic aquifer in IBB basin, Yemen by using remote sensing and GIS tools. *J. Coast. Conserv.* **2020**, *24–27*. [CrossRef]
67. Abuzied, S.; Kaiser, M.; Shendi, E.; Abdel-Fattah, M. Multi-criteria decision support for geothermal resources exploration based on remote sensing, GIS and geophysical techniques along the Gulf of Suez coastal area, Egypt. *Geothermics* **2020**, *88*, 101893. [CrossRef]
68. Abuzied, S.M.; Alrefaee, H.A. Spatial prediction of landslide-susceptible zones in El-Qaá area, Egypt, using an integrated approach based on GIS statistical analysis. *Bull. Int. Assoc. Eng. Geol.* **2018**, *78*, 2169–2195. [CrossRef]
69. Naghibi, S.A.; Pourghasemi, H.R.; Pourtaghi, Z.S.; Rezaei, A. Groundwater qanat potential mapping using frequency ratio and Shannon's entropy models in the Moghan watershed, Iran. *Earth Sci Inf.* **2015**, *8*, 171–186. [CrossRef]
70. Kirubakaran, M.; Johnny, J.C.; Ashokraj, C.; Arivazhagan, S. A geostatistical approach for delineating the potential groundwater recharge zones in the hard rock terrain of Tirunelveli taluk, Tamil Nadu, India. *Arab J. Geosci.* **2016**, *9*, 382. [CrossRef]

71. Arnous, M.O. Groundwater potentiality papping of hard-rock terrain in arid regions using geospatial modelling: Example fromWadi Feiran basin, South Sinai, Egypt. *Hydrogeol. J.* **2016**, *24*, 1375–1392. [CrossRef]

72. Das, S.; Gupta, A.; Ghosh, S. Exploring groundwater potential zones using MIF technique in semi-arid region: A case study of Hingoli district, Maharashtra. *Spat. Inf. Res.* **2018**, *25*, 749–756. [CrossRef]

73. Mekki, O.A.E.; Laftouhi, N. Combination of a geographic system and remote sensing data to map groundwater recharge potential in arid and semi-arid areas: The Haouz plain, Morocco. *Earth Sci. Inf.* **2016**, *9*, 465–479. [CrossRef]

74. Hamimi, Z.; Abd El-Wahed, M.; Gahlan, H.A.; Kamh, S.Z. Tectonics of the Eastern Desert of Egypt: Key to understanding the Neoproterozoic evolution of the Arabian-Nubian Shield (East African Orogen). In *Geology of the Arab World—An Overview*; Bendaoud, A., Hamimi, Z., Hamoudi, M., Djemai, S., Zoheir, B., Eds.; Springer Geology: Berlin/Heidelberg, Germany, 2019; pp. 1–81. [CrossRef]

75. Abdel-Karim, A.M.; El-Mahallawi, M.M.; Finger, F. The ophiolite mélange of Wadi Dunqash and Arayis, Eastern Desert of Egypt: Petrogenesis and tectonic evolution. *Acta Miner. Petrogr. Szeged* **1996**, *36*, 129–141.

76. Ramadan, T.M.; Kontny, A. Mineralogical and structural characterization of alteration zones detected by orbital remote sensing at Shalatein District, SE Desert, Egypt. *J. Afr. Earth Sci.* **2004**, *40*, 89–99. [CrossRef]

77. Ren, D.; Abdelsalam, M.G. Tracing along-strike structural continuity in the Neoproterozoic Allaqi-Heiani Suture, southern Egypt using principal component analysis (PCA), fast Fourier transform (FFT), and redundant wavelet transform (RWT) of ASTER data. *J. Afr. Earth Sci.* **2006**, *44*, 181–195. [CrossRef]

78. Obeid, M.A.; Hussein, A.M.; Abdallah, M.A. Shear zone-hosted goldmineralization in the Late Proterozoic rocks of El Beida area, south Eastern Desert, Egypt. *Trans. Inst. Min. Metall. (Sect. B Appl. Earth Sci.)* **2001**, *110*, 192–204. [CrossRef]

79. Ramadan, T.M. Geological and Geochemical Studies on some Basement Rocks at Wadi Hodein Area, South Eastern Desert, Egypt. Ph.D. Thesis, Geology Department, Faculty of Science, Cairo, Egypt, 1994.

80. Obeid, M.A. The Pan-African arc-related volcanism of the Wadi Hodein area, south Eastern Desert, Egypt: Petrological and geochemical constraints. *J. Afr. Earth Sci.* **2006**, *44*, 383–395. [CrossRef]

81. Hamimi, Z.; Sakran, S. Boudinage structures in Um Teneidbah area, southern Eastern Desert, Egypt: Dynamic and paleostrain analysis. *Egypt J. Geol.* **2001**, *45*, 37–49.

82. Hassan, O.A.; El-Manakhly, M.M. Gold deposits in the southern Eastern Desert, Egypt. In *A Commodity Package*; Egyptian Geological Survey and Mining Authority: Cairo, Egypt, 1986.

83. Klemm, R.; Klemm, D. *Gold and Gold Mining in Ancient Egypt and Nubia, Geoarchaeology of the Ancient Gold Mining Sites in the Egyptian and Sudanese Eastern Deserts*; Springer: Berlin/Heidelberg, Germany, 2013; 663p.

84. Gabra, S.Z. *Gold in Egypt: A Commodity Package*; Geological Survey of Egypt: Cairo, Egypt, 1986; 86p.

85. Takla, M.A.; El Dougdoug, A.A.; Gad, M.A.; Rasmay, A.H.; El Tabbal, H.K. Gold-bearing quartz veins in mafic and ultramafic rocks, Hutite and Um Tenedba, south Eastern Desert, Egypt. *Ann. Geol. Surv. Egypt* **1995**, *20*, 411–432.

86. Ashmawy, M.H. The Ophiolitic Me'lange of the South Eastern Desert of Egypt: Remote Sensing Fieldwork and Petrographic Investigations. Ph.D. Thesis, Berliner Geowiss. Abh., Berlin, Germany, 1987; 135p.

87. Kokaly, R.F.; Clark, R.N.; Swayze, G.A.; Livo, K.E.; Hoefen, T.M.; Pearson, N.C.; Klein, A.J. *USGS Spectral Library Version 7 (No. 1035)*; US Geological Survey: Washington, DC, USA, 2017.

88. Tavarnellia, E.; Holdsworthb, R.E.; Cleggb, P.; Jonesc, R.R.; McCaffrey, K.J.W. The anatomy and evolution of a transpressional imbricate zone, Southern Uplands, Scotland. *J. Struct. Geol.* **2004**, *26*, 1341–1360. [CrossRef]

89. Schreurs, G.; Giese, J.; Berger, A.; Gnos, E. A new perspective on the significance of the Ranotsara shear zone in Madagascar. *Int. J. Earth Sci.* **2009**, *99*, 1827–1847. [CrossRef]

90. Peixoto, E.; Pedrosa-Soares, A.C.; Alkmim, F.F.; Dussin, I.A. A suture–related accretionary wedge formed in the Neoproterozoic Araçuaí orogen (SE Brazil) during Western Gondwanaland assembly. *Gondwana Res.* **2015**, *27*, 878–896. [CrossRef]

MDPI

St. Alban-Anlage 66

4052 Basel

Switzerland

Tel. +41 61 683 77 34

Fax +41 61 302 89 18

www.mdpi.com

Minerals Editorial Office

E-mail: minerals@mdpi.com

www.mdpi.com/journal/minerals

www.ingramcontent.com/pod-product-compliance
Lightning Source LLC
LaVergne TN
LVHW080458180726
843515LV00007BA/1929